BASIC ELECTRICITY

AND AN

INTRODUCTION

TO

ELECTRONICS

BY

The Howard W. Sams
Editorial Staff

Howard W. Sams & Co., Inc.
4300 WEST 62ND ST. INDIANAPOLIS, INDIANA 46268 USA

THIRD EDITION
SEVENTH PRINTING—1981

International Standard Book Number: 0-672-20932-2
Library of Congress Catalog Card Number: 73-75082

Printed in the United States of America.

PREFACE

From beginning to end, this book has been specially written for the student and beginner who sees the bright future in electronics and wants to learn more about it. It begins with a description of the composition of matter and the role of the tiny, but powerful, electron. Progressing in an orderly manner, the text lays the foundation for each new subject, from the structure of atoms to basic circuits, in the important science of electronics.

Discussions move through simply worded explanations of electrical symbols, basic electrical laws, direct current, cells and batteries, magnetism, alternating current, measurement and control, distribution, heating and lighting, radiation, current flow in gases, electrochemistry, and an introduction to the science of electronics.

The concise, down-to-earth text plus the liberal use of diagrams and illustrations provide all the background needed to proceed with advanced study in any one of the many interesting branches of electronics.

This new edition has been prepared with the hundreds of comments and suggestions offered by instructors and students. The logical format used in the presentation of the subject lends itself to student study and also serves as a quick reference.

CONTENTS

CHAPTER 6

CHAPTER 7

CHAPTER 8

CHAPTER 9

CHAPTER 10

CHAPTER 11

1

FUNDAMENTALS

NATURE OF MATTER

STRUCTURE OF MATTER

Matter is anything that has mass and occupies space. Matter exists in three states—solid, liquid, and gaseous—and can be changed from one state to another (by temperature or pressure changes, for example). Scientists once thought that matter could neither be created nor destroyed. But recent developments in atomic physics have shown that matter can be changed into energy. All matter is made of extremely small particles called atoms. An atom is the smallest particle into which matter can be divided by purely chemical means. The next larger division of matter is called a *molecule.* It can be made up of a single atom, or a group of two or more *atoms* of the same *element,* or two or more atoms of different elements.

The size of an atom is somewhere near 10^{-8} centimeter, and of this small size only about one million-millionth of the volume is solid matter, located in the *nucleus* at the center of the atom. More than 99 percent of the mass is contained in the nucleus; the rest is in the electron or electrons circling the nucleus.

The simplest atom is the hydrogen atom, which contains one *proton* and one *electron.* The proton has a positive charge that exactly balances the negative charge of the electron. The atoms of other elements have more than one proton and one electron, but in any stable or electrically balanced atom the protons and electrons are equal in number. Thus, in its stable form, the atom is said to have a *neutral* charge. Atoms, other than the hydrogen atom, also have one or more *neutrons* in the nucleus. These particles have no charge, but they do add to the *atomic weight* of the atom. The *atomic number* is the same as the number of protons or electrons, but the atomic weight can be greater, because of the neutrons.

The Electron

The electron is the basic negative charge of electricity. An atom can have from one electron to over a hundred electrons, depending on the element. For example, oxygen, with the atomic number 8, has 8 electrons; whereas uranium, element number 92, has 92 electrons. The electrons orbit about the nucleus as a center, much as the planets orbit about the sun. The electrons trace paths called shells, and an atom can have as many as seven shells in some elements. Electrons in center shells are, in general, more tightly bound to the nucleus than those in outer shells. Loosely bound electrons are called *free, orbital,* or *planetary* electrons. It is these free electrons that move about as current in an electrical circuit.

An atom that has gained or lost one or more electrons is called an ion. If it has gained electrons, it has a net negative charge; if it has lost electrons, it has a net positive charge.

NATURE OF ELECTRICITY

The nature of electricity cannot be clearly defined because it is not certain whether an electron is a pure charge with no mass, or whether it is a particle of matter with a negative charge. In other words, we do not know exactly what electricity is, only what it does. Electricity can best be defined in terms of its behavior. It is classified as either *static* or *dynamic,* depending upon whether the electrons are at *rest* or in *motion.*

Electrical Charge

Matter that is deficient in electrons is said to be positively charged.

Matter that contains an excess of electrons is said to be negatively charged.

Like charges repel each other; unlike charges attract each other.

Potential Difference (Voltage)

A difference in electrical *potential* exists between two points if they have different amounts of electrical charge. The unit of measurement for potential difference is the *volt.*

Current

The movement of electrons between two points is called current. At first scientists thought that current flowed from a positive point to a point less positive (more negative). We now know that electrons move from negative to positive. However, the older idea of current flowing from positive to negative is still retained for strictly electrical circuits. In electronic circuits, we think of the electron as moving from negative to positive. These two concepts are illustrated in Figs. 1-1 and 1-2. Hereafter, in this book, when current is mentioned without qualification as to direction, it should be taken to mean the direction of electron flow. The rate of electron flow (current) is measured in *amperes.*

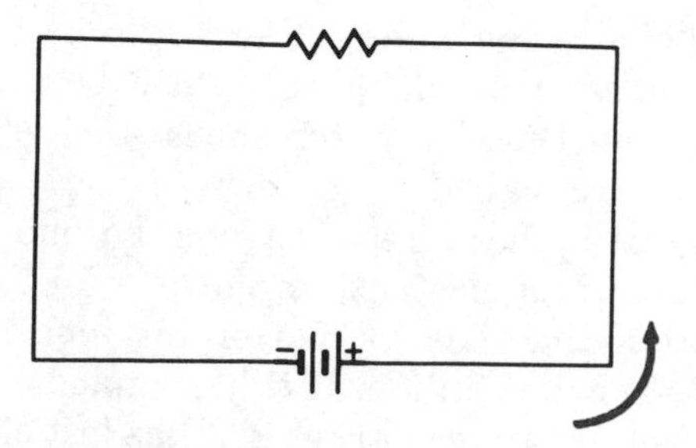

Fig. 1-1. Current direction.

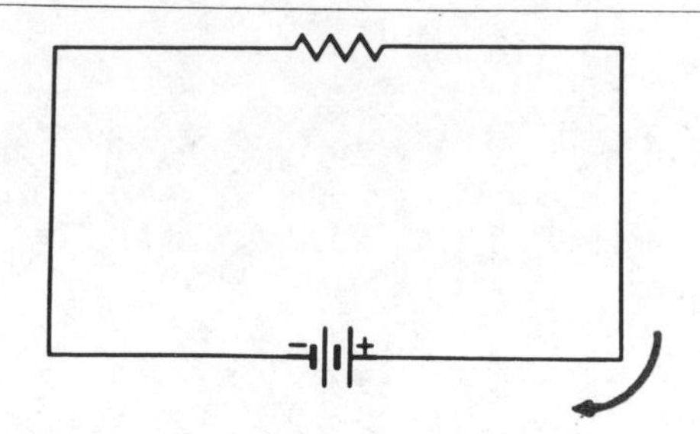

Fig. 1-2. Electron flow.

A *conductor* is a material that has many free electrons and that permits electrons to move through it easily. The following metals are fairly good conductors. They are listed in the order of their *conductivity,* from good to poor: silver, copper, gold, aluminum, tungsten, zinc, brass, platinum, iron (pure), tin, and lead.

An *insulator* is a material that has few free electrons and that resists the flow of electrons. Some common insulators are glass, rubber, mica, Bakelite, paper and silk. Air, mineral oil, and pure water are also good insulators.

A *semiconductor* is a material that has some of the properties of both a conductor and an insulator. Its resistance is less than that of an insulator and more than that of a conductor. Common examples of semiconductor materials are *germanium* and *silicon.*

An electrical impulse travels through a conductor at a speed nearly equal to the speed of light (186,000 miles per second). However, the movement of individual electrons through the conductor is much slower—slower than a speeding automobile, in fact. The following illustration may help to make this clear. Imagine that a long tube, completely filled with billiard balls, reaches from Chicago to San Francisco. A man pushes an extra ball into the tube at Chicago, and a ball immediately pops out of the tube at San Francisco. It takes only an instant to insert the ball in the tube, and the ball moves just a few inches—yet the impulse has traveled the 2,200 miles to the other end of the tube in that same instant. In this hypothetical illustration, the tube represents the electrical conductor and the billiard balls represent the electrons moving through it.

Methods for Generating an Electromotive Force (emf)

Electrons, like matter, cannot be created. Nevertheless, there are several ways to create a potential difference between two points. Such a potential difference tends to force the electrons to move. Therefore, it is termed an electromotive force.

SOURCES OF ELECTROMOTIVE FORCE

Chemical (Cell or Battery)

When two unlike substances (usually metals) are placed in a suitable electrolyte, an emf is generated. That is, a potential difference will exist between the two substances. The two substances are called electrodes. (See Fig. 1-3.)

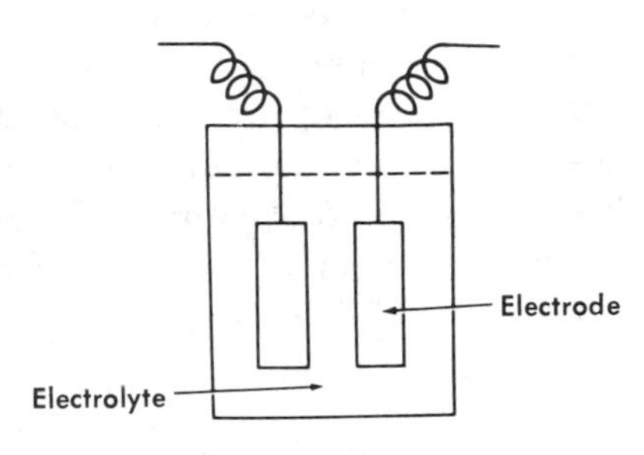

Fig. 1-3. Chemical cell.

Thermoelectric Effect (Heat)

If two different metals are connected in a closed circuit, two junctions are formed as in Fig. 1-4. When one of the junctions is made warmer or colder than the other, current will flow in the circuit. The amount of current depends on the difference in temperature and the choice of metals. Some materials commonly used for *thermojunctions* are bismuth, nickel, platinum, copper, lead, silver and antimony.

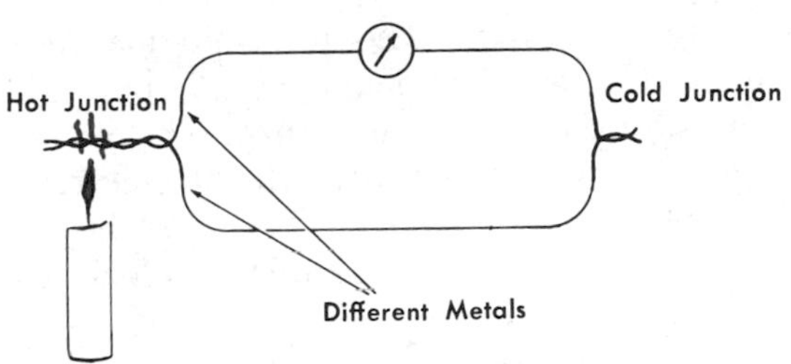

Fig. 1-4. Thermoelectric effect.

Photoelectric Effect (Light)

Certain materials will generate an emf when exposed to light. Some compounds of germanium, selenium, and silicon have this property. (See Fig. 1-5.)

Mechanical—Magnetic (Generator)

If a conductor is moved through a magnetic field, a potential difference is developed in the conductor. This is the basic principle of a generator. (See Fig. 1-6.)

Piezoelectric Effect

Certain types of natural *crystals* and some manufactured *ceramic* materials will generate an emf when subjected to mechanical stress. Some materials commonly used for this purpose are Brazilian quartz, rochelle salts, and tourmaline.

Friction

When certain unlike substances such as glass and wool, silk and hard rubber, etc., are rubbed together and then separated, they become electrically charged. This is static electricity. It is the same kind of electricity one encounters in everyday life when stroking a cat's fur, combing one's hair, or dragging one's feet across a rug. In each case, one of the objects of the pair becomes negatively charged, and the other becomes positively charged.

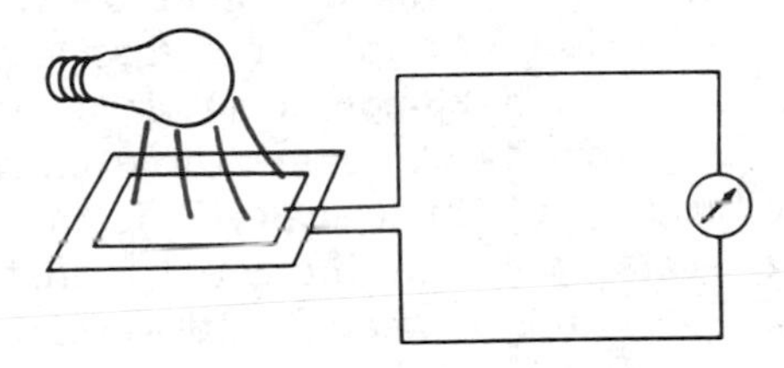

Fig. 1-5. Photoelectric effect.

Basic Facts of Electron Flow (Current)

A current is necessary to operate any electrical equipment. Some effects of current are:

Heating Effect (*Thermal*). Some electrical appliances using this effect are lamps, irons, ranges, toasters and electric heaters. The well-known vacuum tube also uses the heating effect.

Chemical Effect. The chemical effect is used in electroplating, electrolysis, and battery charging.

Magnetic Effect. Electric bells, motors, meters, relays and electromagnets are some examples that use the magnetic effect.

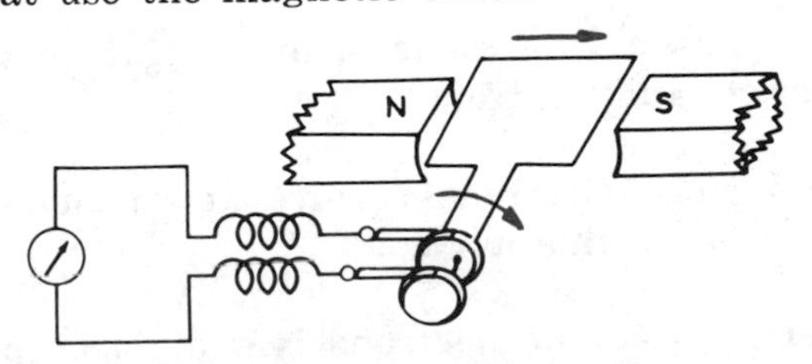

Fig. 1-6. Mechanical-magnetic effect.

Physiological Effect. The physiological effect is the effect of an electrical current upon a human being or other living creature. The physiological effect is commonly used in the electric fence, the electric prod, electrotherapy, the electric needle for removing hair, and the electric chair.

NATURE OF ENERGY

Definition of Energy

Energy is defined as "the ability to do work." Although there are several definitions of energy, this one is most common and describes the different forms of energy most accurately.

Force

Force is defined as "any agent that *produces* or *tends* to produce motion." Force can be *mechanical, electrical, magnetic,* or *thermal.* Note that force does not always produce motion. A small force may fail to move a large body, but it *tends* to do so. For example, if a man exerts a 50-pound force on a 25-pound suitcase, he lifts it easily. If he exerts the same force on a 100-pound suitcase lying on a scale, it is not lifted from the scale; yet the indicator registers a weight of 50 pounds instead of 100—this shows that force tends to lift the suitcase. The word "body" refers to any object—a stone, a building, an automobile, a dust particle—that has mass. Force is usually measured in *pounds,* although it can be measured in *dynes* or in other units.

Work

When we consider energy and force, we are naturally led to consider work. Work is done when a force moves any particle of matter through any distance. The formula for calculating the amount of work done is $W = F \times D$, where W = work, F = force, and D = distance. The unit of work measurement depends upon the units chosen for force and distance. For example, if "F" is in *pounds* and "D" is in *feet,* the work unit would be the foot-pound; if "F" is in *dynes* and "D" is in *centimeters,* the work unit is the dyne-centimeter, more commonly called the *erg.* Since 1 pound equals 444,800 dynes and 1 foot equals 30.48 centimeters, the erg is more than 13 million times smaller than the foot-pound. Therefore, a larger unit, the *joule,* is often used. One joule equals 10^7 ergs, or 0.7376 foot-pound.

Power

Power indicates the *rate* at which work is done. It equals the amount of work done, divided by the time required to do it. The fundamental unit of electrical power is the watt. When the power is one *watt,* work is being done at the rate of one joule per second, or 0.74 foot-pound per second. Note that the watt is not a *quantity,* but a *rate* unit. Larger power units are the *horsepower* and the *kilowatt.* One horsepower equals 746 watts (746 joules per second), or 550 foot-pounds per second.

The following formulas illustrate the Work-Power-Time relationship:

$$\text{Work} = \text{Power} \times \text{Time}$$

$$\text{Power} = \frac{\text{Work}}{\text{Time}}$$

$$\text{Time} = \frac{\text{Work}}{\text{Power}}$$

Summary

An atom is the smallest chemical division of matter.

Atoms consist of a nucleus, or central portion, of protons and neutrons.

One or more electrons revolve at various distances around this nucleus.

Matter deficient in electrons is said to be positively charged. Matter having an excess of electrons is said to be negatively charged. Unlike charges attract each other; like charges repel each other.

Electricity can be obtained by the following common means: Chemical, Thermoelectric, Photoelectric, Mechanical—Magnetic, Piezoelectric and Frictional.

The most common units of electric power are the watt and the horsepower.

Questions and Problems

1. Which contributes most to the weight of an atom, the nucleus or the electrons revolving about it?

2. What *element* has the simplest atom?

3. Describe the difference between static and dynamic electricity.

4. What is the unit of measurement for electrical potential difference? Current?

5. Identify the following materials as insulators, conductors, or semiconductors: Silver, mica, platinum, pure iron, Bakelite, rubber, tugnsten, air, gold, germanium, mineral oil, lead, and silicon.

6. Name all the basic effects of electrical current as used in the various appliances found in the home. Indicate which appliance uses which effect.

7. Which appliance in your home uses electricity at the fastest rate? Express this rate in horsepower.

8. Is the ampere a unit of *quantity* or a unit of *rate?* Explain.

ELECTRON THEORY

In times past, several theories or explanations have been offered by investigators and experimenters to explain the nature of electricity. Some of these theories seemed sound at the time but have since been replaced with ones which can be scientifically proven.

EARLY THEORIES OF ELECTRICITY

One was the early *"two-fluid" theory* which maintained that certain substances possessed two kinds of mysterious invisible fluids that could flow from one place or object to another. If the object had more of one kind of "fluid," it was "charged" (as we now say) ; but if it contained both kinds of fluid in equal amounts or had neither kind, it was neutral. This theory was based entirely on what was known at that time about static electricity. Current electricity was then unknown.

Later, Benjamin Franklin, an early American scientist, became very interested in static electricity and studied it extensively. From the results of his experiments, he concluded that there was only one kind of electric fluid, and he proceeded to develop the "one-fluid" theory. He believed that when an object contained too much of this electric fluid, it was what he chose to call "positive" in charge. If it had lost some or did not have enough, he called it "negative" in charge. But if it had a normal amount, it was "neutral" or not charged at all.

He also believed that the fluid would "flow" from an object or place that had "too much fluid" (positive charge) to one that had "too little fluid" (negative charge). We still use these terms, but their meanings have been changed somewhat. Franklin's one-fluid theory seemed to account for all the facts then known about the behavior of electricity and was so simple and direct that it was generally adopted by the scientists of the time. Nothing was yet explained about the cause or nature of electricity or what it was that flowed. No conclusive proof of his theory or of the older theory could be offered. However, they were workable theories, formulated from experimental observation and study, and constitute an early example of the use of the "scientific method." Man's adoption of the "scientific method" of experiment, study, and evaluation, as opposed to the previous "witchcraft-metaphysical method," marked the dawn of modern science.

ELECTRON DISCOVERED

Finally, about the end of the last century and some 150 years after Franklin's time, the electron was discovered by Sir J. J. Thomson, an English physicist. This was revolutionary—one of the greatest scientific discoveries of all time, an adventure into the unknown. Along with X rays, radioactivity, and related phenomena, all discovered about the same time as the electron, it changed the whole scientific world's way of thinking. His discovery pointed the way to modern chemistry, electronics, atomic energy, and other amazing developments of the last 60 years. Never before had science and industry advanced so rapidly and produced so much.

Franklin might possibly have discovered the electron if an invention used by Thomson had been available in Franklin's time. This was a device invented by Heinrich Geissler, known as a "Geissler" tube. It consists of a glass tube from which the air may be pumped.

This tube is diagrammed in Fig. 1-7. Sealed inside the tube at each end is a metal plate connected by a wire to a source of high-voltage electricity, usually an induction coil. Before the air pump is started, nothing happens when the induction coil is energized, because the air at atmospheric pressure offers such high resistance that no current passes between the plates. As the air is pumped out, the resistance drops until it reaches a point where *current can jump the gap.* An invisible substance actually moves in a stream through the remaining air in the tube from the negative plate to the positive plate. Investigation by Thomson and others proved that this stream was made up of tiny particles, uniform in size and *negative* in charge. He named them electrons and showed that they constituted an electric current when in motion and a static charge when standing still. You may be interested in learning more of the details about the Geissler tube and these epoch-making experiments from other books.

STRUCTURE OF THE ATOM

Fluid theories of electricity were soon discarded in favor of the electron theory. Stating in elementary form and omitting nuclear structure for the sake of simplicity, one may say the electron theory holds that all matter is made up of three kinds of particles within the atom. They are called *electrons, protons,* and *neutrons.* Electrons carry a negative charge and protons, a positive charge, but neutrons have no charge at all. They are neutral particles and, so far as is known, do not enter into ordinary electrical activity such as we are considering; for that reason neutrons will be largely disregarded in this book.

Although all electrons are alike and all protons are alike, electrons and protons are not like each other. Some important facts about them are:

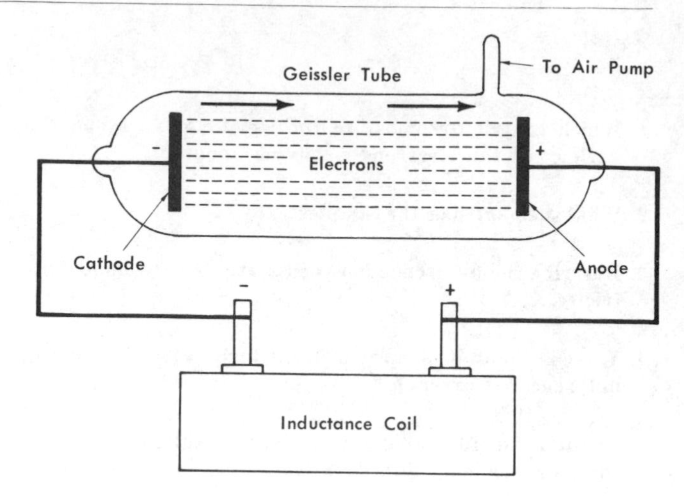

Fig. 1-7. Geissler tube.

1. Electrons and protons (negative and positive charges) attract each other with very strong forces. But electrons repel electrons, and protons repel protons. It also happens that either a positive or a negative charge will attract a neutral (uncharged) object. All this may be summed up in the statement: *Like charges repel, unlike charges attract.* This is one of the fundamental laws of electricity.

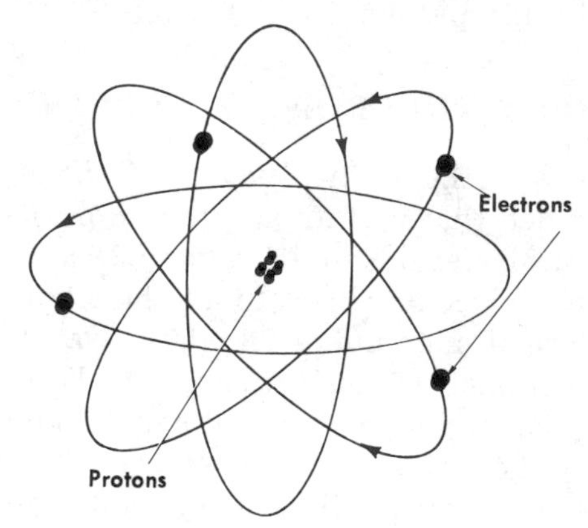

Fig. 1-8. Electrons move in orbits around the nucleus (protons) in a 4-electron, 4-proton atom.

2. The protons of an atom are packed closely together at its center, called the *nucleus,* while the electrons revolve around the nucleus in circular or elliptical paths at terrific speeds. Much as the planets of our solar system revolve in their orbits around the sun, these "electron planets" revolve about a "nucleus sun." Fig. 1-8 shows in a general way how this may occur in a 4-electron, 4-proton atom. We are indebted to Niels

Bohr, famous Danish scientist, for this simple picture of the atom.

3. All electrons possess the *same negative charge,* while all protons possess an equal *positive charge.* Their charges are equal and opposite; therefore, they exactly neutralize each other, so that the normal atom is balanced in charge and neutral. This is the usual condition of atoms.
4. Protons are much smaller and heavier than electrons. Both are very tiny compared to the size of the atom, so tiny that billions of them could be assembled on the point of a pin and never be seen.

There are between 90 and 100 elements (pure substances). There are also many artificial elements, man-made and not occurring in nature, that have been produced in connection with nuclear research. Each element has a kind of atom different from the atom of its nearest neighbor by one electron and one proton. (The variation in neutrons is neglected here.) In a table of the elements hydrogen comes first, with one electron and one proton in each atom. It is the lightest known element. Next is helium, with two electrons and two protons. The next element, lithium, has three of each; beryllium is the fourth element and has four of each. A copper atom has 29 protons and 29 electrons, while nickel, which is next above copper in the table, has 28 of each. Zinc, next below copper, has 30 of each. Thus, it goes down to uranium, the largest and heaviest natural atom, having 92 electrons and 92 protons. Uranium also has 146 neutrons which are useful in atom "splitting" but do not have much significance in ordinary electrical phenomena.

The dense nucleus or central core shown in the drawings of Fig. 1-9 contains all the protons, indicated by plus signs. Around the nucleus, the electrons, indicated by small circles with negative or minus signs, revolve in orbital paths. These drawings may make it appear that all electrons revolve in one plane. That is not true. There may be, and usually are, *as many orbits as there are electrons,* all revolving in different directions (see Fig. 1-8). Since this is difficult to show on flat paper, the circles are usually used. These circles show the number of electrons in each layer or "shell" of the atom. Each shell is spaced at a fixed distance from the nucleus. The number of shells, their distance from the nucleus, and the arrangement of electrons in each shell is always the same in atoms of the same kind. Atoms also differ in ways not mentioned here, but to list them all would take us beyond the scope of this book. You may be interested in studying the subject further in other books. The main point to remember here is that all atoms contain protons that are positively charged and electrons that are negatively charged.

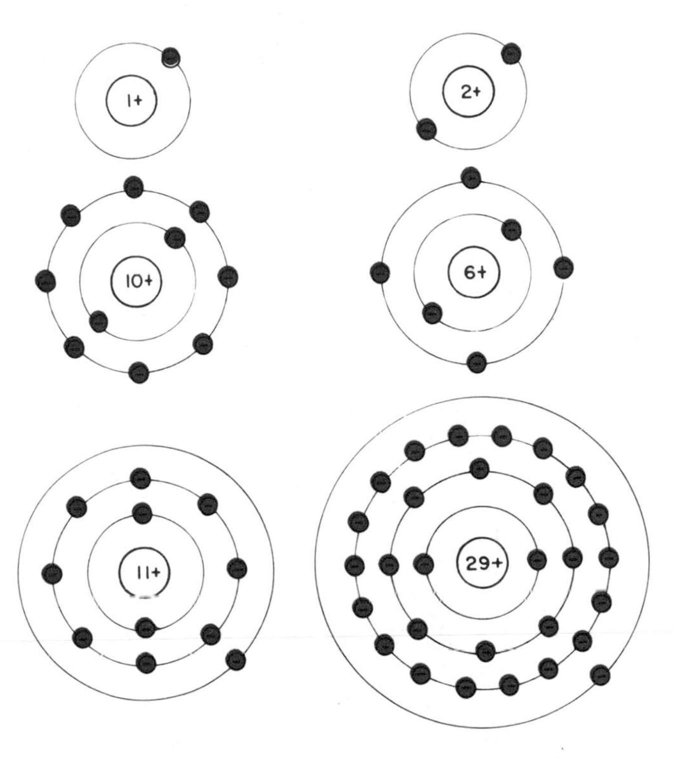

Fig. 1-9. The dense nucleus of the six elements shown contains the protons around which the electrons revolve.

ELECTRICAL NATURE OF MATTER

All substances are either pure elements, mixtures, or compounds. Atoms of each element group together to form *molecules* of that element. Compounds result from combining certain amounts of two or more elements caused by the expenditure of energy in some form. There are thousands of compounds all around us. The molecules of an element are made up of only one kind of atom. Atoms are the smallest particles into which an element can be divided and still retain its original identity. In all compounds *different kinds of atoms* (atoms of different elements) are joined together to form a molecule. The smallest particles into which a compound can be divided are molecules.

If this division is carried further, that is, if these molecules are broken up, the compound is destroyed, leaving only its various atoms. Water and common table salt are examples of compounds. In water two atoms of hydrogen (H) and one of oxygen (O) combine to form a water molecule (H_2O). In common table salt an atom of sodium (Na) and one of chlorine (Cl) combine to form a molecule of sodium chloride (NaCl). Breaking down (decomposing) water molecules results in nothing resembling water, but in two gasses, hydrogen and oxygen; breaking down salt molecules results in two component parts, a solid, sodium, and a gas, chlorine.

Not everything containing two or more elements is a compound. When many of the elements are mixed, they do not combine into a compound but result in a mixture. In a mixture, the original materials do not lose their identity, even though the outward appearance may be different. Air is a mixture of nitrogen, oxygen, water vapor, and other gases.

Some molecules are very complex and contain several kinds of atoms, often in large numbers. Examples of these are many drugs, chemicals, and the tissues of plants. Chemists deal largely with molecules, tearing them down and rebuilding them in different combinations to make new materials or to improve old ones. Thus, many synthetic materials have come into being. Plastics and fabrics such as nylon, orlon, spun glass, and many others, are examples.

Since each molecule is composed of atoms, and atoms in turn contain electrons, it turns out *that all matter is electrical in composition*. That is why the discovery of the electron unlocked one of nature's greatest secrets and brought us untold benefits.

The term *static* means standing still or at rest. A static charge, therefore, is generally at rest, standing still on the surface of an object. Tests show that static charges are never found on the interior of objects, even if hollow, but always on the outer surfaces. They move easily on the surfaces of conductors, such as metals, and become very evenly distributed over the flat, smooth, rounded surfaces of metal objects; but they become concentrated at points, corners, and sharp edges to such an extent that they rapidly leak off into the surrounding air. On insulators, however, static charges usually remain where they are formed until they dissipate into the air, since electrons cannot easily travel over the surface of or through insulating materials.

CREATING A STATIC CHARGE

Normally every atom has the same number of electrons in its orbits as there are protons in the nucleus, and it is electrically balanced (neither negative nor positive in charge, but neutral). If an object composed of such atoms were touched, no shock would be received, because no electrons would flow from the object into the hand, which is also normally neutral. But let every atom in an object, or many of them, gain an electron by contact or otherwise from an outside source, and the *object will have an excess of electrons and be charged negatively*. These excess electrons will collect on the surface of the object and remain at rest there unless conducted away. This charge, standing still, is a static charge, and a shock may now be received by a person touching the object, because the excess electrons will *flow from it to the hand*. If the number of these electrons is great enough, a perceptible shock will be felt.

STATIC CHARGES

A charged condition would also occur if the object were *robbed of electrons*, but it would then have a *positive charge*. Since protons are located at the centers of atoms and are so tightly bound there, they cannot escape unless the atom is split. Every atom that loses an electron or two has more protons than electrons left behind. This *excess of protons* gives every such atom and the object of which it is a part a *positive charge*. If this object is now touched, a shock may be felt because electrons will *flow to it from the hand*.

ELECTRIC CURRENT

Although "static" electricity and "current" electricity may seem to be two different kinds they are really the same. Both consist of electric charges. *Static electricity stands still* and does little work. *Current electricity moves and does work;* it is very beneficial.

As an experiment, place on metal ball A (Fig. 1-10) a negative charge consisting of several billion electrons more than A would have normally. Then, by means of a fine wire connect A to ball B, also made of metal, which has a much smaller neg-

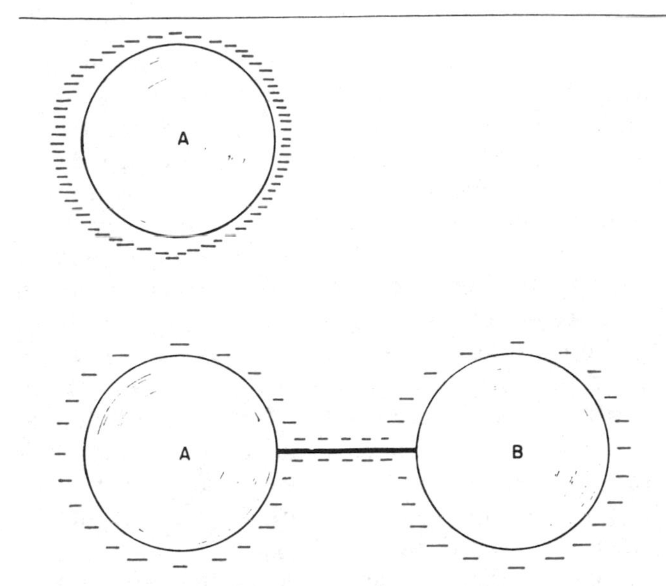

Fig. 1-10. The negative charge is transferred from metal ball A to ball B through the wire.

ative charge than A. Immediately part of the charge will flow from A along the wire B and add to its charge. That is, the charge will divide between A and B, A losing electrons and B gaining electrons. The charge on both will still be negative, but less intense because it now covers and is distributed over a large surface.

On the other hand, if a positive charge is placed on A and the wire is now connected from A to B, electrons will flow the other way from negative B to positive A. In both cases a static charge causes a movement of electrons along the wire. While the electrons are at rest, they are static; when moving, they form current. The method just described does not create much current; the supply of electrons is too limited and the current lasts for only a fraction of a second. Currents of practical magnitude and duration are obtained from such sources as batteries and generators.

Summary

Electrons and protons attract each other with very strong forces, but electrons repel electrons and protons repel protons.

All electrons possess the same negative charge, while all protons possess an equal positive charge.

Static charge is generally at rest on an object. It is never found on the interior of an object but is always on the outer surface.

Questions and Problems

9. **Describe the "two-fluid" theory.**

10. **Describe the "one-fluid" theory which is credited to Benjamin Franklin.**

11. **What is the significance of the Geissler tube, and how does it work?**

12. **What are the three basic parts of an atom?**

13. **State the law of electricity about charged particles.**

14. **Describe the structure of an atom. To whom do we give credit for the picture of the atom?**

15. **What does the word "static" mean, and where are static charges found on all charged objects?**

16. **What is electrical current, and how does it differ from a static charge?**

ELECTROSTATICS AND COULOMB'S LAW

The statement "like charges repel, unlike charges attract" is one of the fundamental laws of electricity. Another of the fundamental laws of electricity is Coulomb's law. Charles Coulomb proved that the force (either attraction or repulsion) between two charges (either like or unlike) is inversely proportional to the square of the distance between them. When this is expressed in mathematical language, we get the following formula:

$$F = \frac{Q_1 Q_2}{d^2}$$

where,

F is the force (in dynes),
Q_1 is the charge on one body (in statcoulombs),
Q_2 is the charge on other body (in statcoulombs),
d is the distance between bodies (in centimeters).

The *statcoulomb* or esu (electrostatic unit) is an electrostatic unit of charge. It is defined as the charge required to repel an identical charge that is one centimeter away with a force of one dyne.

Static electricity ordinarily has little practical value; in fact, it is often a nuisance. It is usually of very high voltage, is difficult to control, and tends to discharge in a fraction of a second. In contrast, current electricity is easily controlled, can be generated at moderate voltages, flows continuously, and delivers energy that can be made to do useful work.

The implication that static electricity is of little value can be qualified somewhat, since capacitors (also called condensers) are often used to store static charges of electricity. A charge of electricity within a capacitor does no useful work while stored. It does useful work only while moving into or out of the capacitor.

Static charges can be produced by friction, by contact, or by induction. We can find many everyday situations where friction produces static electricity. For example, when a cat's back is stroked in dry, cold weather, small sparks of electricity will discharge with a crackling sound. If one's hair is combed briskly when dry, it will fly about as if alive; and the hard rubber comb will then attract bits of lint, paper, or feathers. A glass rod, when rubbed with silk, gives up electrons to the silk; and the rod becomes positively charged. An amber or hard-rubber rod, when rubbed with fur, removes electrons from the fur; and the rod becomes negatively charged. Therefore, it is evident that friction between two unlike substances can produce electrical charges.

We can identify electrical charges and observe their forces of attraction and repulsion with an *electroscope.* (See Fig. 1-11.) The simplest electroscope is a pith ball suspended from a fine silk thread. A small piece of plastic foam like that used to pack fragile items for shipping can be substituted for the pith ball.

The pith ball is uncharged at first. We say it has a neutral charge—the same number of electrons and protons. When an object with either a positive or negative charge nears the ball, the ball is attracted to the object. Now let us see why. First, let us take an object with a negative charge. When it nears the pith ball, its electrons chase the electrons on the ball to the other side of the ball (because like charges repel). Since the protons cannot move, they remain on the side facing the object. As a result, the ball now has a positive charge and will be attracted to the negatively charged object (because unlike charges attract). What if the object has a positive charge? When it nears the pith ball, the electrons on the ball rush to the side of the ball facing the object (because unlike charges attract). The ball now has a negative charge and will be attracted to the positively

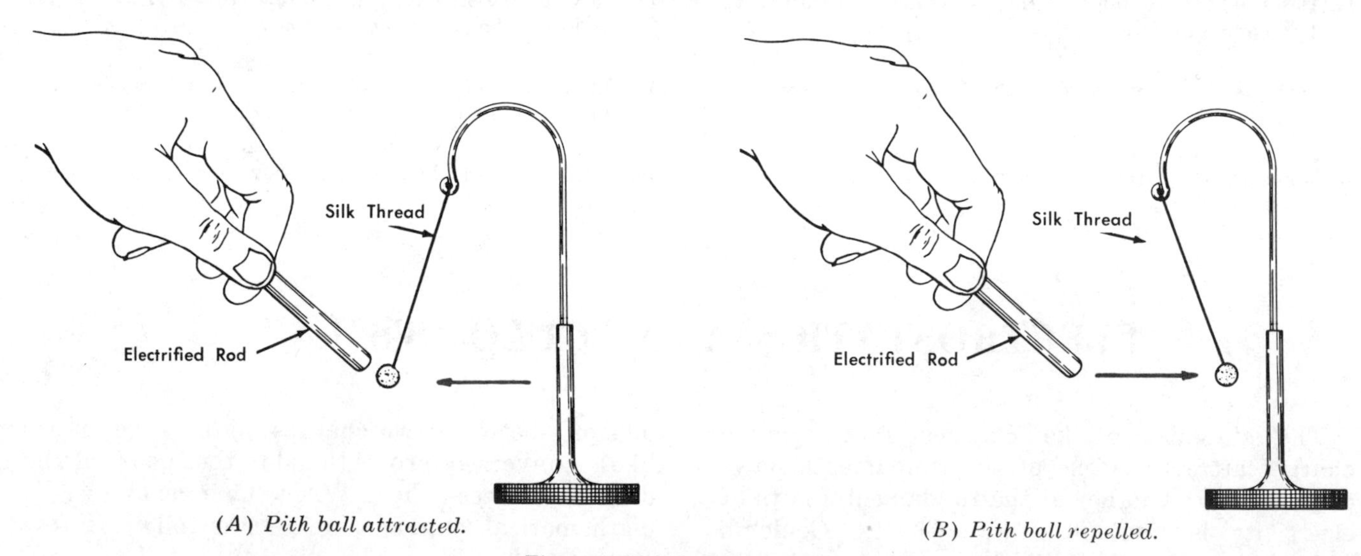

(*A*) *Pith ball attracted.* (*B*) *Pith ball repelled.*

Fig. 1-11. Pith-ball electroscope.

charged object (again, because unlike charges attract).

We can use the electroscope to determine whether an unknown charge is positive or negative. To use a more scientific term, we say that we identify its *polarity*. The electroscope is first given a charge of known polarity. This is easily done by touching the ball with a charged object. Let's assume a glass rod that has been rubbed with a silk cloth is touched to the ball. The rod has a positive charge and thus attracts the electrons on the ball. When the rod is removed, the ball is left with a positive charge. A negative charge could have been placed on the ball by rubbing a hard-rubber rod with a piece of fur and touching the rod to the ball.

A positively charged electroscope will be repelled by a positively charged object and will be attracted to a negatively charged object. The amount of swing is a rough measurement of the amount of charge.

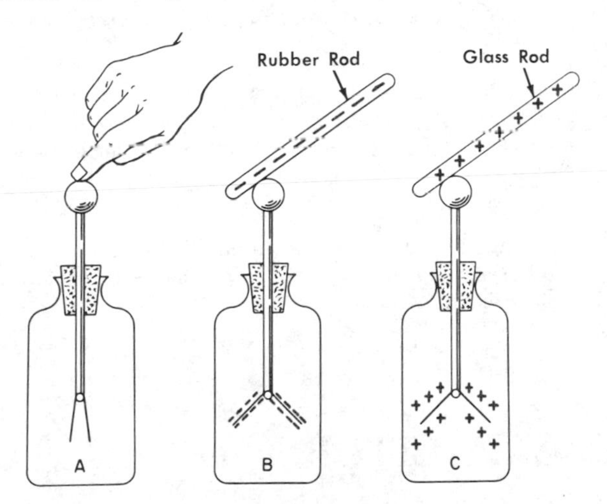

Fig. 1-12. Foil electroscope being charged by contact.

Another form of electroscope is shown in Fig. 1-12. It consists of two strips of very thin metal foil, usually gold leaf suspended from the lower end of a metal rod. The two leaves of the electroscope are approximately one inch long. The top of the metal rod is fastened to a knob or disc. The metal rod runs through a rubber stopper. The entire assembly is placed in a glass jar to prevent air currents from disturbing the foil leaves. Such an electroscope is very sensitive and is more practical than the pith-ball electroscope. When the foil electroscope is charged, the leaves spring apart, because they are charged alike and thus repel each other. Removing the charge, by touching the knob with the hand, causes the leaves to drop.

Charging by Contact

An electroscope can be charged by contact. First, make certain that the electroscope is uncharged, by touching the knob with your hand. Next, stroke a rubber rod with a wool cloth or a piece of fur, and touch the knob with the rod. This puts an excess of electrons (negative charge) on the knob. The knob is connected to the leaves by the metal rod. The electrons flow around and down the rod to the leaves. The leaves receive a negative charge, which causes them to repel each other. When the rubber rod is removed, the charge remains on the leaves.

If the electroscope is touched by a glass rod rubbed with silk, the leaves will be robbed of electrons and will be positively charged. The leaves will again repel each other, because both have been given like charges. Note that an electroscope charged by contact has a positive charge on its leaves if the charging object is positive and a negative charge if the object is negative.

Charging by Induction

An electroscope can be charged by induction. A charged rubber rod is brought *near* the knob of the electroscope, but *not* in contact with it. (See Fig. 1-13.) According to the law of repulsion, a negative charge on the rod pushes the electrons from the knob down to the leaves. As a result, the leaves spread apart. However, if you touch the knob with your finger while the charged rod is near (Fig. 1-13), the electrons will be driven off through your hand instead of down to the leaves, which then become neutral. If you remove your

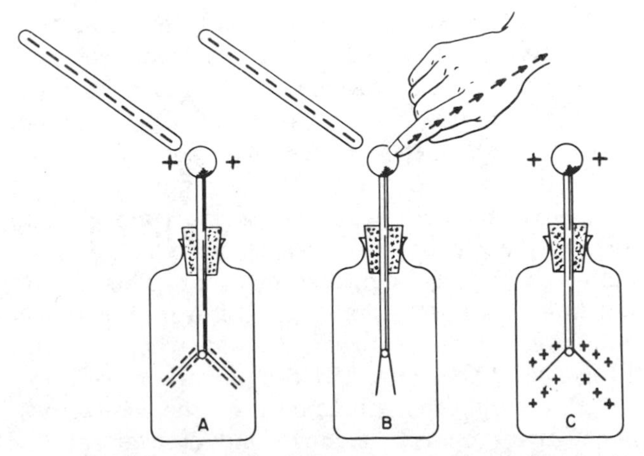

Fig. 1-13. Foil electroscope being charged by induction.

finger before you remove the charged rod, the electrons that were driven off cannot return and the electroscope is now charged. A positive charge will be left on the electroscope because there is a deficiency of electrons, and the leaves will repel each other. Note that an electroscope charged by induction has a positive charge if the charging object is negative, and a negative charge if the charging object is positive. This is just the opposite from an electroscope charged by contact.

Summary

Like charges repel, and unlike charges attract.

The force between two charges is directly proportional to the product of the charges, and it is inversely proportional to the distance between them.

Static electrical charges are on the surface, never in the interior, of an object.

Static electricity can be produced by friction, by contact, or by induction.

Charging an electroscope by contact gives the same charge as that of the charging object.

Charging an electroscope by induction gives the electroscope an opposite charge from that of the charging object.

Questions and Problems

17. **What is the polarity of the static electrical charge obtained on a glass rod by rubbing it with a cloth?**
18. **What is the name of the electrical instrument used to detect an electrostatic charge?**
19. **What type of force (charge) exists between a negatively charged object and a neutral object?**
20. **State Coulomb's law.**
21. **What type of electrical component can "store" static electricity?**
22. **Name two methods of producing static electrical charges.**
23. **Sketch a foil electroscope that has been charged by contact with a glass rod rubbed with a silk cloth. Label the parts of the electroscope, and show the position of all charges.**
24. **How can the amount of electrical charge on an object be estimated with an electroscope? What are the units of measurement?**

CAPACITANCE AND CAPACITORS

A simple *capacitor* can be made by mounting two plates or sheets of conducting material parallel to each other and separated by a very short distance. The plates must be electrically insulated from each other. If a source of dc potential (such as a battery) is connected to the plates as in Fig. 1-14, electrons will leave the plate connected to the positive terminal and accumulate on the plate connected to the negative terminal. This action continues until the potential difference between the plates of the capacitor equals the potential difference of the source. If the power source is removed, the charges will remain on the conductors, the negative charge on one conductor attracting and holding the positive charge on the other. This ability to accept and retain a charge is called capacity or *capacitance.*

The amount of charge that can be stored in a capacitor depends on the capacitance and on the voltage used to apply the charge. This relationship is expressed by the formula:

$$Q = CV$$

where,

Q is the charge in coulombs,
C is the capacitance in farads,
V is the potential difference in volts.

The formula can be rearranged into two other, equivalent forms: $V = Q/C$ and $C = Q/V$. The last form, written out in words, can be considered as a concise definition of capacitance—capacitance equals charge per volt.

The unit of capacitance is the farad, but this unit is so large that smaller, more practical units are used. The *microfarad* (μF) is one millionth of a farad. The *picofarad* (pF) is one millionth of a microfarad. Another unit, the *nanofarad* (nF), is one thousandth of a microfarad.

The insulator between the plates of the simple capacitor shown in Fig. 1-14 is air. Many other materials can be used for this insulator, which is also called a *dielectric*. Some commonly used dielectric materials are air, oil, paper, wax, mica, ceramics, and various oxides. The plates may be aluminum sheets, aluminum foil, copper sheets, copper foil, silver sheets, silver paint, plated steel, and other similar materials. The choice of materials for a capacitor depends on the desired characteristics, both physical and electrical. Some of these characteristics are capacity, voltage rating, operating temperature, polarity, frequency, and temperature coefficient.

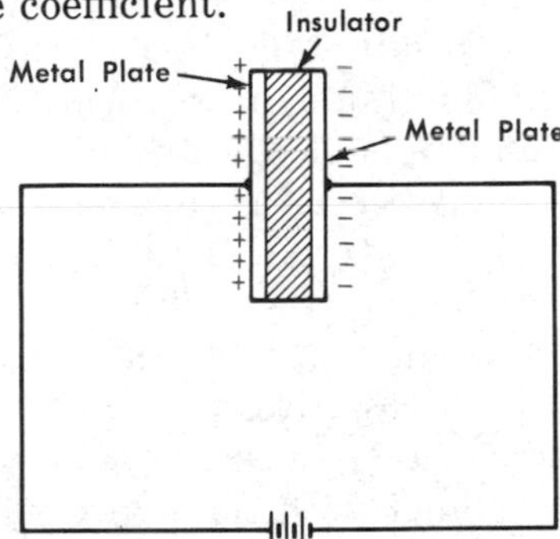

Fig. 1-14. Charging a capacitor.

The capacitance of a simple, parallel-plate capacitor whose plate dimensions are large compared to the distance between them can be calculated by the formula:

$$C = \frac{0.2246KA}{d}$$

where,

C is the capacitance in picofarads,
K is the dielectric constant of the material between the plates,
A is the area of active dielectric in square inches,
d is the distance between plates in inches.

The foregoing formula shows that there are several ways to increase or decrease the capacitance of a capacitor. We can change K by choosing a dielectric material with a different constant; we can change A by using larger or smaller plates (the area of active dielectric changes at the same time); and we can change d by moving the plates closer together or farther apart (thinner or thicker dielectric).

Table 1-1. Dielectric Constants and Dielectric Strengths of Materials

Material	Dielectric Constant	Dielectric Strength (Volts per Mil)
Vacuum	1.0	
Air	1.0	31
Bakelite (paper base)	3.8	250-585
Barium titanate	1250.0	75
Glass (window)	7.0	760
Mica	2.5	2030-5080
Nylon	3.4-22.4	285-470
Paper (paraffin coated)	3.5	1170
Polystyrene	2.4	300-710
Porcelain	6.0	200-400
Rubber (hard)	2.0	470

A perfect vacuum is assigned a dielectric constant of 1, and all other dielectric materials are compared to it. If a material has a dielectric constant of 2, this means that a capacitor constructed with this material would be twice as effective at storing a charge as it would be with a vacuum for a dielectric. Air is also said to have a constant of 1. Actually, the dielectric constant of air is 1.0006, but this can be taken as 1.0 in most cases, without causing any difficulty. Some typical values of dielectric constant for a few electrical insulators are given in Table 1-1. The table also lists dielectric strength, which is the voltage that the material can stand without insulation breakdown.

CAPACITOR TYPES

The importance of capacitors in an electrical circuit can be understood by noticing the number of places in which a capacitor appears. Capacitors are used in coupling circuits, filter networks, tuned circuits, and bypassing circuits. Capacitors are "fixed," meaning that the capacity cannot be changed after manufacture, or "variable," meaning that the capacity can be changed or "varied" to a desired value. Capacitors are classified by a descriptive term which usually designates one or more of the materials used. The commonly used types are paper-wax tubular, molded-plastic paper tubular, molded mica, ceramic tubular, ceramic disc, electrolytic, variable air, and trimmer.

Paper Capacitors—Tubular

The fixed capacitor consists of two tin-foil or aluminum foil conducting plates separated by a dielectric, such as linen paper or wood-pulp paper. The two plates and the dielectric are rolled into a small cylinder. During manufacture, the residual air is withdrawn or evacuated from the capacitor. The capacitor is then made moistureproof by being soaked in hot wax or oil, or in a special insulating compound that melts under high temperatures. The terminals are usually wires which connect internally to each conducting plate "roll." The entire assembly is placed in a waxed paper tube or is molded into a plastic case. The capacity of the paper capacitors ranges from approximately .001 μF to 2 μF, with voltage ratings from 200 volts to 6000 volts. The voltage at which the capacitor can be safely operated depends on the material in the dielectric and its thickness. The capacity and the voltage rating are generally designated on the capacitor.

Mica Capacitors

The better grades of fixed capacitors use mica as the dielectric, because mica has a higher breakdown voltage and a smaller dielectric loss than other common materials. Mica is more expensive than the paper dielectric. Therefore, it is not used in the larger fixed capacitors. The mica capacitor is constructed of alternate layers of tin foil and mica molded in a plastic casing to make it moistureproof. The mica capacitors range in capacity from approximately .000005 μF to .02 μF. Precision units can be made from mica sheets coated with silver paint, which becomes the conducting plate. These units are known as "silver micas."

Ceramic Capacitors

These capacitors are popular because of their small size and their wide range of values and voltage ratings. They consist of discs or tubes of ceramic material on which a silver paste is painted. Heat is applied to the unit to dry the paste (conductor) and cure the ceramic (dielectric). Leads or terminals are fastened in place, and the entire assembly is coated with or molded into an insulating material.

Electrolytic Capacitors

These units have extremely high capacity yet take up very little space. For this reason they are commonly employed as starting capacitors on ac motors and in the filter circuits of dc power supplies.

On polarized capacitors one terminal will be marked positive (+) and the other will be negative (−). This polarity must be observed when the capacitor is connected in a dc circuit. If connected in reverse polarity, there will be a high dc leakage current. Properly connected, there will be some dc leakage, even with the best electrolytics.

Electrolytic capacitors are usually of the "polarized" type, since they are usually employed in dc circuits. Capacitors used in ac circuits are of the "nonpolarized" type.

The plates in a nonpolarized electrolytic are either strips of aluminum which are "etched" or indented to provide a larger active surface or strips of gauze which have been sprayed with molten aluminum. These plates are then "formed" by the action of an electric current. The forming action leaves a thin film of aluminum oxide on each plate. This oxide coating becomes the dielectric of the capacitor. The polarized electrolytic also has an aluminum oxide positive plate, but the negative plate is of pure aluminum. The negative plate usually serves as the container.

An electrolyte is used to establish contact between the oxides on the plates. Years ago this electrolyte was liquid, and the capacitors were known as "wet" capacitors. Newer units contain a moist paste which saturates a porous strip of material in contact with the plates. These capacitors are known as "dry" electrolytic capacitors.

Because of its versatility, the electrolytic capacitor is available in many voltage ratings and capacities, and in multisection units. Capacities usually range from 1 μF up to 2000 μF, with voltage ratings from 6 volts to 600 volts.

Tantalum Capacitors

This is an electrolytic capacitor of more recent development than the aluminum electrolytic. It has approximately the same capacity range as the latter capacitor but is smaller in size. The anode is made of tantalum. The working temperature range extends to lower and higher extremes in the tantalum capacitor than in the aluminum capacitor.

Mylar Capacitors

These units have exceptionally high insulation resistance—low dielectric absorption and a high power factor. They are also extremely stable, having a relatively small capacitance change with temperature variations over a range of 0° to 85°C. The mylar capacitor has an excellent resistance to humidity and moisture due to the casing material which will not burn, soften or melt at any operation temperature.

CHARACTERISTICS OF A CAPACITOR

To show that a capacitor can store electrical energy, the electrical action of a capacitor can be compared to the mechanical action of a piston and diaphragm. (See Fig. 1-15).

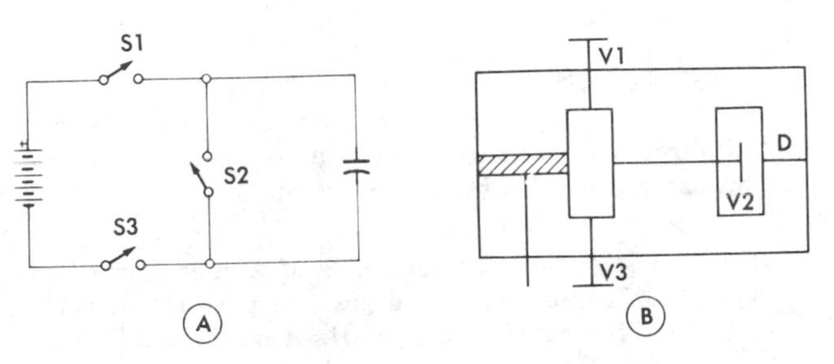

Fig. 1-15. Zero energy at both A and B.

When the mechanical pressure on either side of the diaphragm is the same, the diaphragm is in an "unstressed" condition. When the electrical pressure on either side of the dielectric is the same, the dielectric is in an "unstressed" condition. Under such conditions, no energy is stored in either system.

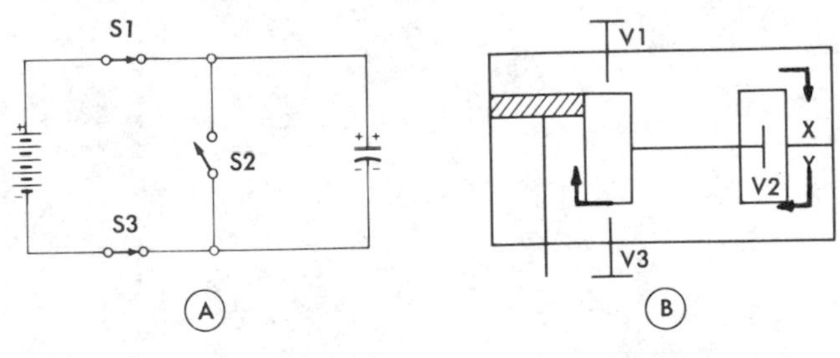

Fig. 1-16. Stored energy at both A and B.

When the piston moves up, the pressure at X increases and creates a difference in pressure that stresses the diaphragm. When a difference in electrical pressure is applied to the capacitor, the dielectric is similarly stressed. Both systems now store energy. (See Fig. 1-16.)

Providing a path through C (as in Fig. 1-17) relieves the mechanical stress in the diaphragm and the electrical stress in the dielectric. The momentum of the liquid causes the diaphragm to reverse its position. Thus, oscillations are set up in both systems that finally reduce the stored energy to zero.

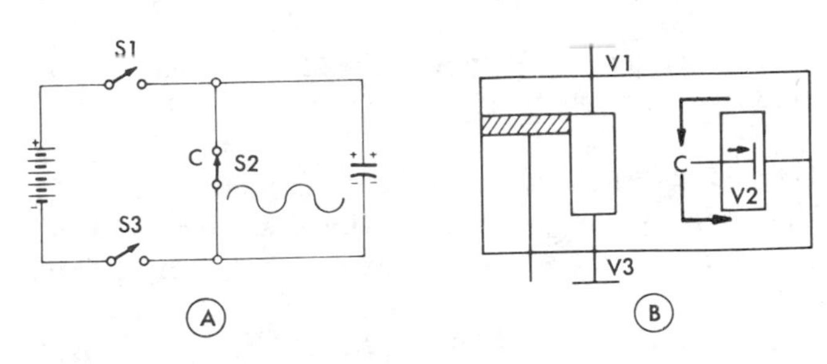

Fig. 1-17. Oscillating energy gradually drops to zero.

When the difference in pressure is reversed, the diaphragm is stressed in the opposite direction. The stress in the dielectric is also reversed, and the capacitor receives a charge of the opposite polarity. Both systems again store energy. (See Fig. 1-18.)

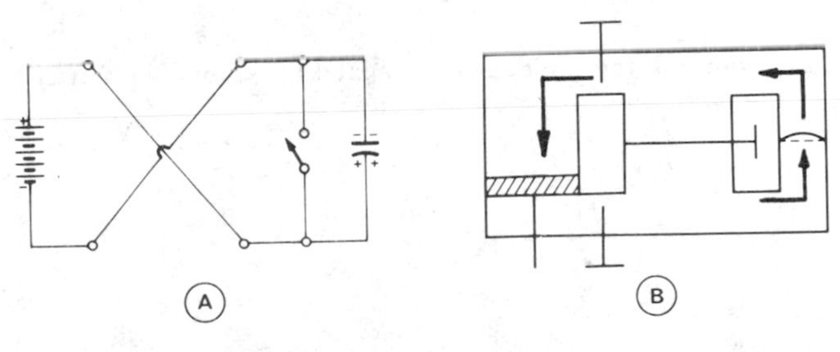

Fig. 1-18. Stored energy at both A and B opposite of that shown in Fig. 1-16.

Moving resistor arm "A" (Fig. 1-19) upward will charge the capacitor; moving the arm downward will discharge the capacitor. Moving the arm upward and downward will thus charge and

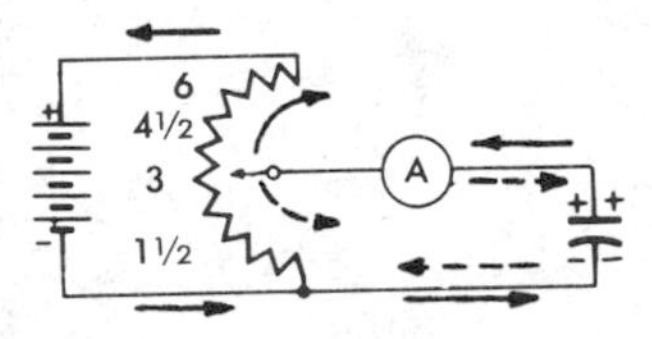

Fig. 1-19. Charging and discharging a capacitor by applying pulsating dc.

discharge the capacitor repeatedly. Note that the input voltage of the capacitor is dc, while its current is ac. Thus, although the current appears to be flowing through the capacitor, it is really pulsating in and out of the capacitor.

Summary

Capacitance is that property which allows electrical energy to be stored on two conducting surfaces by a dielectric.

The electrical energy is stored by *electrostatic* stress of the dielectric.

Important factors controlling capacity are: (1) the area of the plates, (2) the distance between the plates, and (3) the nature of the dielectric.

The unit of capacity is the *farad*. Since this is a very large unit, capacitance is usually stated in *microfarads* (μF).

Capacitors are used in coupling circuits, filter circuits, and tuned circuits, and for bypassing applications.

A capacitor blocks passage of dc voltage and readily passes ac voltage.

Capacitors of more than 5 microfarads are generally electrolytic capacitors which employ a chemical dielectric.

Questions and Problems

25. Describe the construction of a simple capacitor.

26. List six materials useful as dielectrics.

27. Does the type of dielectric affect the value of capacitance? Explain.

28. If the voltage applied to a capacitor is doubled, how is the charge on the capacitor affected?

29. Will direct current pass through a good mica capacitor? Through a good electrolytic? Explain.

30. Does it matter how the terminals of a nonpolarized capacitor are connected to the positive portions of a circuit? Does it matter with a polarized capacitor? Explain.

31. What types of capacitors are used for high-capacity applications?

2

DIRECT CURRENT

CIRCUITS AND SYMBOLS

An electric current is the motion of electrons. This motion is called a current flow. To enable these electrons to flow and to confine them in a particular path, an electrical *circuit* must be provided. (The word "circuit" means *to go around.*) A complete electrical circuit provides a continuous path for the passage of current (electron flow). See Fig. 2-1.

A practical electrical circuit has at least four parts: (1) a source of electromotive force (*emf*), (2) a set of conductors, (3) a load, and (4) a means of control.

Electromotive force is defined as "the force that can move electrons." It can come from a cell or battery, a dc or ac generator, an electronic power supply, or any equipment that can provide a difference of electrical pressure.

Although wires of many sizes can be used as *conductors,* the term "conductor" actually refers to any material which offers low resistance to a current. Conductors may be good or poor; however, poor conductors are usually referred to as resistors or as *insulators* if the conductivity is very low. There are no sharp lines of distinction separating conductors, resistors and insulators; for instance, *semiconductors* fit into the areas between conductors and resistors and between resistors and insulators.

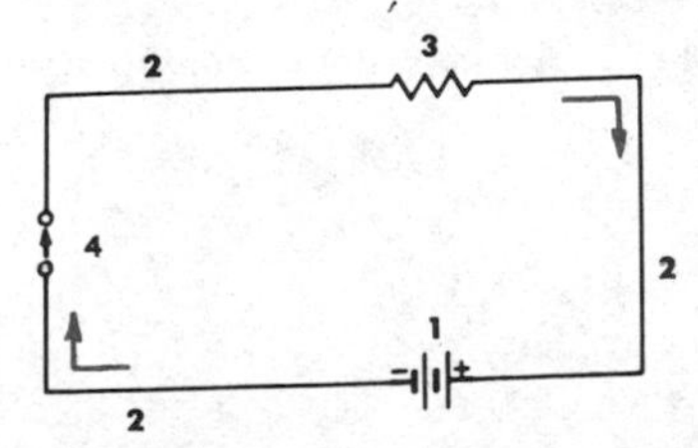

Fig. 2-1. Simple electrical circuit.

The *load* of an electrical circuit may be any device that uses electrical energy, such as a lamp, a bell or buzzer, a toaster, a radio, or a motor. The load is usually considered as being separate from the conductors that connect it to a current source.

The current flowing in an electrical circuit is stopped or started by means of *switches.* Further control is provided by variable resistances, such as *rheostats* and *potentiometers.* Fuses, circuit breakers, or relays can be used as controls.

Series and Parallel Circuits

In Fig. 2-2 three resistors are connected in *series.* Note that there is only one path for the current and that the same current flows through all parts. Assuming the resistors are lamps, all lamps will stop glowing if any lamp is turned off. This type of connection is called a *series* connection. It is used with lamps, bells, or other appliances.

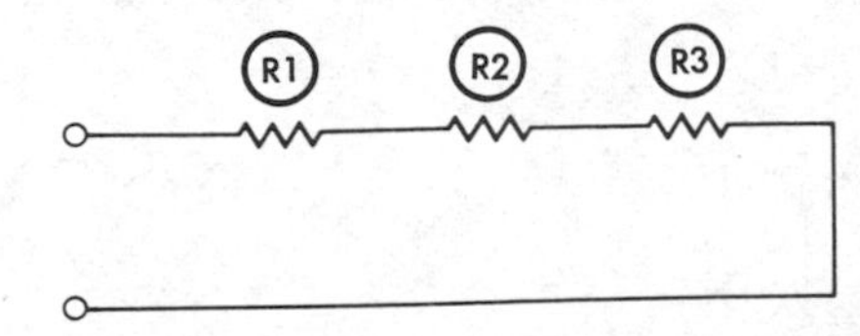

Fig. 2-2. Resistors connected in series.

Fig. 2-3 shows the same three resistors connected in shunt or *parallel*. Note that there is more than one path for the current to take. The current is not necessarily the same in each resistor. The current through any resistor is less than the total current in the entire circuit. If the resistors are lamps, turning off one lamp will not turn off the others. The parallel circuit is quite useful because the other lamps can remain lit when one lamp goes out. The lamps in a home are wired in parallel because one lamp can be turned off without plunging the house into total darkness.

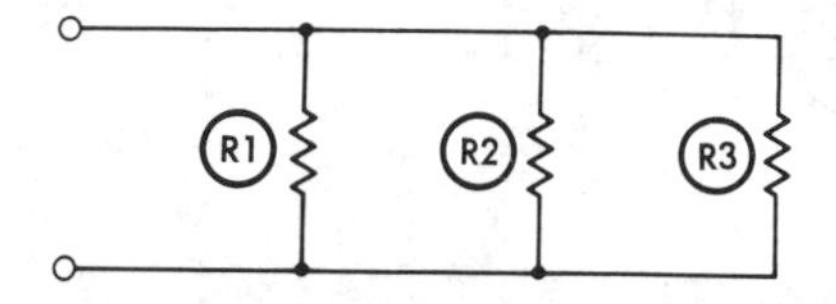

Fig. 2-3. Resistors connected in parallel.

Not all parallel connections are alike. (See Fig. 2-4.)

Fig. 2-5 shows the three resistors connected in still another way. They are not all in series, nor are they in parallel. Two of the resistors are in series, and the two together are in parallel with the third resistor. Since both a series and a parallel circuit are found here, this is known as a combination or *series-parallel* circuit.

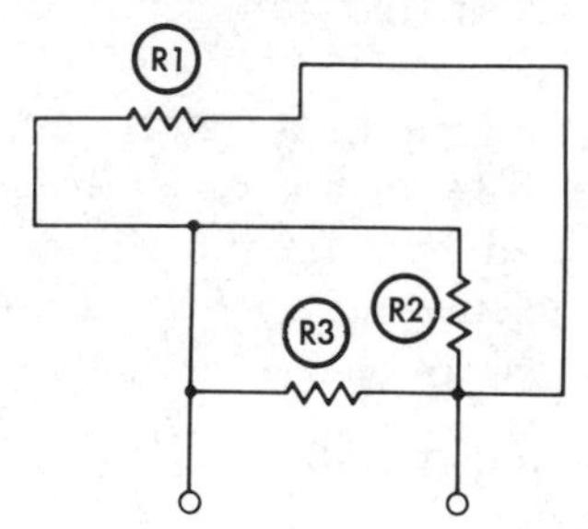

Fig. 2-4. Another example of resistors in parallel.

Fig. 2-6 shows other combination (series-parallel) circuits.

Common Units and Symbols

Instead of drawing pictures to represent the parts of an electrical circuit (pictorial method), electrical symbols (schematic method) are used. These universally recognized symbols make it easy for anyone to draw electrical diagrams that can be understood by any electrician or other technically trained person.

The symbols used in this lesson conform to the graphic symbols and electronic diagrams approved by the American National Standards Institute (ANSI). More and more manufacturers are switching from their individual pet symbols and conforming to the ANSI standards. You may find variations of the symbols in Fig. 2-7, but after some experience you will be able to recognize them.

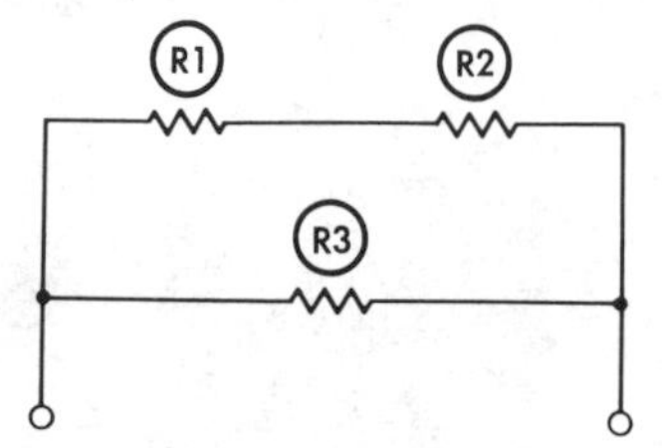

Fig. 2-5. Series-parallel or combination circuit.

ANSI standards allow some variations of certain symbols, so you may find more than one symbol for the same component.

In electrical and electronic theory, abbreviations are used to represent most electrical units. The four most common abbreviations are (1) *R* and *r*, symbols for *resistance*, (2) *E*, symbol for *emf* (pressure), (3) *I*, symbol for electrical current, and (4) *P*, symbol for electrical power.

For example: R = 10 ohms; E = 25 volts; I = 2.5 amperes; P = 62.5 watts.

Table 2-1 lists many common electrical abbreviations.

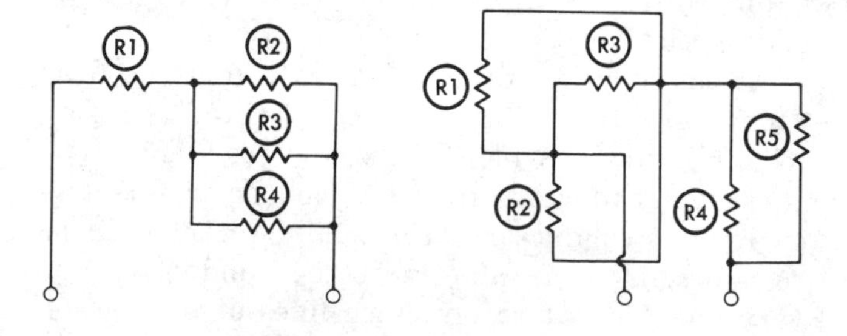

Fig. 2-6. Other series-parallel or combination circuits.

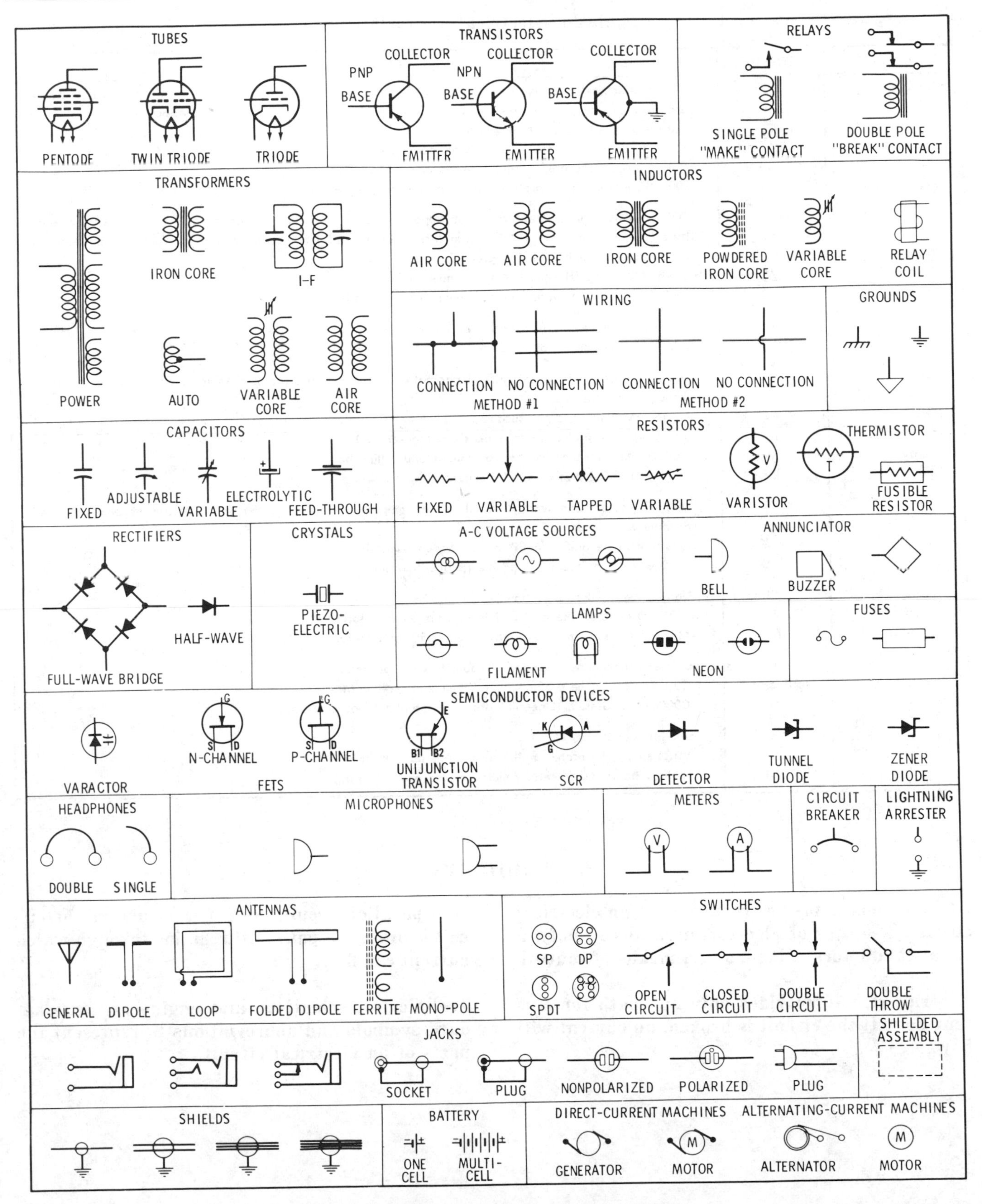

Fig. 2-7. Common electrical and electronic symbols.

Table 2-1. Electrical Abbreviations

UNITS	SYMBOLS	DESCRIPTION
Coulomb	C	Unit of electrical quantity. The number of electrons which must pass a point in one second to produce a current of one ampere. The quantity which will deposit .0000116 ounce of copper from one plate to the other in a copper sulfate solution.
Ampere	A or amp	Unit of current. One coulomb flowing per second.
Milliampere	mA	.001 ampere. (The prefix "milli" means one-thousandth.)
Microampere	μA	.000001 ampere. (The prefix "micro" means one-millionth.)
Ohm	ohm or Ω	Unit of resistance (R). Measure of the opposition offered to the flow of current. The resistance offered by a column of mercury 106.3 centimeters in length and 1 square millimeter in cross-sectional area, at 32 degrees Fahrenheit or 0 degrees Celsius.
Megohm	M	1,000,000 ohms. (The prefix "meg" means million.)
Microhm		.000001 ohm. (The prefix "micro" means one-millionth.)
Mho	g	Unit of conductance (g). Measure of the ease with which a conductor will permit current to flow. A mho is the reciprocal of an ohm.
Volt	V	Unit of pressure difference (emf—electromotive force). Pressure required to force one ampere of current through a resistance of 1Ω.
Hertz	Hz	Frequency (one cycle per second.)
Millivolt	mV	.001 volt. (The prefix "milli" means one-thousandth.)
Microvolt	μV	.000001 volt. (The prefix "micro" means one-millionth.)
Kilovolt	kV	1000 volts. (The prefix "kilo" means one-thousand.)
Watt	W	Unit of power. One watt is equal to one ampere of current under the pressure of one volt. The formula for power is $P = I \times E$.
Milliwatt	mW	.001 watt. (The prefix "milli" means one-thousandth.)
Kilowatt	kW	1000 watts. (The prefix "kilo" means one-thousand.)
Watthour	Wh	Unit of work. (Power × time.)
Kilowatt hour	kWh	1000 watt-hours. (The prefix "kilo" means one-thousand.)
Horsepower	hp	746 watts. The power required to raise 550 lbs one foot in one sec.
Farad	F	Unit of capacitance. Capacity of capacitors (condensers).
Microfarad	mfd or μF	.000001 farad. (The prefix "micro" means one-millionth.)
Picofarad	pF	.000001 microfarad. (One-millionth of one-millionth of a farad.)
Henry	H	Unit of inductance (L).
Millihenry	mH	.001 henry. (The prefix "milli" means one-thousandth.)
Microhenry	μH	.000001 henry. (The prefix "micro" means one-millionth.)

Summary

At least four items are required for an electrical circuit: a source of electromotive force (emf), a set of conductors, a load, and a means of control.

A series circuit provides only one path for current flow. If the circuit is broken, no current will flow.

In a parallel circuit, even if a branch of the circuit is broken, a path still remains through which current can flow.

Technicians, scientists, and engineers use standard symbols and abbreviations to represent the parts of an electrical circuit.

Questions and Problems

1. What will happen in a circuit when a very poor conductor is used?

2. Name two pieces of electrical equipment that can produce an emf.

3. If there are three lamps in a series circuit, what happens to the other two lamps when one lamp is turned off? Why?

4. If there are two resistors of equal resistivity in a parallel circuit, does equal current flow through each resistor?

5. If resistors R1, R2, and R5 were removed from the circuit in Fig. 2-6B, would current continue to flow in the rest of the circuit? Explain.

6. If resistors R4 and R5 were removed from the circuit in Fig 2-6B, would current continue to flow in the rest of the circuit? Explain.

7. What is the electrical abbreviation for current? Emf? Power?

8. What is the unit of measurement for emf? Resistance? Current?

VOLTS, AMPERES, OHMS

AMPERES

Current is measured in amperes. The term "ampere" refers to the number of electrons passing a given point in one second. This number is unbelievably large. If one could count the individual electrons, he would see approximately 6243 quadrillion electrons go by during the one second that one ampere was flowing.

This number of electrons, 6243 quadrillion, is a *coulomb*. This is a measurement of quantity (like saying there are eight pints in a gallon of water). When the electrons are moving, there is current. Current can be measured in amperes, which is a measurement of quantity divided by time. This would be similar to saying that we could fill so many gallon buckets with water at the rate of so many pints per minute.

One ampere is equal to one coulomb per second. An instrument called an ammeter will measure electron flow in coulombs per second. The ammeter is calibrated in amperes, which we always use instead of coulombs per second when speaking of the amount of current. Fig. 2-8 is a schematic showing an ammeter connected in a circuit to measure the current in amperes.

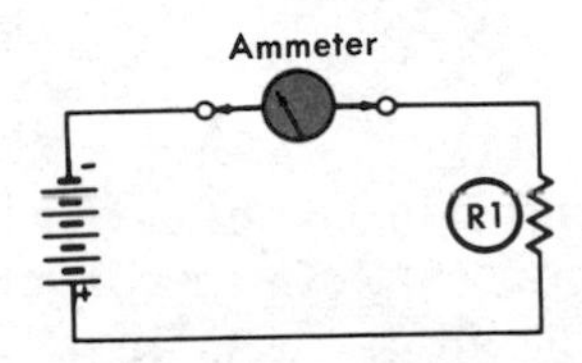

Fig. 2-8. An ammeter connected into a simple circuit.

VOLTS

Electromotive force is measured in *volts*. This is the amount of pressure difference between points in a circuit. It is this pressure or difference of potential that forces current to flow in a circuit. For example, suppose we have two automobile tires, one inflated to a pressure of 30 pounds per square inch, the other to a pressure of 10 pounds per square inch. If we connect a hose to the valves of the tires, the difference in pressure will send air from the 30-psi tire to the 10-psi tire. Air will continue to flow until the pressure is the same in both tires.

One volt (potential difference) is required to force one ampere of current through one ohm of resistance. This is similar to a water pump that forces water to flow through a pipe. The water pump can be compared to the potential difference. The number of pounds of pressure produced by the water pump corresponds to the number of volts produced by a current source. The action of the pump pushing a number of gallons of water per second past a certain point in a water system could be compared to the action of a current source sending a number of amperes of current. A valve in the pipe offers a resistance to the flow of water; the amount of resistance offered is comparable to the ohm. Keep this water-pump analogy in mind as you study electricity. It will help you to better understand the action of the "Big Three"—voltage, amperage, and resistance—in an electrical circuit.

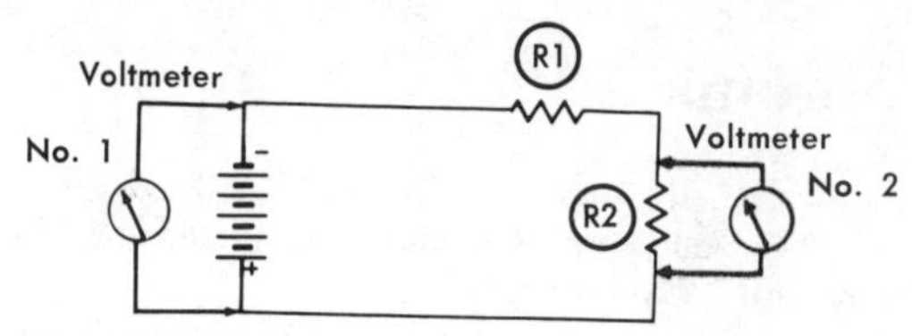

Fig. 2-9. Voltmeter connected in a simple circuit.

An instrument that will measure voltage is known as a *voltmeter*. Fig. 2-9 is a schematic showing voltmeters connected in the circuit to measure the voltage. Voltmeter No. 1 is connected to read the applied or source voltage. Voltmeter No. 2 is connected to measure the voltage drop, or potential difference, across R2.

OHMS

Resistance is measured in *ohms*. Resistance opposes the flow of electrons (current). The amount of opposition to the flow is stated in ohms.

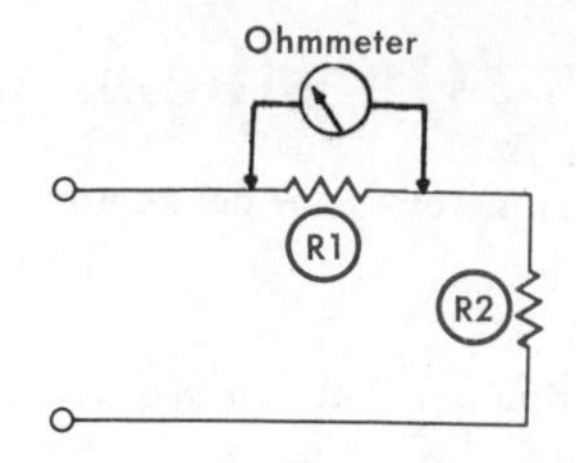

Fig. 2-10. An ohmmeter connected to indicate resistance.

If a glass tube 106.3 centimeters (approximately 41 inches) in length and one square millimeter in cross-sectional area is filled with mercury and maintained at zero degrees Celsius (32 degrees Fahrenheit), the tube will offer a resistance of one ohm.

An instrument that will measure ohms is known as an *ohmmeter*. Fig. 2-10 is a schematic showing an ohmmeter connected to read the resistance of R1. The resistance of any material depends on the type, size, and temperature of the material. Even the best conductor offers some opposition to the flow of electrons.

Summary

Current is measured in a unit called an ampere, which is equivalent to a certain number of electrons passing a given point in one second.

Electrons are driven through a circuit by a force called electromotive force. The unit of force is called a volt.

Anything that offers opposition to current flow is said to have resistance. The unit of resistance is called an ohm.

Questions and Problems

9. In measuring current, what meter is used?

10. How long does it take 12,486 quadrillion electrons to pass a given point when one ampere of current is flowing through that point?

11. If one ampere of current is flowing through a one-ohm resistance, what is the voltage?

12. What instrument is used to measure voltage?

13. What is the unit of measurement of resistance?

14. Sketch the circuit symbol for a voltmeter. An ammeter. A resistance unit.

15. What is the unit of measurement for electric current?

16. What is the unit of measurement of electromotive force?

17. What instrument is used to measure resistance?

OHM'S LAW

The law which governs most simple and many complex electrical phenomena is known as Ohm's law. It is the most important law in electricity. In 1827, a German physicist, Dr. Georg Simon Ohm (1787-1854), introduced the law which bears his name. His many years of experimenting with electricity brought out the fact that the amount of current which flowed in a circuit was directly proportional to the applied voltage. In other words, when the voltage increases, the current increases; when the voltage decreases, the current decreases.

If the voltage is held constant, the current will change as the resistance changes, but in the opposite direction. The current will decrease as the resistance increases and will increase as the resistance decreases.

Ohm's law states: The current which flows in a circuit is directly proportional to the applied voltage and inversely proportional to the resistance.

There are three ways of stating this fundamental law.

(1) The current in amperes is equal to the pressure in volts divided by the resistance in ohms:

$$I = \frac{E}{R} \quad (1)$$

(2) The resistance in ohms is equal to the pressure in volts divided by the current in amperes:

$$R = \frac{E}{I} \quad (2)$$

(3) The pressure in volts is equal to the current in amperes multiplied by the resistance in ohms:

$$E = I \times R \quad (3)$$

Equation (1) states that the current is equal to the voltage divided by the resistance. Fig. 2-11 shows a circuit with a lamp that has a resistance of 240 ohms and operates on 120 volts. Let us find out how much current is flowing through the lamp.

$$\text{Using Ohm's law: } I = \frac{E}{R} = \frac{120}{240} = 0.5 \text{ ampere}$$

Equation (2) states that the resistance is equal to the voltage divided by the current. In Fig. 2-12 the resistor allows 4 amperes of current to flow through it when connected to a 12-volt battery. Let us find its resistance.

$$\text{Using Ohm's law: } R = \frac{E}{I} = \frac{12}{4} = 3 \text{ ohms}$$

Equation (3) states that the voltage is equal to the current multiplied by the resistance. Fig. 2-13 shows a door bell that has a resistance of 36 ohms

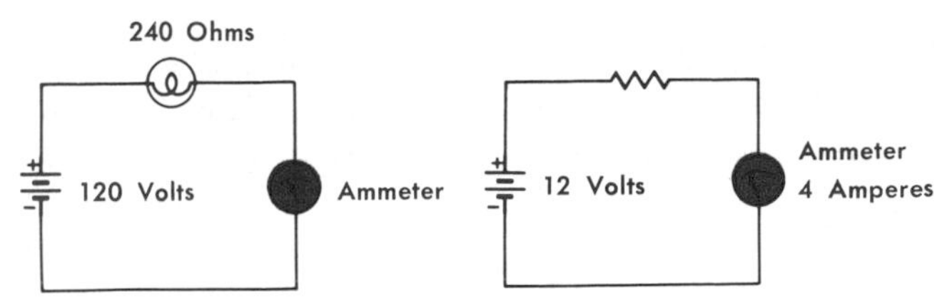

Fig. 2-11. Ammeter connected to measure current through a lamp.

Fig. 2-12. Circuit illustrating use of current and voltage to find resistance.

and requires an average current of 0.5 ampere. Let us find the voltage required to operate it.

Using Ohm's law:

$$E = I \times R = 0.5 \times 36 = 18 \text{ volts}$$

From these three Ohm's law expressions, any one of the quantities can be obtained if the other two are known.

From the time Dr. Ohm published his findings until the present, the Ohm's law formulas have been used by electrical, electronics, and other scientific men in the world as their fundamental rule. You may wonder: "why use Ohm's law when a lamp or a motor or a resistor is to be measured? Why not use a meter that would give us the values we need, whether they be voltage, current, or resistance?" Meters cannot always provide the information or values required. For example, the important resistance value of a lamp is its resistance when lit. (Electrical circuits have *two* resistances—a "hot" resistance and a "cold" resistance.) It is not advisable to connect an ohmmeter in a circuit when current is flowing, because the meter may be burned out. However, we can connect an ammeter to find the amperage, and then connect a voltmeter to find the voltage. When the amperage and voltage are known, Ohm's law can be used to determine the resistance. Quite often it is necessary to know in advance the "hot" resistance of some resistors, appliances, and other electrical equipment.

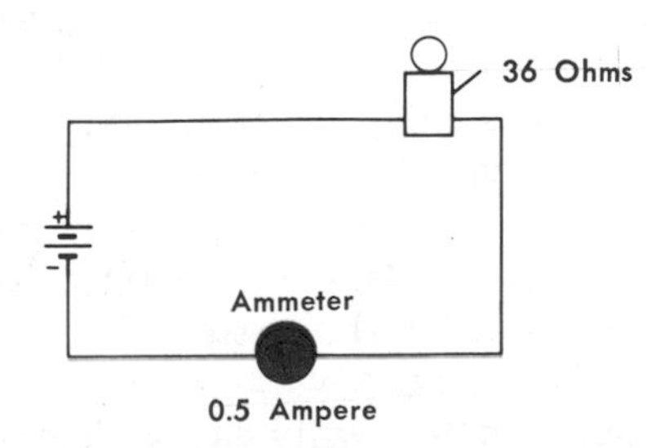

Fig. 2-13. Circuit illustrating use of resistance and current to find voltage.

In designing or engineering components, circuits, or equipment, it is necessary to solve for the proper values of voltage, current, or resistance prior to production.

There are many more applications where Ohm's law can be used. As stated at the beginning of this section, it is a fundamental law of electricity and electronics.

Summary

From the three expressions of Ohm's law ($I = \frac{E}{R}$, $R = \frac{E}{I}$, and $E = I \times R$) any one of the quantities can be obtained if the other two are known.

Ohm's law is applied to most electrical problems in solving for resistance, voltage, and current.

Questions and Problems

18. When the pressure in volts and the resistance in ohms are known, what Ohm's law formula should be used?

19. What voltage will produce a current of 6 amperes through a resistance of 18 ohms?

20. A relay having a resistance of 3200 ohms requires .02 ampere to operate. What voltage is required to energize it?

21. What current is produced by 30 volts acting across 0.735 ohm?

22. What is the resistance of an electric oven if it draws 35 amperes from a 110-volt line?

23. An electric soldering iron draws 1.05 amperes when used on a 110-volt circuit. What is the resistance of the soldering iron?

24. What Ohm's law formula should be used when the pressure in volts and the current in amperes are known?

25. Why is Ohm's law so fundamental in the study of electricity and electronics?

26. What current will flow in a 110-volt circuit if the total resistance of the circuit is 100 ohms?

RESISTANCE AND RESISTORS

The term resistance refers to the opposition that all materials offer to the passage of electrical current. All materials offer some resistance. Silver has the least resistance of the well-known metals, and copper has a little more resistance than silver. A material offering no resistance to current flow is known as a *superconductor*. Temperatures close to absolute zero (−459.6° F.) tend to make a few materials act as superconductors. For instance, suppose we have a ring of lead that is at a temperature of absolute zero. If we start a current flowing in the ring (from a battery, for example) and then disconnect the battery, current will continue to flow indefinitely.

Materials which readily carry electrical current at normal temperatures are called conductors. Conductors are the "highways" that connect the parts of electrical circuits. These can be in the form of wires, metal ribbons, plated areas of a circuit board, or even sections of a metal chassis.

WIRE RESISTANCE

A wire is a thin cylinder or filament of drawn metal. Wires are made in many different sizes to carry different current loads. The resistance of a wire depends upon its temperature, its size, and its material.

Temperature

When the temperature of a *metal* conductor rises, the resistance of the conductor increases proportionately. On the other hand, when the temperature of conductors like carbon, a gas, or an electrolyte rises, the resistance of such conductors decreases proportionately.

Length

The longer the conductor, the greater the resistance. Let us imagine we have a number of small cubes, one behind the other, like the ones in Fig. 2-14. Assume that each cube offers a resistance of one ohm. If two of these cubes are placed end to end, we will find that the resistance is twice as much, or two ohms. A line of four cubes will have a resistance of four ohms, and a line of 100 cubes will have a resistance of 100 ohms.

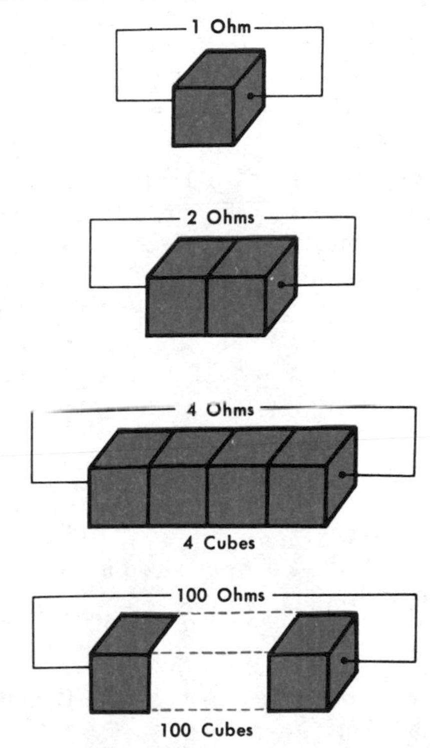

Fig. 2-14. The resistance of a conductor varies with its length.

Therefore, resistance varies directly with the length of the conductor. The reason is simple. The cubes act as resistors in a series, and there is a voltage drop across each "cube resistor."

Area

Now, let us place two of these cubes side by side, as in Fig. 2-15, only this time we will pass the current broadside through the blocks. In reality, we are passing the current through two one-ohm resistors in parallel. The current divides equally between the two equal resistors. The total resistance equals the value of one resistor (one ohm) divided by the number of paths (two), or one-half ohm. In other words, when the area is doubled, the resistance is cut in half. Likewise, if four of these cubes are placed together to make an area of four square inches, the total resistance will be one-fourth ohm. Therefore, the resistance of a conductor varies inversely with its cross-sectional area. The size of wire selected is a compromise between the least resistance to the current and the least cost of the wire.

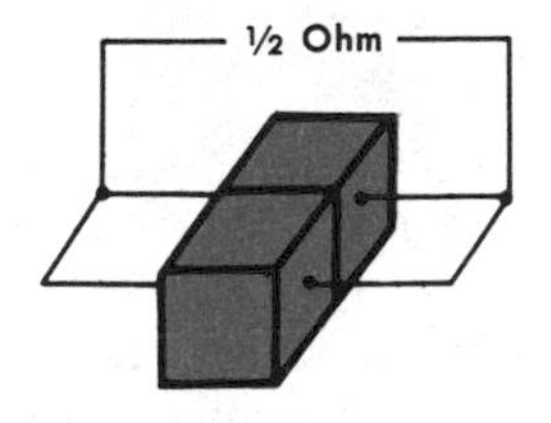

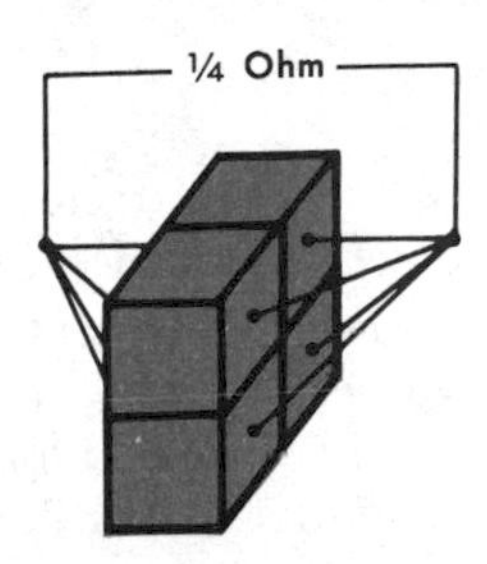

Fig. 2-15. The resistance of a conductor varies with its cross-sectional area.

Material

Specific resistance is the resistance of a piece of material one foot long and with a cross-sectional area of one circular *mil*. A piece this size is known as a *mil foot* or *circular mil foot*. A piece of copper one foot long and with a cross-sectional area of one circular mil has a resistance of 10.4 ohms. Therefore, copper has a specific resistance of 10.4. The specific resistances for several common metals are listed as follows:

Silver	9.8
Iron	63.4
German Silver	128.3
Copper	10.4
Aluminum	17.2

Obviously, resistance varies directly with the kind of material. A formula for finding the resistance of a conductor should include temperature, length, area and material. However, ordinary temperature changes have such a small effect that they can be ignored. The following formula is used:

$$R = \frac{KL}{A}$$

where,

R is the resistance of the conductor in ohms,
K is the specific resistance of the material,
L is the length of the conductor,
A is the cross-sectional area of the conductor.

RESISTORS

The resistances that are necessarily used in electrical and electronic circuits are units called *resistors.* These are graded according to the exact value of their resistance. Resistors are of many types, including insulated, noninsulated, wirewound, carbon, deposited-carbon, fusible, temperature-sensitive, and variable.

Wirewound

Wirewound resistors are made of wire that is wound onto a core of fireproof material, such as a ceramic. These resistors usually have rather low resistances and carry high currents. In other words, wirewound resistors normally have high *wattage* ratings.

Carbon

Carbon resistors are made of finely ground carbon, mixed with a binder. The compressed mixture offers a high resistance to a rather low current flow. Such resistors usually have wattage ratings of 2 watts or less.

Deposited-Carbon

Deposited-carbon resistors consist of carbon vapor deposits on a glass or a ceramic form. Spiral paths are scraped into the carbon until the desired resistance is obtained. Such resistors are usually high-precision types.

Fusible

Fusible resistors are a special type of wirewound resistors made to burn out and open the circuit if the current flow becomes greater than the resistor can carry.

Temperature-Sensitive

Temperature-sensitive resistors change their resistance values as their temperature changes. Since current passing through the temperature-sensitive resistor will generate heat, such a resistor might properly be called a *current-sensitive* resistor. In another type of sensitive resistor, the resistance changes as the applied voltage changes.

Variable

Variable resistors are carbon or wirewound resistors with a movable metal arm that will contact the resistor at any point to which it is set. When the arm is moved, the resistance between it and either end of the resistor changes. Such resistors are used in applications where an easily varied resistance is required. The variable resistors in electronic circuits are usually called controls. The volume control of a radio is a good example of a variable resistor.

Resistors are used in electrical circuits to limit the current flow or to create voltage drops. Suppose we have a circuit like the one in Fig. 2-16A, in which a 5-ohm resistor is connected to a battery. We would like to have a current of only 1 ampere flowing through the resistor, but a 20-volt battery is the only power source available. By using Ohm's law, we find that 4 amperes will flow in this circuit. Therefore, we must add another resistor in series with the 5-ohm resistor to limit the current flow. Again using Ohm's law, we find that a total resistance of 20 ohms is needed to limit the current flow to 1 ampere. Since we already have a 5-ohm resistor in the circuit, a 15-ohm resistor must be added, as in Fig. 2-16B.

Consider the circuit in Fig. 2-17. Suppose a 10-volt difference of potential (or voltage drop) is desired across R1. What value of resistance must be added to the circuit? The final current in the circuit must be:

$$I = \frac{E_{R1}}{R1} = \frac{10}{10} = 1 \text{ ampere}$$

where,

- I is the current in amperes,
- E_{R1} is the 10-volt drop desired across the known resistor R1,
- R1 is the resistance in ohms of the known resistor.

The voltage to be dropped across the added resistor is 90 volts, therefore:

$$R_2 = \frac{E_{R2}}{I} = \frac{90}{1} = 90 \text{ ohms}$$

where,

- R2 is the resistance in ohms of the unknown resistor,
- E_{R2} is the voltage drop across the unknown resistor R2 (the voltage of the battery minus the desired 10-volt drop E_{R1}),
- I is the current in amperes, as calculated from the previous formula.

Variable resistors are used whenever the voltage must be divided easily and quickly. Consider the circuit in Fig. 2-18, in which a 1000-ohm variable resistor is connected across a battery. Obviously, there is no resistance between point A and the positive terminal of the battery; thus, they will be at the same voltage. If the movable contact, represented by the arrowhead, is moved to point A, there will still be no resistance between point A and the positive terminal of the battery, and they will still be at the same voltage. Therefore, voltage E will equal the battery voltage, or 10 volts.

What will happen if the movable contact is moved to the center of the 1000-ohm resistor, toward point B? Now there is a resistance of 500 ohms above the contact and 500 ohms below it. Since the same current will flow through both resistances and since both resistances are equal, the voltage drops across them will be equal. Two equal parts of 10 volts are 5 volts apiece. The output voltage will therefore be 5 volts. The reason is that the voltage across two resistors in series will divide in the same ratio as the resistances.

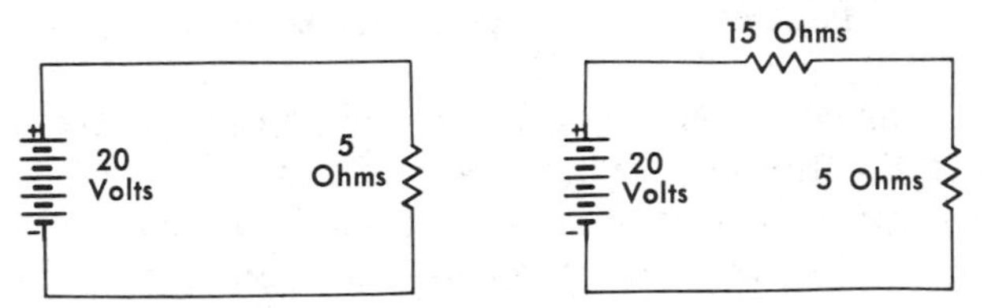

Fig. 2-16. Circuit illustrating use of resistance to limit current flow.

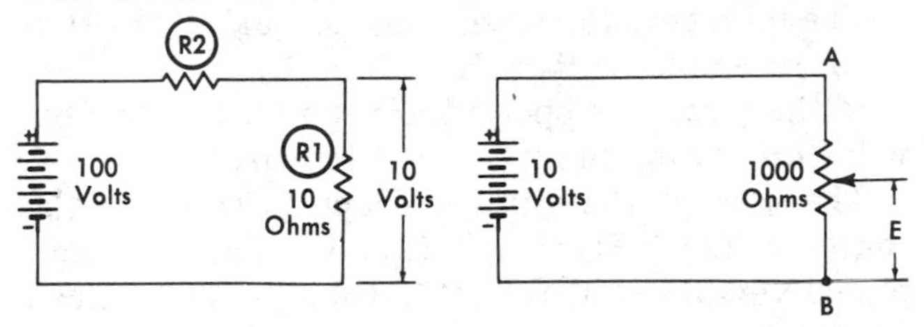

Fig. 2-17. Circuit illustrating use of resistance to create a voltage drop.

Fig. 2-18. Circuit illustrating use of variable resistance.

Summary

Resistance is the opposition to the flow of electrical current.

A conductor is any material that will easily carry electrical current at normal temperatures.

Resistance is directly proportional to the length and specific resistance of the conductor. Resistance is inversely proportional to the cross-sectional area of the conductor.

A circular mil foot of wire is a piece of wire one foot long and with a cross-sectional area of one circular mil.

The resistances in electrical and electronic circuits are called *resistors*.

Resistors are used in electrical and electronic circuits to limit the flow of current or to create voltage drops.

Questions and Problems

27. Name four types of resistors, according to their construction characteristics.

28. What is the specific resistance for copper? For silver? For aluminum?

29. Does the resistance of a conductor vary directly or inversely with the temperature?

30. When the diameter of a conductor is doubled, what happens to its resistance?

31. What resistance must be added to a circuit consisting of a 13-ohm resistor and a 100-volt battery, if the current is to be limited to one ampere?

32. Define the term *conductor*.

KIRCHHOFF'S VOLTAGE LAW

Series Circuit

A simple electrical circuit is shown in Fig. 2-19. It is called a *series circuit* because each component is connected directly to the next component in a continuous line or series—there are no branch circuits.

Note that the current can follow only one path through the entire circuit of Fig. 2-19. The electrons move from the negative terminal of the battery, go first through R_1, then through R_2 and R_3 in succession, and finally complete the circuit by arriving at the positive terminal of the battery.

The current (I) in amperes will be the same in all parts of the circuit. Let I_T equal total battery current, I_1 equal current through R_1, I_2 equal current through R_2, and I_3 equal current through R_3. Then $I_T = I_1 = I_2 = I_3$.

The total resistance in ohms is equal to the sum of all resistances: $R_T = R_1 + R_2 + R_3$.

If the circuit is open at any point, the current will stop flowing in all parts of the circuit.

The sum of the voltage drops is equal to the source voltage: $E_B = E_1 + E_2 + E_3$. This is a simplified version of Kirchhoff's voltage law and can be stated in the following manner: *The sum of the voltage drops around a dc series circuit equals the source or applied voltage.*

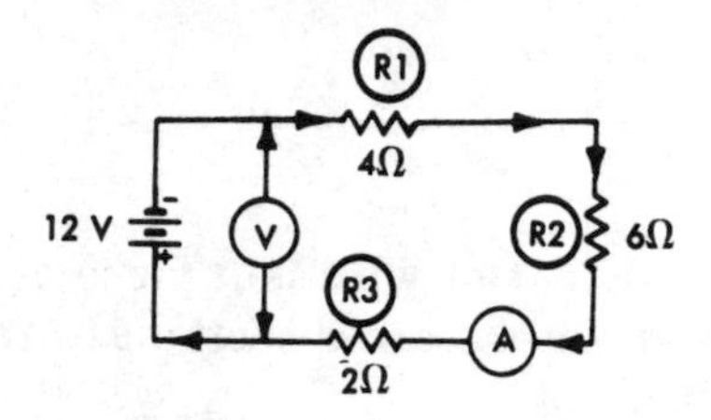

Fig. 2-19. A simple series circuit.

Let us see how the principles mentioned in the preceding paragraphs can be applied to Fig. 2-19 to confirm Kirchhoff's voltage law.

First find the total resistance: $R_T = R_1 + R_2 + R_3 = 4 + 6 + 2 = 12$ ohms. Using Ohm's law, find the current:

$$I_T = \frac{E_T}{R_T} = \frac{12}{12} = 1 \text{ ampere}$$

Next, find the voltage drops across each resistor, again using Ohm's law:

$$E_1 = I \times R_1 = 1 \times 4 = 4 \text{ volts}$$
$$E_2 = I \times R_2 = 1 \times 6 = 6 \text{ volts}$$
$$E_3 = I \times R_3 = 1 \times 2 = 2 \text{ volts}$$
$$E_1 + E_2 + E_3 = 12 \text{ volts}$$

(Sum of voltage drops) = (Source voltage)

Voltage Drop and Potential Difference

Fig. 2-20 illustrates an electrical circuit in which there are drops of potential. We will call terminal D our reference point. If we connect one lead of a voltmeter to terminal D, we can move the other lead to any point in the circuit to show the potential difference between that point and D. If both leads are at D, the meter will read zero volts. If one lead is moved to A, the meter will show the source voltage (6 volts). When the lead is moved to B, the reading will be 5 volts, indicating that 1

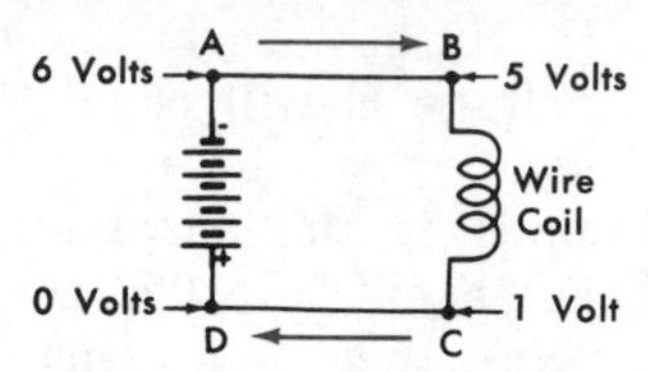

Fig. 2-20. A circuit having potential drops.

volt has been lost, or *dropped,* across the line from A to B. Similarly, when the lead is moved to C, a reading of 1 volt will be obtained, indicating a drop of 4 volts across the coil. Also, since the voltmeter is now connected across the line section C-D, the 1-volt reading is the voltage drop across C-D.

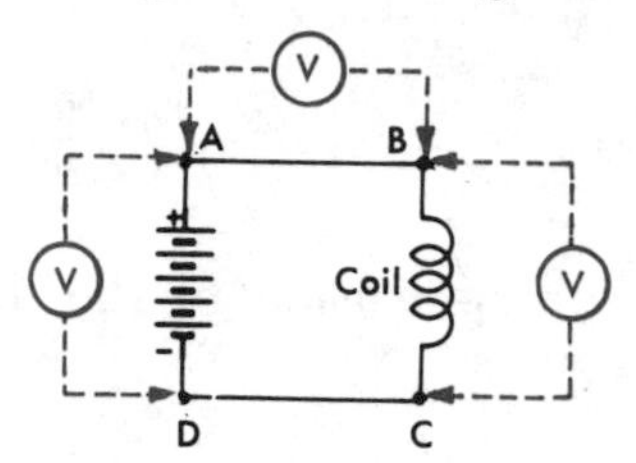

Fig. 2-21. Use of voltmeters to measure potential drops.

The voltage drop can be read across any section of a series circuit if a voltmeter is connected directly across that section, as shown in Fig. 2-21.

Water Analogy

We can refer to a water supply system as an aid in understanding the behavior of electrical systems. The two systems can be compared as follows: *water pressure* compares to *electrical pressure* or voltage, *water flow* compares to *electrical current,* and the *resistance* of the interior walls of the pipe compares to the resistance of the electrical conductor.

Fig. 2-22 shows a closed-circuit water system in which a centrifugal pump is used to maintain water pressure. Pressure gauges, indicating pounds per square inch, are placed at regular intervals along the pipeline.

Each gauge shows the difference in pressure between that point and the inlet to the pump. The pressure is greatest at the outlet of the pump and decreases as the distance from the outlet increases, because of frictional losses in the pipe. This loss compares to the voltage loss that occurs

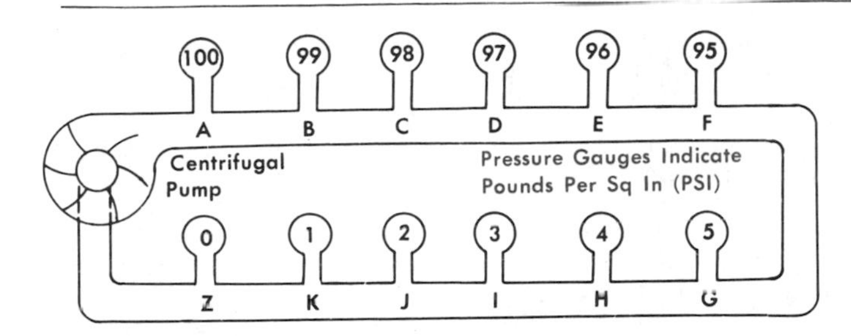

Fig. 2-22. A closed-circuit water system.

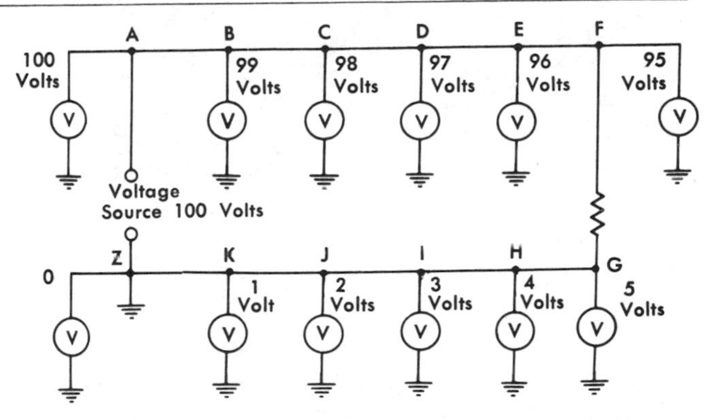

Fig. 2-23. An electrical version of the closed-circuit water system.

when electrical current flows through a resistance in an electrical circuit.

Fig. 2-23 is the electrical version of Fig. 2-22. The circuit from A to B, B to C, etc., has resistance, although it is not shown by the symbol for resistance. Between F and G there is a much greater amount of resistance, and it is shown as the resistor R.

Summary

There is only one current path in a series circuit.

The current in a series circuit is the same in all parts of the circuit.

The total resistance in ohms is the sum of all the series resistance.

If a series circuit is open at any point, the current will stop flowing in all parts of the circuit.

The sum of the voltage drops in a dc series circuit equals the source or applied voltage.

Questions and Problems

Refer to Fig. 2-22 for Questions 33 through 37.

33. What is the total pressure developed by the pump?

34. Z is the point of lowest pressure, or reference point. Which point is at the higher pressure—A or B; J or I; B or D; K or A?

35. What is the difference in pressure between A and B; C and F; F and G?

36. Is the difference in pressure between two points the same as the drop in pressure between those points?

37. Does the sum of all the pressure drops equal the applied pressure?

Refer to Fig. 2-23 for Questions 38 through 40. The voltage reference point is Z.

38. Which point has the higher electrical pressure—A or B; J or I; B or D; K or A?

39. What is the difference in electrical pressure between points B and D; D and H; F and G?

40. Does the sum of all voltage drops around the circuit equal the applied voltage?

Refer to Fig. 2-24 for Questions 41 through 46.

41. What is the rate of current flow in amperes?

42. What is the voltage drop per 10 feet of line? Note: Resistance of 10 feet is one ohm.

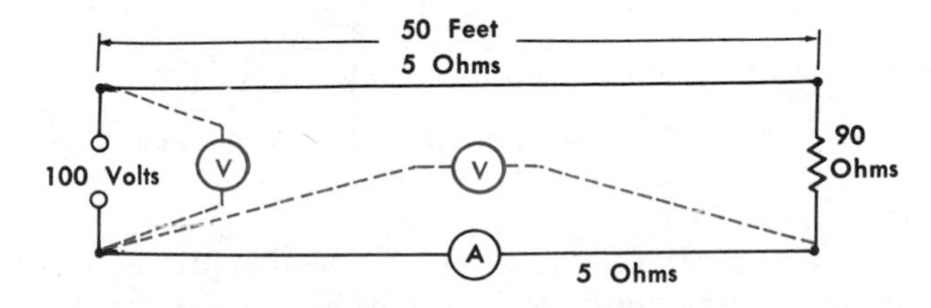

Fig. 2-24. Circuit for Questions 41 through 46.

43. What is the total line voltage drop?

44. What voltage is applied to the load?

45. Does the load voltage plus the line drop equal the applied voltage?

46. Sketch the circuit shown above and mark in the readings on the different meters shown.

KIRCHHOFF'S CURRENT LAW

When series and parallel circuits are combined, it becomes harder to determine the amount of current flowing in the various branches. Fortunately, there is a law that provides help in determining the value of branch currents. Kirchhoff's current law states that the current flowing *toward* a point in a circuit must equal the current flowing *away* from that point. Now, if the point divides into two or more paths, note what must

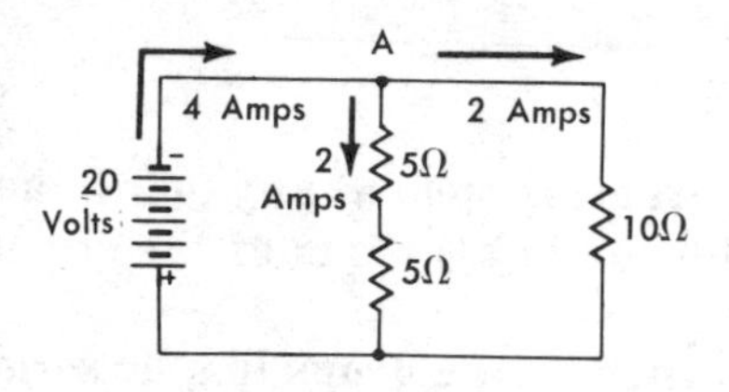

Fig. 2-25. Circuit illustrating use of Kirchhoff's current law.

occur: the current flowing toward the point must equal the sum of all the lesser currents flowing away from that point. If the two were not equal, some current would remain with no place to go—this is obviously impossible.

The circuit in Fig. 2-25 illustrates Kirchhoff's current law. Note that 4 amperes of current flow toward point A, with 2 amperes flowing away through the two 5-ohm resistors and 2 amperes through the 10-ohm resistor. Thus, there are 4 amperes flowing toward point A and 4 amperes flowing away from point A.

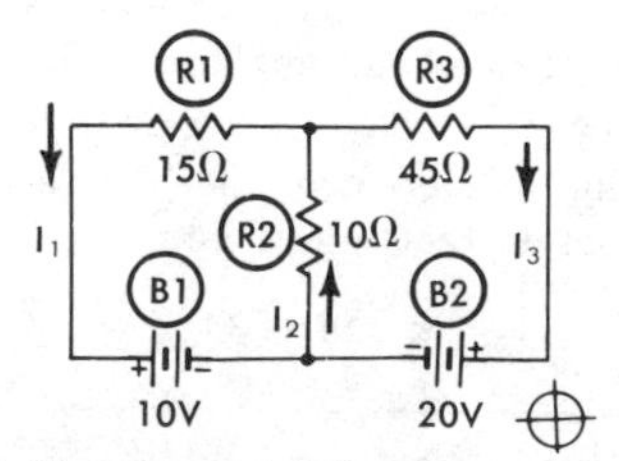

Fig. 2-26. Circuit containing both series and parallel branches.

When a circuit contains both series and parallel branches, the solution is much more involved. Fig. 2-26 illustrates such a circuit. Suppose we want to find the amount of current flowing through R2. We might think that battery B_2, with a higher voltage than battery B_1, would overpower B_1 and force current through the left-hand circuit. This, however, is not true.

The circuit is actually considered as two circuits linked together by the same resistor, R2. The current through R2 is made up of two currents, one through R1 from B_1, and the other through R3 from B_2.

The formula for Kirchhoff's current law for Fig. 2-26 is:

$$I_2 = I_1 + I_3 \qquad (1)$$

where,

I_1 is the current in amperes through R_1,
I_3 is the current in amperes through R_3,
I_2 is the current in amperes through R_2.

Kirchhoff's voltage law states that:

$$VB_1 = R_1I_1 + R_2I_2 = 15I_1 + 10I_2$$
$$\text{or } 10 = 15I_1 + 10I_2 \qquad (2)$$

where,

VB_1 is the voltage of battery B_1,
R_1 is the resistance of R_1 in ohms,
I_1 is the current in amperes through R_1,
R_2 is the resistance of R_2 in ohms,
I_2 is the current in amperes through R_2.

The right-hand circuit is considered in a similar manner:

$$VB_2 = R_2I_2 + R_3I_3 = 10I_2 + 45I_3$$
$$\text{or } 20 = 10I_2 + 45I_3 \qquad (3)$$

where,

VB_2 is the voltage of battery B_2,
R_2 is the resistance of R_2 in ohms,
I_2 is the current in amperes through R_2,
R_3 is the resistance of R_3 in ohms,
I_3 is the current in amperes through R_3.

There are now three equations and three unknowns. To solve for the unknowns, we subtract I_3 from both sides of equation (1):

$$I_1 = I_2 - I_3$$

Substitute this value of I_1 into equation (2):

$$10 = 15\,(I_2 - I_3) + 10I_2$$
$$10 = 15I_2 - 15I_3 + 10I_2$$
$$10 = 25I_2 - 15I_3 \qquad (4)$$

From equation (3):

$$20 = 10I_2 + 45I_3 \qquad (3)$$

Multiply both sides of equation (4) by three:

$$30 = 75I_2 - 45I_3 \qquad (5)$$

Add equations (3) and (5):

$$\begin{array}{l} 20 = 10I_2 + 45I_3 \\ \underline{30 = 75I_2 - 45I_3} \\ 50 = 85I_2 + 0 \end{array}$$

or

$$I_2 = \frac{50}{85}$$
$$I_2 = .5882 \text{ ampere}$$

Substitute this value of I_2 into equation (3):

$$20 = 10\,(.5882) + 45I_3$$
$$20 = 5.882 + 45I_3$$
$$45I_3 = 20 - 5.882$$
$$I_3 = \frac{14.188}{45}$$
$$I_3 = .3138 \text{ ampere}$$

From equation (1):

$$I_1 = I_2 - I_3$$
$$I_1 = .5882 - .3138$$
$$I_1 = .2744 \text{ ampere}$$

Therefore,

$$I_1 = .2744,$$
$$I_2 = .5882,$$
$$I_3 = .3138 \text{ ampere}$$

Study the circuit in Fig. 2-27. We see that a 2-ohm resistor, a 5-ohm resistor, and a 10-ohm resistor are connected in parallel and that a 20-volt battery is connected to the combination circuit. Our problem is to find the resistance of the combination, so we can use Ohm's law to find the amount of current in the circuit.

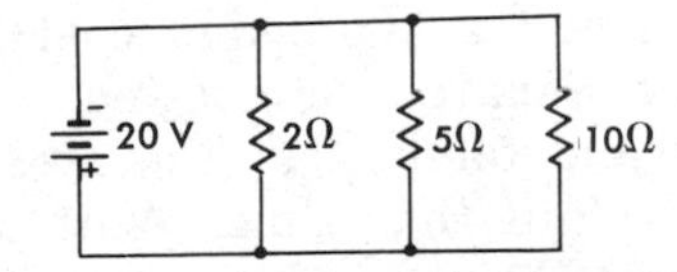

Fig. 2-27. A parallel circuit.

A glance at Fig. 2-27 reveals no easy way to find the resistance in the circuit. However, we can find the resistance indirectly by using Ohm's law. Since the 2-ohm resistor is connected directly across the 20-volt battery:

$$I = \frac{E}{R}$$
$$I = \frac{20}{2}$$
$$I = 10 \text{ amperes}$$

Since the 5-ohm resistor is also connected directly across the battery:

$$I = \frac{20}{5}$$
$$I = 4 \text{ amperes}$$

The 10-ohm resistor is also connected across the battery, and the current in it is:

$$I = \frac{20}{10}$$
$$I = 2 \text{ amperes}$$

The resistance of the combination can be computed with Ohm's law:

$$R = \frac{E}{I}$$
$$R = \frac{20}{10 + 4 + 2} = \frac{20}{16}$$
$$R = 1.25 \text{ ohms}$$

If the problem had involved only the three resistors and if no voltage value had been given, we could have used a voltage value of one volt in the formula to obtain an answer in terms of "current per volt." For example,

Current per volt through

$$2 \text{ ohms} = \frac{1}{2} = .5 \text{ ampere per volt,}$$
$$5 \text{ ohms} = \frac{1}{5} = .2 \text{ ampere per volt,}$$
$$10 \text{ ohms} = \frac{1}{10} = .1 \text{ ampere per volt,}$$
$$\text{combination} = .8 \text{ ampere per volt.}$$

Resistance of combination =

$$\frac{\text{Volts across combination}}{\text{Current through combination}} = \frac{1}{.8} = 1.25 \text{ ohms.}$$

Note that the current through each resistor is the reciprocal of its resistance. (The reciprocal of any number is a fraction expressed by 1 divided by that number; for example, the reciprocal of 2 is ½.) The reciprocal of a resistance is called its conductance. Conductances can be used when we want to find the resistance of a parallel circuit.

To find the conductance of the combination, add the conductances of the separate branches. Invert the conductance of the combination to find the resistance.

Expressed as an equation, this reads:

$$R\ total = \frac{1}{\frac{1}{R_1} + \frac{1}{R_2} + \frac{1}{R_3}}$$

This method of finding the total resistance in a parallel circuit is known as the "Reciprocal of the Sum of the Reciprocals," and is sometimes called the "Reciprocal Method."

Another method of finding paralleled resistances is used when there are only two resistors. The formula is:

$$R\ total = \frac{R_1 \times R_2}{R_1 + R_2}$$

This formula can also be used for more than two resistors if the formula is applied to two resistors at a time. The equivalent resistance of the two resistors then is used as one of the resistances in the next solution. This can be repeated as often as necessary.

There is one special case of paralleled resistors in which the equivalent resistance can be found by simple mental arithmetic. If a number of equal resistors are in parallel, the equivalent resistance is always equal to the value of one resistor divided by the number of resistors.

Summary

Kirchhoff's current law states that the current flowing toward any point in a circuit must equal the current flowing away from that point.

The conductance of a circuit is the reciprocal of the resistance of that circuit.

To find the equivalent resistance of paralleled resistances, add the conductances of the separate branches to get the total conductance. Then invert this conductance to get the equivalent resistance of the paralleled resistances.

Questions and Problems

47. What is the equivalent resistance of three 15-ohm resistors connected in parallel?

48. A 5-, 10-, and 15-ohm resistor are connected in parallel. What is the total resistance of this combination?

49. A 33-, 330-, and 4700-ohm resistor are connected in series. What is the total resistance of this combination?

50. A 3370-, 5280-, and 2000-ohm resistor are connected in parallel. What is the total resistance of this combination?

51. One 6-ohm resistor and two 12-ohm resistors are connected in parallel. What is the total resistance of this combination?

52. If a 3-ohm resistor is connected to a 4.5-volt battery, what current will flow? What current will flow if a 6-ohm resistor is added in parallel with the 3-ohm resistor?

WATT'S LAW

After James Watt (1763-1819) invented the reciprocating steam engine, he found it was difficult to sell the engine unless its power could be compared with that of a horse. Consequently, he performed a series of experiments to determine the power of the average English draft horse.

In his experiments, he had a number of horses, one at a time, pull a 165-pound coal bucket up through a mine shaft. He allowed one minute for each trial and found that the average horse covered 200 feet in that time. Watt called this new measurement *horsepower* and saw that it was a means by which mechanical power could be measured. He determined the average horsepower as follows:

$$P\,(\text{Work}) = F\,(\text{Force}) \times D\,(\text{Distance})$$
$$P = F \times D$$
$$= 165 \text{ pounds} \times 200 \text{ feet}$$
$$= 33{,}000 \text{ foot-pounds}$$

Since *power* is equal to work divided by time:

$$P = \frac{\text{work}}{\text{time}} = \frac{33{,}000 \text{ foot-pounds}}{60 \text{ seconds}}$$
$$= 550 \text{ foot-pounds per second}$$

Electrical power is measured in much the same manner as mechanical power. An electric motor can be substituted for a horse. If the motor did the same amount of work as the horse in the same amount of time, it would be rated as a one-horsepower motor. The electrical energy input of a one-horsepower electric motor is always equal to 746. This figure represents the pressure difference in volts multiplied by the current in amperes. Although the term *volt-amperes* is often referred to, the unit *watt* was coined in honor of the man who had first defined horsepower. The power (watts) in a dc circuit is equal to the current (I) multiplied by the voltage (E).

The Watt's law formula is:

$$P = I \times E$$

There are three common forms of the Watt's law formula:

(1) Power in watts is equal to the current multiplied by the voltage:

$$P = I \times E$$

(2) Current in amperes is equal to the power in watts divided by voltage:

$$I = \frac{P}{E}$$

(3) Voltage is equal to the power in watts divided by the current in amperes:

$$E = \frac{P}{I}$$

Let us see how much power is required to heat an electric iron (Fig. 2-28) operating on 120 volts and drawing 6 amperes.

Using Watt's law:

$$P = I \times E = 6 \times 120 = 720 \text{ watts}$$

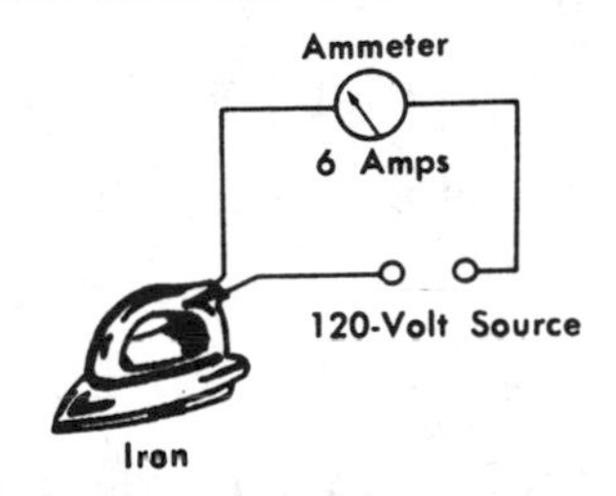

Fig. 2-28. Circuit illustrating use of current and voltage to find power.

Referring to Fig. 2-29 we can find the voltage required to light a 150-watt lamp that normally draws 1.5 amperes.

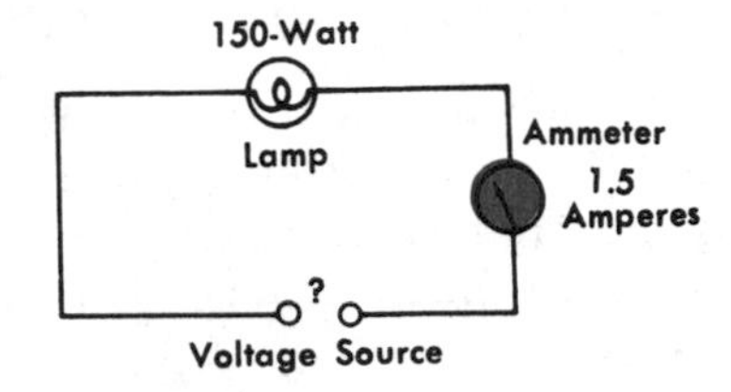

Fig. 2-29. Circuit illustrating use of power and current to find voltage.

Using the formula:

$$E = \frac{W}{I} = \frac{150}{1.5} = 100 \text{ volts}$$

How much current will be drawn if the same 150-watt lamp is operated on 120 volts, as shown in Fig. 2-30?

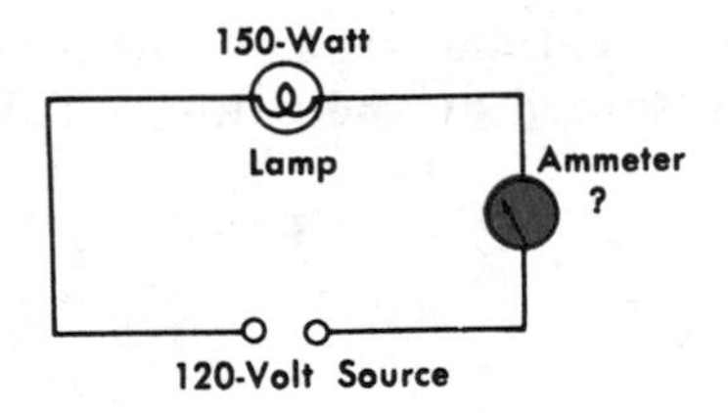

Fig. 2-30. Circuit illustrating use of power and voltage to find current.

The Watt's law formula is:

$$I = \frac{W}{E} = \frac{150}{120} = 1.25 \text{ amperes}$$

In the Watt's law formula—$P = I \times E$—power in watts cannot be found unless both the current and the voltage are known. Therefore, if we know the resistance and current but do not know the voltage, we must perform two steps. Here is a problem: "How much power is consumed in a resistance of 10 ohms when the current is 5 amperes?" To use the Watt's law formula:

$$P = I \times E,$$

we must find the voltage, using Ohm's law:

$$E = I \times R = 5 \times 10 = 50 \text{ volts}$$

Then find the power:

$$P = I \times E = 5 \times 50 = 250 \text{ watts}$$

Instead of having to use two formulas when the voltage is unknown, we can combine them by substituting $(I \times R)$ in the place of E:

$$P = I \times (I \times R) = I^2R =$$
$$5^2 \times 10 = 25 \times 10 = 250 \text{ watts}$$

This gives us the same answer, 250 watts.

There is a total of twelve formulas concerning Ohm's law and Watt's law. All twelve formulas can be conveniently arranged in the wheel shown in Fig. 2-31.

The watt is a small unit of electrical power. When larger amounts of electrical power are measured, the unit of measurement is the kilowatt, which is 1000 watts. A motor rated at one horsepower of mechanical power will use 746 watts, or 0.746 kilowatt, of electrical power.

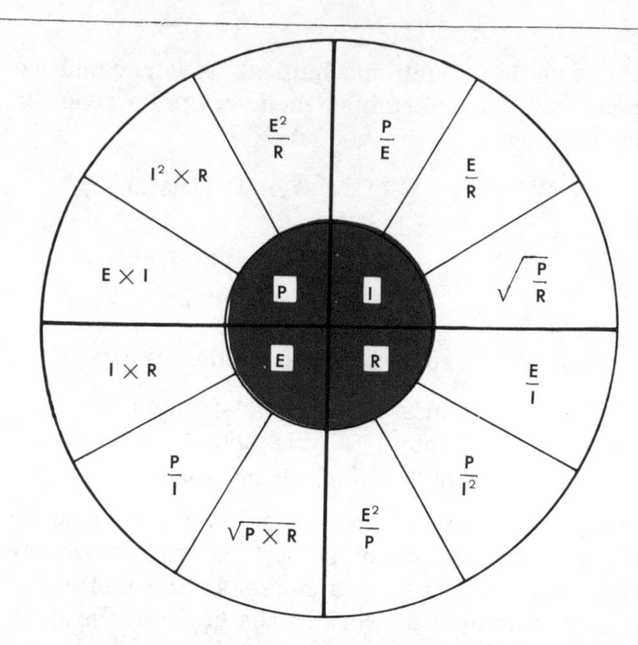

Fig. 2-31. Memory circle for formulas using Ohm's and Watt's laws.

The power companies sell electricity, but in our homes this electricity is converted into electrical energy. The electrical power of one kilowatt used for one hour consumes one *kilowatt hour* of energy.

Summary

The power in watts in a dc circuit is equal to the current multiplied by the voltage. This is Watt's law:

$$P = I \times E$$

A kilowatt is equal to 1000 watts.

A motor rated at one horsepower of mechanical power will use 746 watts, or 0.746 kilowatt, of electrical power.

Questions and Problems

53. What is the wattage of a tv receiver connected to 110 volts and drawing 1.25 amperes?

54. How much current will a 200-watt lamp draw when connected to a 120-volt power supply?

55. What is the resistance of a 200-watt lamp when connected to a 120-volt power supply?

56. An electric iron operating from a 120-volt power supply draws 4 amperes. How much power does it consume?

57. What is the current drawn by a one-horsepower motor connected to a 220-volt power supply?

58. What voltage is required to light a 200-watt lamp drawing 2 amperes of current?

59. If a soldering iron has 200 ohms resistance and draws 0.6 ampere of current, how much power does it use?

3

CELLS AND BATTERIES

FUNDAMENTALS OF CELLS AND BATTERIES

Dry cells and storage batteries are widely used for producing low-voltage direct current. Therefore, we will now learn how chemical action can produce an electrical pressure that will cause current to flow.

The first person to record the production of electrical current by chemical action was Luigi Galvani in 1785. In 1800 Alessandra Volta (for whom the volt was named) devised a voltaic pile. A voltaic pile is a "Dagwood sandwich" consisting of alternate layers of copper and zinc. Each layer is separated by acid-soaked paper. This discovery —that two unlike conductors touching a suitable *electrolyte* will generate a voltage—was the origin of all present-day cells and batteries.

We can make a simple voltaic cell by submerging a strip of copper and a strip of zinc into sulphuric acid, as shown in Fig. 3-1. The zinc—or negative—electrode is called the *cathode,* and the copper—or positive—electrode is called the *anode.* The chemical action between the anode, cathode and electrolyte develops a voltage between the anode and the cathode. If a conductor is connected across the cell terminals, current will flow. Chemical action, while current is flowing, eventually "eats away" the negative electrode.

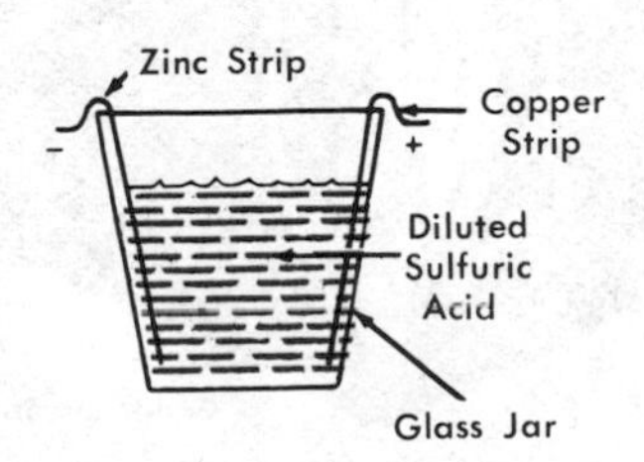

Fig. 3-1. A simple voltaic cell.

TYPES OF CELLS AND BATTERIES

There are many kinds of cells, all used for various purposes. Some examples are the Leclanche, Weston, lead-acid, nickel-iron-alkaline, mercury, Edison, and the Daniell gravity (Crowfoot). The dry cell (a modification of the Leclanche cell) and the mercury cell are the types encountered most often as portable radio power supplies.

Primary Cell

A primary cell is a voltaic cell in which the electrolyte and at least one electrode are used up as the cell is working. This process is not reversible; in other words, a dead cell cannot be brought back to life. A flashlight cell (everyone mistakenly calls it a battery) is a good example. When it no longer works, we throw it away and buy a new one.

Secondary Cell

Secondary cells also produce a current by electrochemical action. However, their parts are not used up, as with the primary cell. The secondary cell can be "charged," that is, restored to its original, useful condition. It is charged by passing a current through it from some external source, but in a reverse direction to that in which the cell delivers current when in use.

Batteries

Batteries are groups of cells of the same type, placed within a container or carton. The cells can be connected in series, parallel, and series-parallel. The term "battery" is often used, even when the battery consists of just one cell.

Dry Cell

A dry cell is actually a "moist" cell in which the electrolyte is a jelly or is absorbed in a porous medium. Because dry cells are light-weight and contain an electrolyte that will not spill, they can be used in places where a heavier cell or one with a spillable electrolyte would be inconvenient. The best-known dry cell is the common flashlight cell (Fig. 3-2).

Wet Cell

The electrolyte of a wet cell is liquid. So that the liquid will not spill, wet cells for portable use are placed in nonspillable containers. Wet cells are used in the automobile battery. Six-volt ignition systems have three lead-acid cells, and 12-volt systems have six lead-acid cells. The cells are connected in series and placed in a compact case. Automobile batteries are storage batteries and are rechargeable. The construction of a typical wet cell is shown in Fig. 3-3.

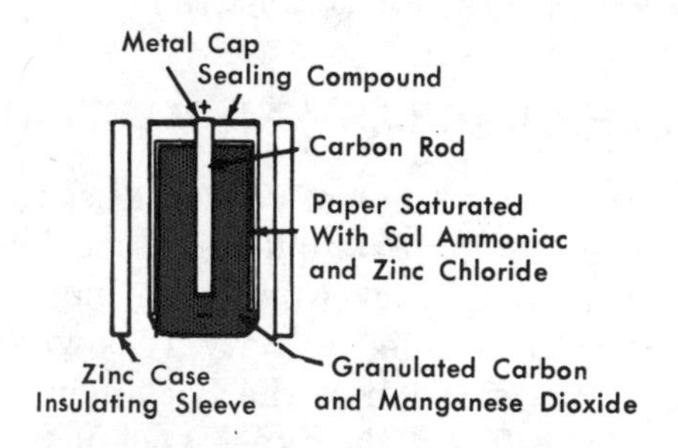

Fig. 3-2. Construction of a dry cell.

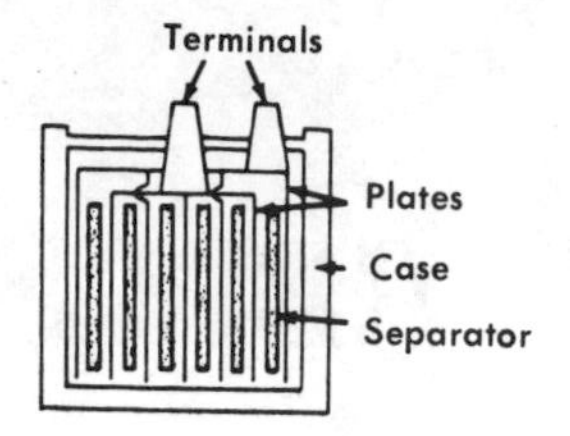

Fig. 3-3. Construction of a wet cell.

Lead-Acid Cell

The construction details of a lead-acid cell are shown in Fig. 3-4. This cell is well known because it is widely used in automobiles. The electrolyte is a solution of sulfuric acid. Each cell develops approximately 2.1 volts.

Nickel-Iron Cell

The nickel-iron cell (Edison) has an alkaline solution electrolyte, and develops approximately 1.2 volts.

Nickel-Cadmium Cell

The nickel-cadmium cell also develops 1.2 volts, and, like the Edison cell, has an alkaline electrolyte. This cell is useful where light weight, rugged-

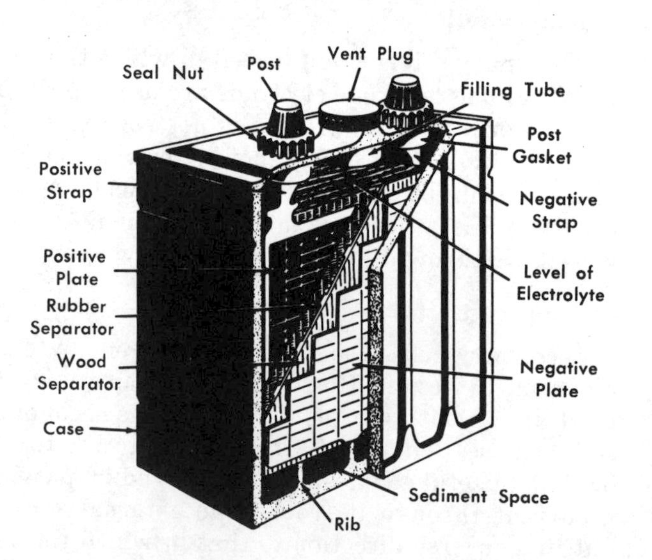

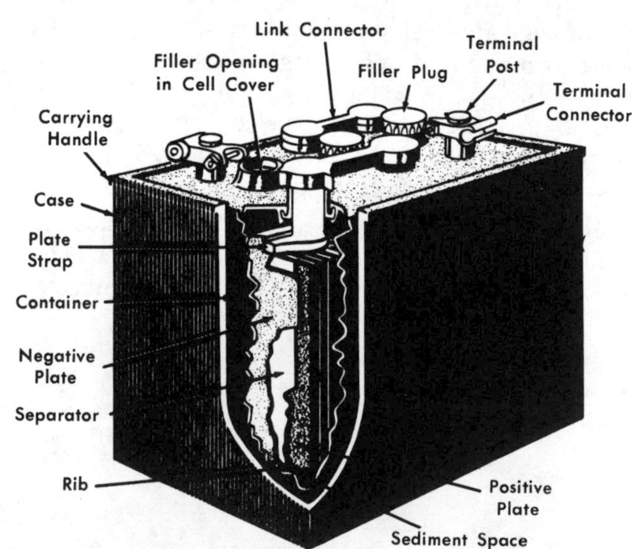

Fig. 3-4. Construction of a lead-acid cell.

ness, and the ability to withstand electrical abuse are required.

Mercury Cell

The mercury cell consists essentially of a depolarizing mercuric oxide cathode, an anode of pure amalgamated zinc and a concentrated aqueous electrolyte of potassium hydroxide saturated with zincate. The cathode and anode are press-shaped structures which are assembled into a sealed steel container.

The fundamental components of the mercury cell are a pressed mercuric oxide cathode in sleeve or pellet form, and pressed cylinders, or pellets of powdered zinc with steel enclosures. These provide precise mechanical assemblies having maximum dimensional stability and marked improvement in performance over the dry cell bateries.

Characteristics and features of the mercury cell include higher sustained voltages under load; relatively constant ampere-hour capacity; low and substantially constant internal impedance; excellent shelf life; good resistance to shock; and no recuperation required, which means that the current capacity is obtained in either intermittent or continuous usage.

Basic cells are available with two formulations designed for different field usage. In general, types having numbers not followed by the letter "E" are recommended for use in those applications when higher than normal temperatures may be encountered, and also where the cells are to be used as voltage reference sources. For example, type "E1" rather than "E1E" would be intended for higher temperature and voltage reference application.

Mercury batteries have been widely used as secondary standards of voltage because of the higher order of voltage maintenance and ability to withstand mechanical and electrical abuses. For use as reference sources in regulated power supplies, radiation detection meters, portable potentiometers, electronic computers, voltage recorders and similar equipment, the desirable features are voltage stability (momentary short circuits will cause no permanent damage) and high voltage drain without permanent damage.

CELL CHARACTERISTICS

The voltage produced by a primary or secondary cell depends on the materials used in the construction of the cell. For example, an ordinary dry cell will deliver approximately 1.5 volts. The lead-acid cell, used in automobile batteries, supplies 2.1 volts, and a mercury cell—often used in transistorized hearing aids—will develop 1.35 volts. The amount of current which a cell produces is determined by the condition of the electrolyte and the size of the anode and cathode. For instance, there are several sizes of common flashlight cells. Since all are made of the same materials, they will deliver about 1.5 volts; however, the larger cells will produce more current for a longer period of time.

The total current-producing capacity of a cell is measured in *ampere-hours.* A 10 ampere-hour cell theoretically can supply a current of one ampere for 10 hours or one-half ampere for 20 hours. However, there are limits to how fast the current can be drawn from a cell. For example, the cell just mentioned might be damaged if we tried to draw 20 amperes in a half hour. Generally speaking, the *total* current will be greater if it is not drawn continuously and if frequent "rest periods" are allowed.

BATTERY CONNECTIONS

Cells or batteries are connected in combinations to increase the voltage or the current, or both.

Series

Cells are connected in series by connecting the positive terminal of each cell to the negative terminal of the adjacent cell, as shown in Fig. 3-5. The total voltage of a series-connected battery is the *sum* of the separate cell voltages. However, the capacity in ampere-hours is equal to the capacity of just one cell in the battery. On the other hand, the power in watt-hours (volts × ampere-hours) is the sum of the watt-hours for each cell in the battery. Let's suppose we have three cells con-

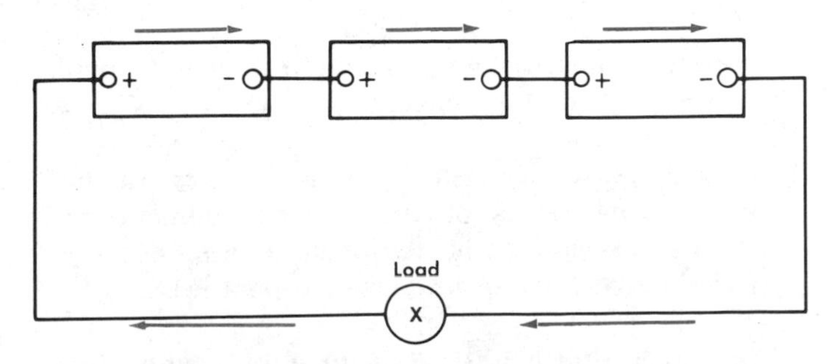

Fig. 3-5. Connection of cells in series.

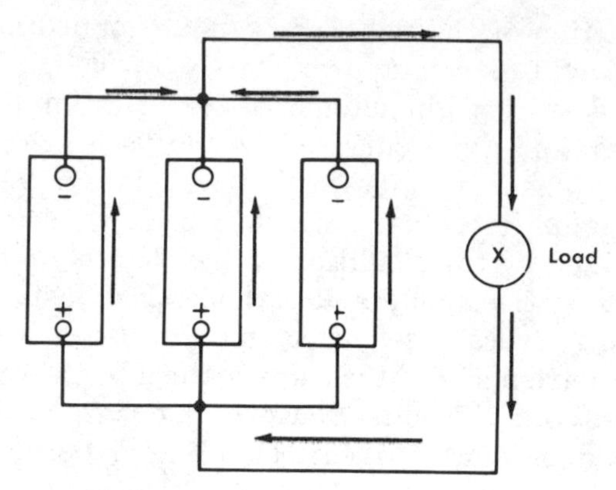

Fig. 3-6. Connection of cells in parallel.

nected in series. For easier figuring, we will say that the voltage of each cell is one volt and the capacity of each cell is one ampere-hour. Therefore, the total voltage of the battery is three volts, the total current capacity is one ampere-hour, and the total power is three watt-hours (3 volts × 1 ampere-hour = 3 watt-hours).

Parallel

Cells are connected in parallel by connecting all positive terminals together and all negative terminals together (Fig. 3-6). The total voltage of a parallel-connected battery is the same as the voltage of any individual cell in the battery.

The capacity in ampere-hours and the watt-hours are the totals of the ampere-hours or watt-hours of all the cells. All cells should have the same voltage; otherwise, wasteful circulating currents will occur within the battery.

Series-Parallel

The series-parallel battery connection is a combination of the series and the parallel connections just mentioned. It is used when both the voltage and the current must be increased. The series-parallel connection is shown in Fig. 3-7. First, the

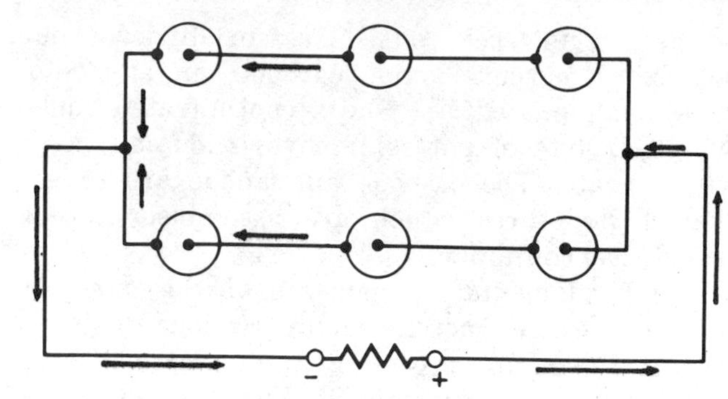

Fig. 3-7. Cells connected in series-parallel.

desired voltage is obtained by connecting the correct number of cells in series. Then one or more identical groups of cells are connected in parallel with the first group of cells until the desired current is obtained.

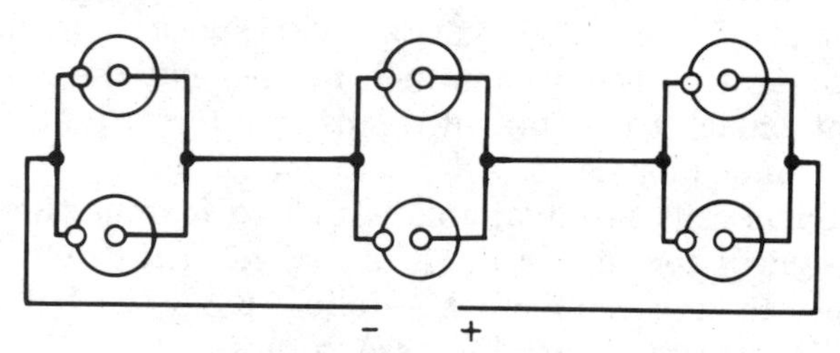

Fig. 3-8. Cells connected in parallel-series.

The same effect can also be obtained by first connecting a group of cells in parallel for the desired current, and then adding similar groups of cells in series for the voltage. A group of cells with this type of parallel-series connection is shown in Fig. 3-8.

Summary

In a chemical cell, chemical action is used to produce an electrical potential.

Cells are classed as primary or secondary. In the primary type, parts of the cell are consumed; thus, the cell cannot be restored. The secondary cell can be recharged and used repeatedly.

A *cell* is a single unit; two or more connected cells make up a *battery*.

The voltage developed by a cell depends on the metals and chemicals in the cell. The quantity of electricity that it can produce (ampere-hours) depends on the size of the electrodes and the condition of the electrolyte.

The common dry cell is a modification of the Leclanche cell. The dry cell develops 1.5 volts; the lead-acid cell, 2.1 volts; and the mercury cell, 1.35 volts.

Batteries or cells can be connected in series, in parallel, or in series-parallel combinations.

A series connection increases the voltage but not the ampere-hours; a parallel connection increases the ampere-hours, but not the voltage; a series-parallel connection increases both the voltage and the ampere-hours.

Questions and Problems

1. **If an automobile is equipped with 12-volt headlights, how many lead-acid cells must the car battery have?**

2. **Explain the difference between: (a) a primary cell and a secondary cell and (b) a cell and a battery.**

3. **When a storage battery is being charged with a dc generator, should the positive terminal of the generator be connected to the positive terminal of the battery or to the negative terminal? Why?**

4. **If six dry cells in *series* deliver a total current of six amperes and six other dry cells in *parallel* on another circuit also deliver six amperes, compare the rate at which zinc is used up in the first group to that of the second group.**

5. **A dry cell delivers 15 amperes at 1.5 volts. How many cells are needed to deliver 45 amperes at 3 volts? Draw the connection diagram for the group.**

6. **Which supplies the greater potential difference (emf)—the small "penlight" dry cell used in many portable transistor radios, or the larger dry cell used in an ordinary-sized flashlight?**

7. **Name two features of a dry cell which make it more convenient than a wet cell.**

INTERNAL RESISTANCE AND EFFICIENCY

An electrical circuit consists of a source of voltage and a device that converts the voltage into a usable effect. The voltage source is a battery or generator, and the usable effect may be heat, light, sound, or mechanical work.

There are two kinds of resistance in a circuit. The resistance in the voltage source (battery or generator) is called *internal resistance.* The resistance anywhere else in the circuit is called *external resistance.* Both the internal and external resistance are measured in *ohms.*

The electrical symbol for external resistance is the capital letter "R"; the symbol for internal resistance is the small letter "r."

Internal resistance has a considerable effect on the efficiency of the voltage source, as we will soon show.

INTERNAL RESISTANCE IN A STORAGE BATTERY

In a storage battery, the electrolyte offers almost all of the internal resistance, although the plates also have some resistance. This internal resistance reduces the voltage (emf) of the storage battery when current is taken from the battery.

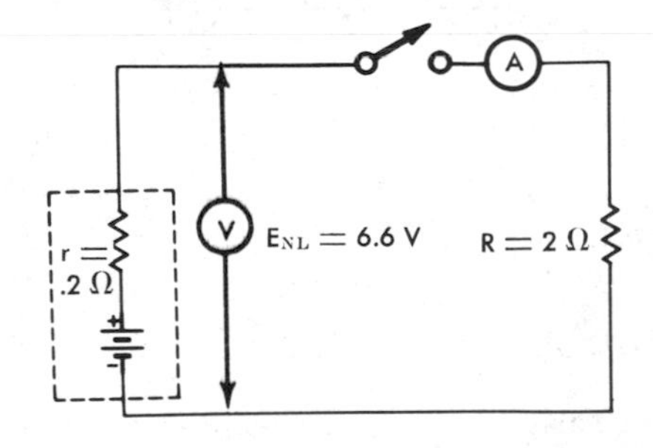

Fig. 3-9. Simple circuit showing both external resistance (R) and internal resistance (r). The voltmeter reads the no-load voltage (E_{NL}).

Fig. 3-9 shows a simple circuit with an external resistance (R) of two ohms and an internal resistance (r) of 0.2 ohm.

Effect of Temperature

When the heating coils of a toaster are cold, more current will flow through them than will flow after they become hot. The hotter the coils, the less current. This is true of any metallic conductor—its resistance *increases* as the temperature rises.

However, the resistance of nonmetallic conductors—carbon, gases, and electrolytes—is just the opposite. When they are hot, their resistance is low. When they are cold, their resistance is high.

This is one reason why an automobile is hard to start on a cold day. The low temperature of the electrolyte in the battery increases its internal resistance. As a result, the higher resistance cuts down the voltage, which—in turn—takes its toll on the current. (Ohm's law, remember!) Therefore, the current-starved spark plugs and starter must do the best they can with what little current they can get.

Variations in Source Voltage

Variations in the voltage of the emf (electromotive force) source are referred to as the *no-load voltage* (E_{NL}) and the internal voltage drop (E_r). Fig. 3-9 shows an ammeter connected in the circuit; but, since the switch is open, there is no reading. However, the voltmeter indicates that the open-circuit, or no-load, voltage of the battery is 6.6 volts.

In Fig. 3-10 the switch has been closed. The *ammeter* now reads "three amperes," and the voltmeter reads "6 volts" (instead of 6.6 volts). This is the voltage under load, or the full-load voltage (E_{FL}).

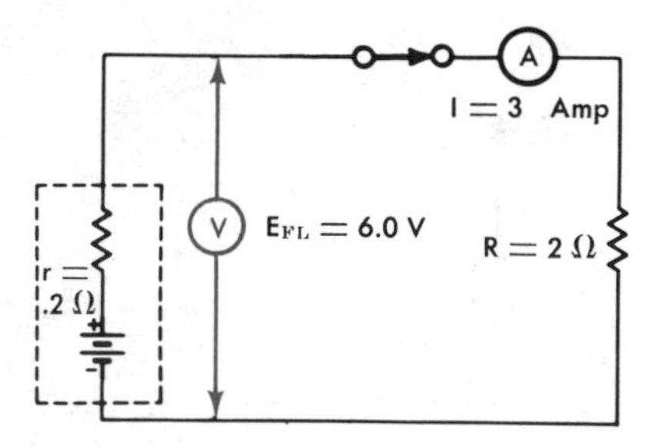

Fig. 3-10. Same circuit as Fig. 3-9, with switch closed, causing current of 3 amperes. Voltmeter now reads the full-load voltage (E_{FL}).

Determining the Internal Resistance of a Battery

Operate a storage battery at its rated load for approximately ½ hour. Then take a voltmeter reading. This reading is the full-load voltage (E_{FL}). Open the switch and quickly take another voltmeter reading before the effect of the polarization current disappears. This is the no-load voltage (E_{NL}).

The difference between these two readings (E_{NL}-E_{FL}) is the internal voltage drop. Substitute this voltage and the load current in the Ohm's law formula. Your answer will be the internal resistance of the battery.

The change in voltmeter reading from 6.6 to 6.0 volts shows that there has been a loss of 0.6 volts in the internal part of the circuit (the battery). This 0.6-volt loss is the internal voltage drop of the battery. Therefore, three voltages are involved in this simple circuit: the no-load voltage (6.6 volts), the full-load voltage (6.0 volts), and the internal voltage drop (0.6 volt). Any one of these values can be found if the other two are known.

The internal resistance of a cell or battery can easily be determined by Ohm's law. If the current through a resistance and the voltage drop across the resistance are known, the resistance is the voltage drop divided by the current. In the example in Fig. 3-10, the current is 3 amperes and the voltage drop is 0.6 volt. Dividing 0.6 by 3 gives a value of 0.2 ohm. This is the internal resistance of the battery.

VOLTAGE OF A STORAGE CELL

The voltage rating of a storage cell varies somewhat, depending on how much the cell is charged, and whether it is in the process of charging or discharging. For example, the voltage of a typical lead-acid storage cell may vary from 2.0 to 2.5 volts while charging and from 2.0 to 1.7 volts while discharging. The voltage of an Edison storage cell will vary between 1.5 and 1.8 volts while charging and between 1.0 and 1.5 while discharging, and will have an *average voltage* of 1.2 volts.

Rates of Discharge

If a cell is discharged too rapidly, its voltage will drop below average. Therefore, certain rates of discharge are recommended by the manufacturer. The minimum discharge period of a lead-acid storage battery is eight hours; however, the type used in vehicles is rated at only six hours. The Edison storage battery has a recommended five-hour discharge rate.

These recommendations are for continuous rates of discharge. If the batteries are used for short intervals only, with rest periods in between, their ampere-hour capacity will be increased. When used continuously at high discharge rates, however, the batteries will have a lower ampere-hour capacity. A high discharge rate will not damage a battery, provided its plates and conductors can endure the heat generated by the rapid discharge.

When measuring the efficiency of a battery, it is wise to use the recommended rate of discharge. A lower current will result in a higher efficiency value, and a higher current will have the opposite effect.

EFFICIENCY

The *efficiency* of an electrical appliance is its ability to perform with the least waste of energy. Mathematically, efficiency is the ratio of *output* to *input,* expressed as a fraction, a decimal, or a percentage. Efficiency as a percentage can be determined by the formula:

$$\text{Efficiency} = \frac{\text{output} \times 100}{\text{input}}$$

Both output and input must be expressed alike, usually in watts.

If the output and input were the same, the efficiency would be 100%. However, the efficiency is *never* 100%, because we never get as much work out of an appliance as the energy we put into it. Therefore, 100% efficiency is impossible, although many large substation power transformers do approach that point.

The following efficiencies are typical for electrical equipment working under load:

Transformers (power)	93% to 98%
Generators (medium and large)	80% to 96%
Motors (medium and large)	75% to 95%
Motors (small fan)	25% to 60%
Storage batteries (Edison)	approximately 60%
Storage batteries (lead-acid)	68% to 80%

Let's determine the efficiency of the storage battery represented in Figs. 3-9 and 3-10. Beginning with the formula:

$$\text{Efficiency in } \% = \frac{\text{output} \times 100}{\text{input}}$$

The output is represented by the number of watts in the load, and since:

$$P = I \times E_{FL}$$
$$= 3 \times 6$$
$$= 18 \text{ watts}$$

where,
I is the current in amperes,
E_{FL} is the full-load voltage.

Therefore, the output is 18 watts. The input equals the output plus its losses. We now know the output. To find the input, we must determine the losses and add them to the output.

We find the losses (in watts) by using Watt's law,

$$P = 3 \times 0.6 = 1.8 \text{ watts}$$

Adding the losses of 1.8 watts to the output of 18 watts gives us 19.8 watts for the input. Therefore,

$$\text{Efficiency in } \% = \frac{\text{output} \times 100}{\text{input}}$$
$$= \frac{18 \times 100}{19.8}$$
$$= 91\%$$

The efficiency value of our storage battery is rather high. This is due to its light load (three amperes). With a 10-ampere load, the internal voltage drop would have been greater, indicating a lower efficiency.

We can solve the problem by using two Watt's law formulas:

$$P = I^2 R \text{ (for output)}$$
$$P = I^2 r \text{ (for losses)}$$

We could obtain the same result by using resistance and voltage $\left(\frac{E^2}{R} \text{ and } \frac{E^2}{r}\right)$ instead of current and resistance (I^2r and I^2R).

Let's suppose that both the *current* and the *resistance* are unknown. Nevertheless, we can still find the efficiency. Because this is a series circuit, the same amount of current is traveling through the internal resistance and through the external resistance. Therefore:

$$\text{Efficiency in } \% = \frac{IE_r \times 100}{IE_r + IE_R}$$

where,
IE_r is the watts lost in the internal resistance,
IE_R is the watts used by the external resistance.

By factoring the I in the denominator, we get:

$$\text{Efficiency in } \% = \frac{IE_r \times 100}{I(E_r + E_R)}$$

By canceling the two "I's," our equation becomes:

$$\text{Efficiency in } \% = \frac{E_r \times 100}{E_r + E_R}$$

For the input in the denominator, we may substitute the no-load voltage, E_{NL} (because the input voltage has not yet encountered the load). For the output in the numerator, we may substitute the full-load voltage, E_{FL} (because the output voltage will have had to "buck" the load). Therefore, our new formula becomes:

$$\text{Efficiency in } \% = \frac{E_{FL} \times 100}{E_{NL}}$$

If the internal and external resistances are known, but not the current and voltage, the efficiency can still be determined as follows:

$$\text{Efficiency in } \% = \frac{I^2R \times 100}{I^2R + I^2r}$$
$$= \frac{I^2(R \times 100)}{I^2(R + r)}$$
$$= \frac{R \times 100}{R + r}$$

where,

I^2R is the power in watts used by the external resistance,
I^2r is the power in watts lost in the internal resistance.

Once again we simplify the equation by factoring the I^2R terms from the numerator and denominator, and canceling them.

For example, if the external resistance (R) is six ohms and the internal resistance (r) is also six ohms, the efficiency in percent equals:

$$\text{Efficiency in } \% = \frac{6 \times 100}{6 + 6}$$
$$= \frac{600}{12}$$
$$= 50\%$$

Even if the internal and external resistance were both seven ohms, the efficiency would still be 50%. In other words, no matter what their value, whenever the external and internal resistance are equal, the efficiency of the current source will always be 50%.

Parallel Sources

Connecting power sources in parallel increases the efficiency. This is quite understandable, if we consider first a single battery and then several batteries connected in parallel. Let's assume the single battery has an internal resistance (r) of 2 ohms and an external resistance of two ohms, also. The efficiency is therefore 50%. If an identical battery is connected in parallel with the first battery, resistance of the combination is one ohm (equal resistors in a parallel circuit). We substitute the resistance values as follows:

$$\text{Efficiency in } \% = \frac{R \times 100}{r + M}$$
$$= \frac{2 \times 100}{2 + 1}$$
$$= \frac{200}{3}$$
$$= 66\tfrac{2}{3} \text{ or } 67\%$$

where,

R is the external resistance in ohms at the output of each battery,
r is the internal resistance in ohms at the input of each battery,
M is the resistance in ohms of the two batteries connected in parallel.

Therefore, the addition of an identical battery, connected in parallel, has increased the efficiency of the power source from 50% to 67%, a gain of a little more than one-third.

Maximum Power From a Source

In calculating the efficiency of the storage battery in Fig. 3-10, it was found that the power output was 18 watts, with a 2-ohm load. What change in output could be expected if the load were increased, or decreased? According to the formula $P = I^2R$, power would increase if R were increased, but this would be true only if current I did not change. The current does change, however, because in the formula $I = E/R$ we have changed R without changing E. Therefore, with increased R, the current I will decrease. Thus, we have an increased R, tending to increase the power, and a decreased I, tending to reduce the power. Will the net change in power be an increase or a decrease? Let us work an example to find out. Suppose that we allow R to increase to 3.1 ohms (this value simplifies the calculations). With the internal resistance of the battery at .2 ohm, the total series resistance is now 3.3 ohms, and the new current is 6.6 volts divided by 3.3 ohms, or 2 amperes. Power is I^2R or $2^2 \times 3.1$, or 12.4 watts. So increasing R has reduced the power output from the original 18 watts.

What happens if, instead, we decrease R to 1 ohm, for example? Now the total resistance is 1.2 ohms, and the current will be 6.6 divided by 1.2, or 5.5 amperes. The power is $5.5^2 \times 1.2$, or 36.6 watts, more than double the original value.

You can see that all this naturally leads up to the question, is there a value for R that will give maximum output and what is its value? We could go on, trying new values for R and calculating each new value of power, but that might involve a lot of calculation. Fortunately, mathematicians have found the solution to this problem, using calculus to do so. The answer is, *maximum power output is obtained when the output load equals the internal resistance of the source.* This is an important point and one that is of great practical use in electrical and electronic devices. If we apply this principle and refigure the output using R = .2 ohm, we get I = 6.6 volts divided by .4 ohm, or 16.5 amperes. Power output is $16.5^2 \times 2$, or 54.45 watts. If you want to satisfy yourself that this is the maximum power obtainable from this circuit, you might calculate the power, using values for R just a little above and below .2 ohm, say, .22 ohm and .18 ohm.

Notice that maximum efficiency and maximum output are not obtained at the same value of R. When R = r, the formula, efficiency = R × 100/(R + r), gives a value of 50%.

Summary

A working electrical circuit consists of at least two parts: (1) a source of voltage, and (2) an external circuit connected to the source. The resistance of the source is the *internal* resistance; and the resistance of the rest of the circuit is the *external* resistance.

The internal resistance of cells and storage batteries becomes greater as the temperature falls.

In an electrical circuit, useful work is accomplished, but voltage and power are lost because of internal resistance.

Cells and batteries will deliver the maximum number of ampere hours if they are not operated above their rated discharge currents. Rest periods will also prolong their life.

The efficiency ratio of any power device is equal to the output (watts or other units) divided by the input. Multiplying this ratio by 100, gives the percentage.

A working device can never be 100% efficient, because the amount of work we get from it will always be less than the amount of energy we put into it.

Questions and Problems

8. Why is the no-load voltage of a cell greater than its full-load voltage?

9. Does the voltage loss due to the drop across an internal resistance perform any useful work? Explain.

10. Does a high discharge rate injure a cell if the discharge occurs within the rated limits of the cell? Explain.

11. On what does the internal resistance of a lead-acid battery depend?

12. When measuring the internal resistance of a cell, why can't we take the ammeter and voltmeter E_{NL} readings at the same time?

13. When does more current flow through the heating coils of an electric stove—when they are cool or when they are hot?

14. If power sources in parallel increase the efficiency, what would a series connection do to the efficiency? Explain.

15. Two dry cells of 1.5 volts each are connected in series. Each cell has an internal resistance of 0.4 ohm. What current will flow if the cells are connected to an external resistance of 2.2 ohms?

16. If the dry cells in Problems 15 are connected in parallel and are then connected to the 2.2-ohm external resistance, what current will flow?

17. The emf of a cell is 1.4 volts. In supplying current to a circuit, the voltage drop across the internal resistance of the cell is 0.3 volt. What is the terminal voltage?

4

MAGNETISM

NATURE OF MAGNETISM

Most electrical and electronic equipment depends directly or indirectly upon magnetism. It has been said that without magnetism, the entire electrical world would collapse. There are few electrical devices in use today that do not make use of magnetism.

NATURAL MAGNETS

Many ancient peoples have been credited with the discovery of magnetism. The Chinese claim the invention of the magnetic compass around 2637 B. C. The magnets in their primitive compasses were called *lodestones* or *leading stones.* It is now known that lodestones were crude pieces of iron ore known as *magnetite.* (See Fig. 4-1.) Since magnetite has magnetic properties in its natural state, lodestones are classified as *natural* magnets. The only other natural magnet is the earth itself—all other magnets are man-made and are known as *artificial* magnets.

THE EARTH'S MAGNETIC FIELD

The reaction between the earth and a magnet (or compass) led to the conclusion that the earth itself is a huge magnet—but a very weak one, considering its size. Like a bar magnet, the earth has two magnetic poles, one at each end (if a ball can be thought of as having "ends"). In fact, we might say the earth is a ball-shaped bar magnet. The magnetic poles of the earth are not at the geographical poles. Hence, a compass points to true north at certain longitudes only. These longitudes are called *agonic* lines. At other points on the earth's surface, the compass needle does not point to true geographic north. This variation from true geographic north is called *declination* (as distinguished from *deviation* due to nearby masses of metal). At the *magnetic* equator (known as the *aclinic line*), a compass needle mounted on a horizontal axis will not "dip" toward the earth. North of the magnetic equator, however, the needle does dip or incline. Directly over the north magnetic pole, the inclination will be 90 degrees (straight down). The polarity of the north magnetic pole is assumed to be south, since it attracts the north-seeking pole of a compass. (See Fig. 4-2.)

Fig. 4-1. Natural magnet known as magnetite.

A magnet is identified by its ability to attract small pieces of iron or steel (nails, tacks, needles, iron filings, etc.) and other magnetic materials. The term "magnetic material" generally refers to iron and various alloys of iron, such as *steel, Alnico,* and *Permalloy.* However, three other metals—*nickel, cobalt,* and *gadolinium*—are also magnetic materials. In all other materials, magnetic effects are so weak that the materials are usually regarded as nonmagnetic.

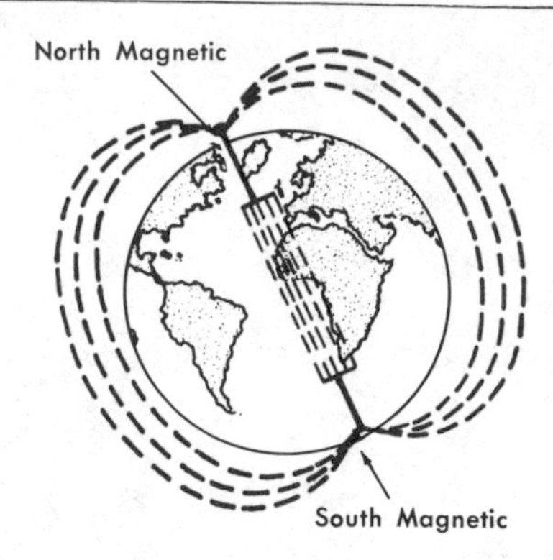

Fig. 4-2. The earth is a huge magnet.

In all magnetic materials, the magnetization is caused by a *magnetic field.* If the magnetization is in the same direction as the field, the material is said to be *paramagnetic;* if it is in the opposite direction, the material is said to be *diamagnetic.* In *ferromagnetic* materials, the magnetization is in the same direction as the magnetic field, just as with paramagnetic materials, but the magnetic effects are many times stronger, as much as a million times with some ferromagnetic materials. Diamagnetic materials and paramagnetic materials are very weak in magnetic effect; therefore, practical magnets are made using ferromagnetic material.

An explanation of magnetism takes us down to the level of the atom. In the atom, the electron is orbiting around the nucleus and at the same time is spinning about its own axis. This motion is accompanied by a magnetic field. In most materials, although there may be billions of electrons, their magnetic fields are in a more or less haphazard relation to each other; therefore, they cancel each other out so that there is little or no net magnetic field.

With ferromagnetic materials it is different; a group of atoms may spontaneously orient themselves so that their magnetic fields are parallel and aiding each other. This is called a magnetic *domain,* and it may contain several million atoms. A material such as iron has six directions of easy magnetization and the domains tend to orient themselves in one or another of these directions, but there still may be very little net magnetization in the piece of iron. Now, if the iron is subjected to some external magnetic field, such as another magnet, or a current-carrying wire, many of the domains will align themselves with the applied field and the iron becomes magnetized. If the majority of the domains keep their alignment after the external field is removed, the iron is said to be permanently magnetized.

A permanent magnet can be weakened by mechanical shock or by heat. If enough heat is applied, the domains themselves are demagnetized and ferromagnetism disappears. The temperature at which this happens is called the Curie point for that material. For iron, the Curie point is about 770°C.

Magnetic Lines of Force

To the eye, a piece of magnetized iron looks like an unmagnetized piece; there is no visible difference. But they can be easily told by their action on other ferromagnetic material; one attracts it, the other does not. The attraction, in the one case, is due to an external *magnetic field.* The field is said to be made up of *lines of force.* Really, there are no lines; the field is continuous. As in many other scientific explanations, this is merely a convention adopted to make the explanation easier.

Although the field of a magnet is invisible, its effects can be made visible by dipping the magnet in a pile of iron filings. Fig. 4-3 shows how the filings are picked up by the magnet. The magnetic field is strongest in the areas where the clusters of iron filings are the heaviest.

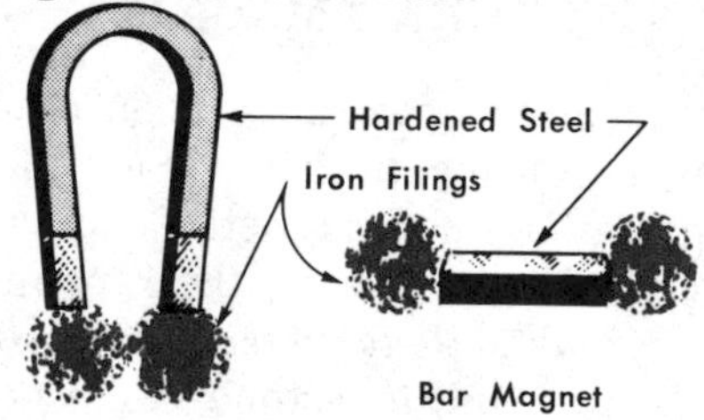

Fig. 4-3. The iron filings show where the magnetic field is strongest.

We can demonstrate even more clearly that a magnetic field actually exists in the space surrounding the magnet, by placing a piece of cardboard or glass between a horseshoe magnet and the loose iron filings, as shown in Fig. 4-4. The cardboard or glass prevents the magnet from touching the iron filings; yet the filings can still be picked up. Now we know that there is an attraction, in the space surrounding the magnet, between the filings and the magnet.

Although the lines of force in the magnetic field are invisible, we can demonstrate just what kind of pattern these lines assume around the magnet. This we can do by using the materials in the

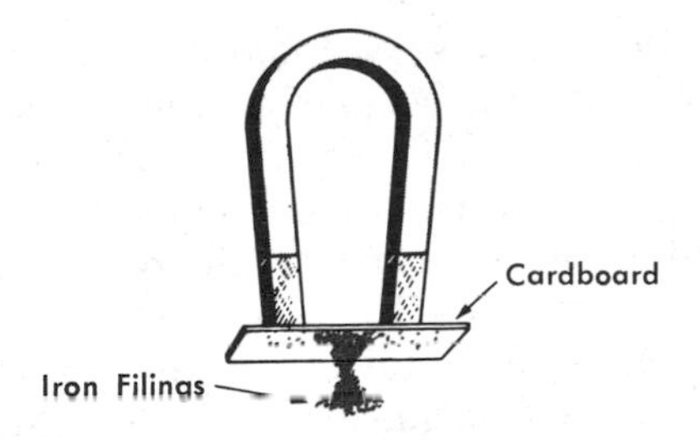

Fig. 4-4. The iron filings are attracted, even though the cardboard prevents the magnet from touching them.

previous illustration. Point the poles of the magnet upward. Lay the cardboard across the ends of the poles, as shown in Fig. 4-5. Then sprinkle the iron filings on the cardboard. (You do not have to use cardboard for this demonstration. A piece of glass, rubber, or Bakelite—or just plain writing paper—will work just as well.)

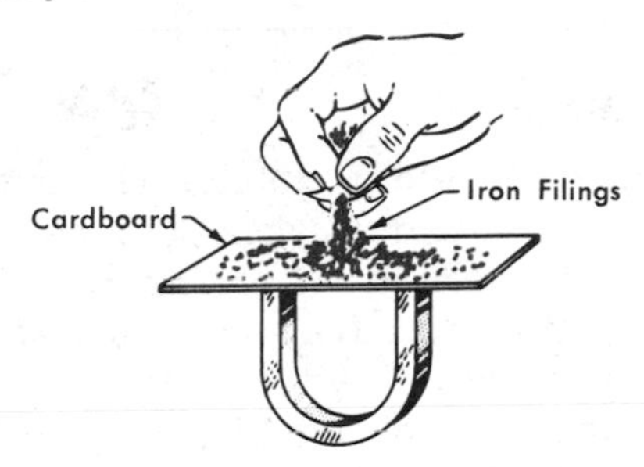

Fig. 4-5. How to demonstrate the pattern of the magnetic field.

Watch the cardboard as you sprinkle the iron filings on it. They will form a pattern on its surface. This pattern occurs because the iron filings align themselves with the lines of force in the magnetic field around the poles. Notice that the filings have arranged themselves into curved lines extending outward from the ends of the magnet. Fig. 4-6 shows the patterns of the filings.

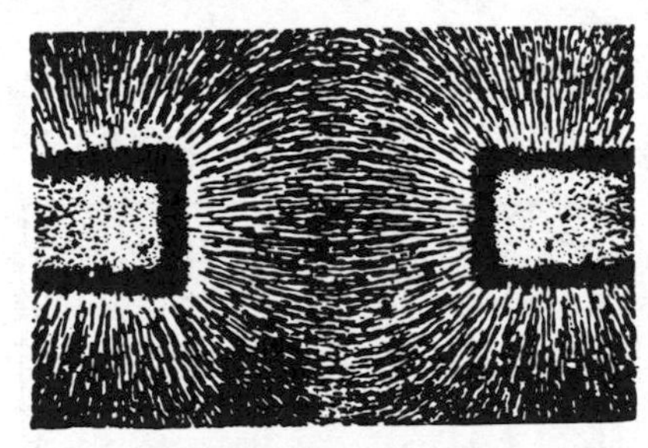

Fig. 4-6. The pattern of the lines of force.

The direction of this peculiar force in the space around the poles of the magnet takes definite lines. Therefore, it is referred to as *lines of force*. The electrical symbol which represents lines of force in a magnetic field is shown in Fig. 4-7.

Scientists tell us that the lines of force of a magnet emerge from the north pole of the magnet and travel in curved lines to the south pole of the magnet. This assumption cannot be proven in the laboratory, but it is accepted as being the most logical. Once a line of force leaves the north pole and moves through space to the south pole, it does not stray from its path unless something changes the magnetic arrangement.

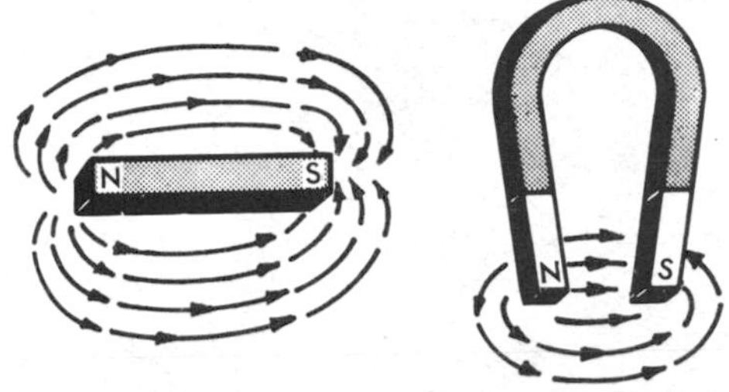

Fig. 4-7. Electrical symbol of lines of force in a magnetic field.

The lines of force travel in a circular or elliptical path outward from the north pole and through the space surrounding the metal of the magnet. They re-enter the magnet at the south pole and pass up the inside of the magnet to the north pole, forming a continuous, closed loop. We might think of the lines of force as rubber bands which can pass lengthwise through the metal of the magnet and then fan out in all directions at each pole to form a field near the magnet.

All of the lines of force which go into the space between the poles also pass through the metal of the magnet. Therefore, the magnetic lines of force are more concentrated within the metal of the magnet than in the space surrounding the magnet (because the lines of force are farther apart in space). The lines of force in and around a magnet are shown in Fig. 4-8. The illustration shows the concentration and distribution of the magnetic forces about the metal of the magnet.

Magnetic lines of force can pass through iron and other *ferrous* metals more easily than through air. The ease with which a material accepts magnetic lines of force is called its *permeability*. The permeability of a perfect vacuum is one, or unity. Air, paper, oil, and other nonmagnetic materials usually have a permeability of unity, also. Substances such as nickel and cobalt, with a permeability slightly greater than one, are *paramagnetic;* substances similar to iron and its alloys, with a

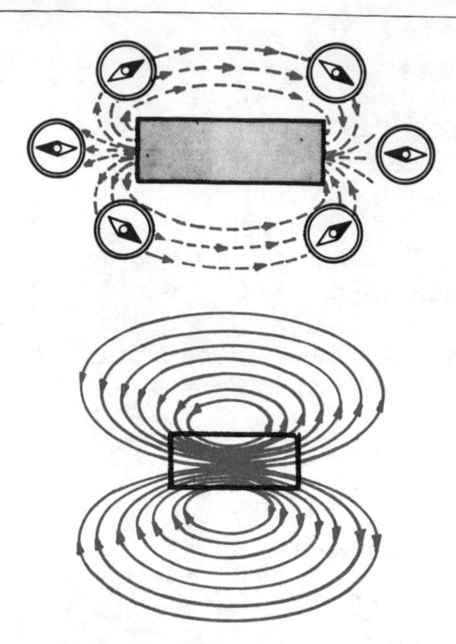

Fig. 4-8. The lines of force in and around a magnet.

much higher permeability, are *ferromagnetic*. Antimony and bismuth, with a permeability of less than one, are *diamagnetic*.

ARTIFICIAL MAGNETS

All man-made magnets are *artificial* magnets. Artificial magnets are made by one of two methods. The older method consists of stroking a *natural* magnet against a piece of iron or steel—stroking many times in only one direction. This is the *contact* method.

A piece of iron or steel placed near (or in contact with) a magnet will become magnetized. This is the *induction* method. The process can be speeded up by heating the iron or steel and allowing it to cool while in the magnetic field. Originally, the magnetic field of the earth was used to make magnets by induction. Today, an electric current supplies the magnetic field.

Permanent Magnets

A *permanent* magnet will retain its magnetism long after the external magnetizing force has been removed. Hard steel makes an excellent permanent magnet. Better permanent magnets are now made from various alloys, the best permanent magnets being made of *Alnico* (an alloy of iron, aluminum, nickel and cobalt). Other modern magnetic materials are Permendur, Remalloy, Perminvar, and Vicalloy.

The ability of a magnetic material to retain its magnetism over long periods of time is called its *retentivity*. (Retentivity should not be confused with *residual* magnetism, which is unwanted magnetism remaining in a magnetic material after the magnetizing force has been removed.) Soft iron is a magnetic material but is a very poor permanent magnet because of its low retentivity.

Permanent magnets are made in many shapes; the most common are the *bar* and the *horseshoe*.

Bar Magnets

These magnets have a pole at each end. They are used for experimental purposes and are of little commercial value. (See Fig. 4-9.)

Fig. 4-9. A bar magnet.

Horseshoe Magnets

These types are used in motors, telephones, generators, magnetos, and speakers. Horseshoe magnets are usually made in one piece and have a pole at each end. However, they may consist of two bar magnets joined by a *yoke*. Actually, even if a horseshoe magnet is in one piece, it can be considered as being two bar magnets joined by the curved section (yoke). (See Figs. 4-10 and 4-11.)

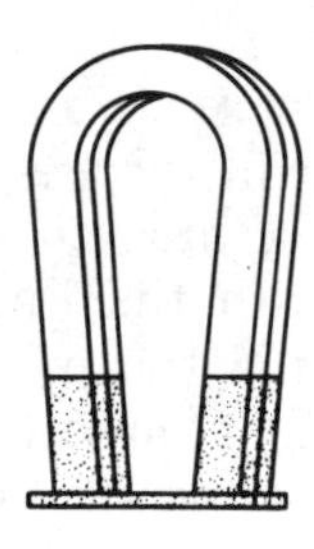

Fig. 4-10. A horseshoe magnet.

A *yoke* must not be confused with a *keeper*. A keeper is a piece of soft iron placed across the

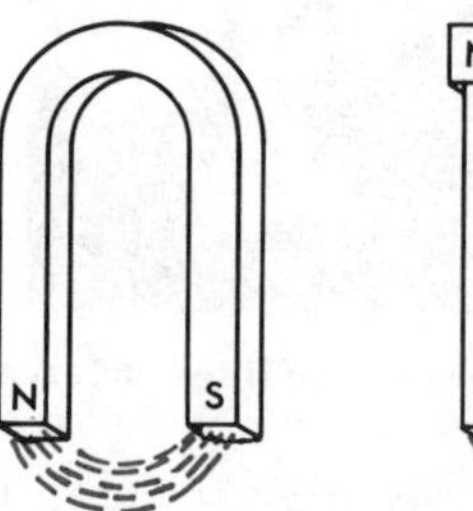

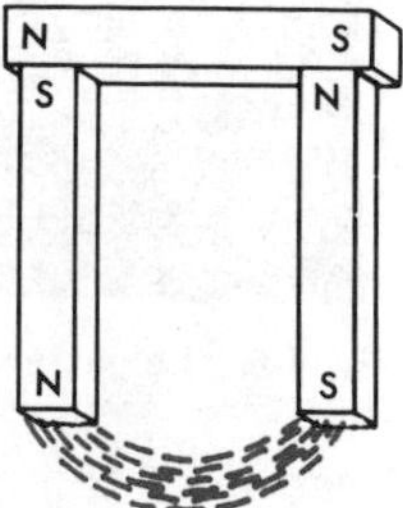

Fig. 4-11. Two forms of horseshoe magnets.

ends of a horseshoe magnet when it is not in use. The keeper provides a completely closed circuit of magnetic material, so that air-gap reluctance is eliminated. As a result, the magnet will last longer. (See Fig. 4-12.)

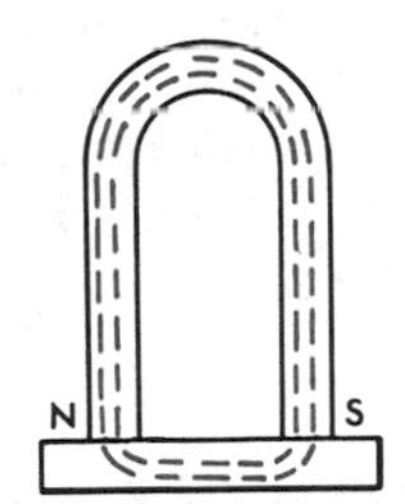

Fig. 4-12. A keeper placed across the poles of a horseshoe magnet.

Compound Magnets

Special permanent magnets can be made by joining two or more permanent magnets with their poles clamped together. The combination is called a *magnetic battery* or *compound magnet.* (See Fig. 4-13.)

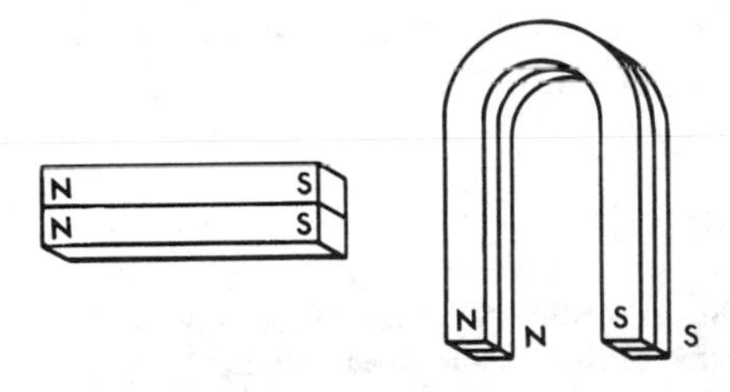

Fig. 4-13. Two forms of compound magnets.

Occasionally a permanent magnet will have more than two poles (points of attraction). This occurs when adjoining sections of magnetic material are also magnetized. Additional poles—called *consequent* poles—are formed in the magnet. (See Fig. 4-14.) Consequent poles can be induced in a temporary magnet by passing a strong magnetic force near the opposite ends of the temporary magnet.

Fig. 4-14. Consequent poles in a permanent magnet.

Flat Magnets

Flat magnets are magnetized throughout the thickness, rather than the length, of the material. They are used primarily as pole pieces in motors and generators.

Ring Magnets

A ring magnet, which looks like a washer, has no poles. These magnets are used in certain types of instruments. The closed core of a transformer is a good example of a *temporary* ring magnet.

POLES OF A PERMANENT MAGNET

The poles of a permanent magnet are usually at the ends (points) of the magnetic material. They represent the greatest concentration of magnetic force. Every permanent magnet (except a ring magnet) has a north pole and a south pole.

When a compass needle is held near a magnet, the needle will line up parallel with the lines of force between the poles and will point toward the poles.

MAGNETIC UNITS OF MEASUREMENT

The strength of a magnet—that is, the effort required to produce the lines of magnetic force—is known as *magnetomotive force.* It is measured in *gilberts.*

Reluctance

Reluctance is the opposition to the passage of the magnetic flux (the total number of lines of force). It is measured in *CGS* units (formerly called *oersteds*). One cubic centimeter of air has a *reluctance* of one *CGS* unit, or one oersted.

Permeability

The ease with which magnetic lines of force pass through a material depends upon the *permeability* of the material. Iron and steel have high permeability, the permeability of iron being greater than the permeability of steel.

Permeance

Permeance is the ease with which the flux passes through a material. Note that permeance is the opposite of reluctance (which is the opposition to

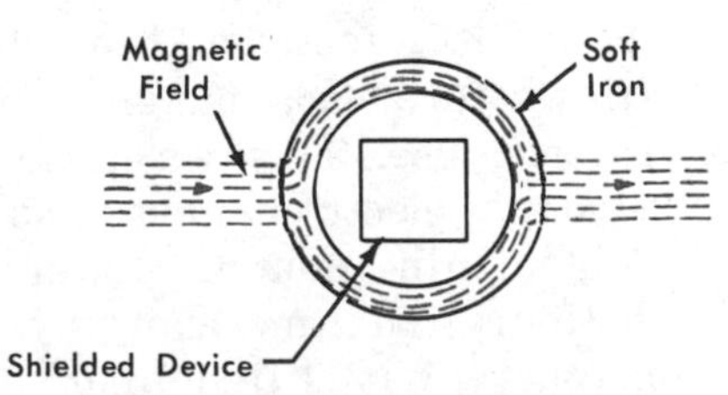

Fig. 4-15. Shielding by providing an easy path for the magnetic lines of force.

the passage of flux). Permeance is the reciprocal of reluctance.

INSULATION AND SHIELDING

There is no insulation for magnetism. Magnetic lines of force will penetrate any material. However, it is often desirable to keep magnetic lines of force out of certain places. For example, sensitive instruments must be shielded to prevent their being magnetized. When shielding is necessary, we take advantage of the fact that some materials allow magnetic lines of force to pass more easily than others. Therefore, a good magnetic conductor is placed between the magnet and the place where the magnetic lines of force are to be kept out. The lines of force follow the easier magnetic path around the conductor. As a result, the object is "shielded." (See Fig. 4-15 Page 55.)

Summary

Most electrical and electronic equipment depends directly or indirectly upon magnetism.

The original, natural magnets were pieces of an iron ore called *magnetite.* The magnets were called *lodestones.*

The earth is a huge magnet with north and south magnetic poles.

The field of a magnet consists of magnetic lines of force concentrated around the poles of the magnet.

Lines of force emerge from the north pole of a magnet, re-enter the magnet at the south pole, and travel inside the magnet to the north pole again.

The ease with which a material accepts magnetic lines of force is its *permeability.* Substances with a permeability *greater* than *one* are called *paramagnetic.* Those with a permeability of *less* than *one* are called *diamagnetic.*

Objects can be shielded—but never insulated—against magnetism.

Questions and Problems

1. Describe a simple experiment for demonstrating the presence of a magnetic field and the distribution of magnetic lines of force about a bar magnet.
2. Explain why low permeability of a magnetic material is associated with high retentivity.
3. What is the distinction between ferromagnetism, paramagnetism, and diamagnetism?
4. Why would an ordinary magnetic compass be of little value to a navigator inside a submarine?
5. How could you magnetize one of the blades of a pocket knife? Demagnetize a screwdriver?
6. Do magnetic lines of force enter into, or emerge from, the earth's Northern hemisphere? Explain.
7. What magnetic property corresponds to electrical conductance? To resistance?

ELECTROMAGNETISM

Magnetism and its effects had been observed for many centuries before electromagnetism was first put into practical use. This was in 1830, when the first "strong" electromagnet was made. It lifted a weight of nine pounds! Later that same year, Joseph Henry, an American physicist, built an electromagnet that lifted more than 700 pounds. A year later, Henry made another electromagnet that lifted almost a ton.

Our modern electric industry really began when men first used an electric current to produce a magnet and then used the magnetic "field" to produce an electric current. Electromagnets are employed in such signal devices as door bells, telegraph sounders, buzzers and chimes. They are also used in relays, annunciators, motors, motor controllers, generators, telephone receivers, radio receivers, broadcast transmitters, meters, transformers, induction coils, amusement devices, hoists, elevators, electric railways, arc lamps, and circuit breakers—plus many other applications.

Lines of Force

A magnetic field is established around a conductor whenever electrons move through the conductor. Such a conductor, which is normally nonmagnetic, does not develop this field but provides a path for the flow of electrons. The movement of electrons creates the magnetic field.

The effect or *strength* of the magnetic field is proportional to the current flowing in the conductor. The direction of the field is perpendicular to the direction of the current flow.

But how can we *prove* that a magnetic field exists around a conductor carrying a current? A simple compass (see Fig. 4-16) is all we need. If the compass is held first on one side of the wire and then on the other, we can readily "see" that magnetic lines of force are present.

If we investigate further with the compass, we will see that the magnetic lines of force reverse their direction whenever the current through the wire reverses its direction. This tells us that the direction of the magnetic lines of force in a conductor is determined by the direction in which the current is flowing. You can easily remember the direction of the magnetic lines of force by using the "left hand rule," which states: *Grasp the wire with the left hand so that the thumb points in the direction of the current flow, and the fingers will then point in the direction of the magnetic lines of force.* This rule is illustrated in Fig. 4-17.

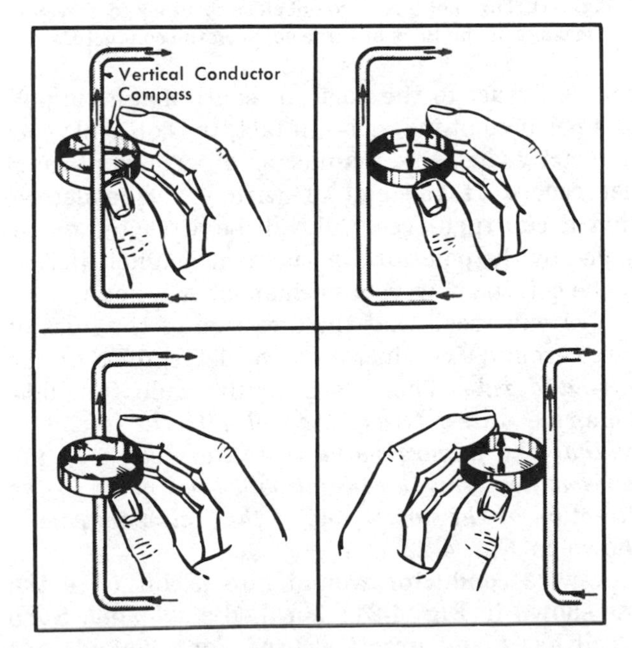

Fig. 4-16. A compass detects the presence of electromagnetic lines of force around a conductor carrying an electric current.

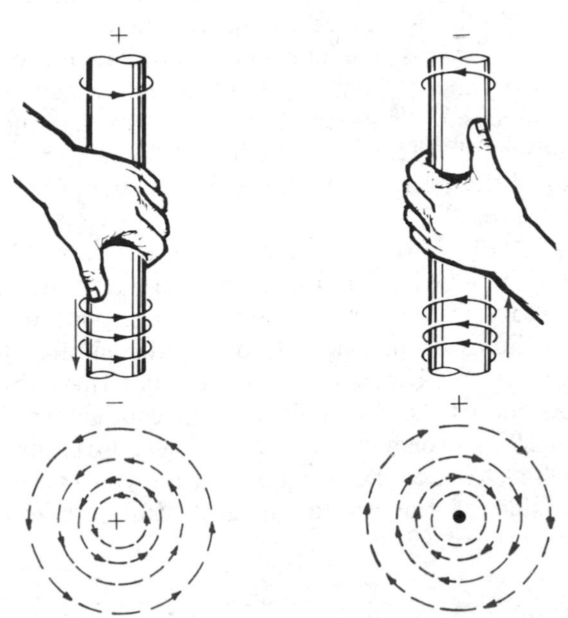

Fig. 4-17. Left-hand rules for a wire carrying a current.

We have shown that magnetic lines of force are produced when current is flowing through a conductor, and we have shown the path taken by the magnetic field. Now let us see what takes place when current flows through two parallel conductors—that is, when current flow through two conductors is in the same direction. Fig. 4-18 shows two parallel conductors, both carrying current in the same direction. A plus sign (+) was placed inside the circles of Fig. 4-18 to indicate that the conductor carried current into the page—away from you, the reader. If the current flow is reversed, so that it is coming out of the page—toward you—a dot (·) will be put inside the circle. These symbols indicate either the feathered end

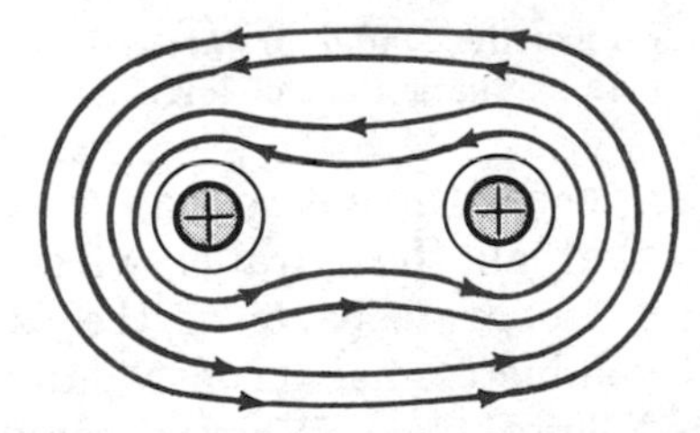

Fig. 4-18. Lines of force are joined around two parallel conductors carrying current in the same direction.

(rear) or the point of an arrow, which is used to symbolize the current flow. Note how the magnetic lines of force have arranged themselves about the two conductors shown in Fig. 4-18. The lines of force do not form separate patterns around each wire. Instead, they have *partially* combined to form a single, but larger, magnetic field.

If three wires are parallel, a similar pattern will also result. That is, the magnetic field of each conductor will combine to form a single magnetic field. This single field will be larger and stronger than any of the individual fields of magnetic force.

Fig. 4-19 shows six adjacent conductors. The current is flowing in the same direction in all conductors. Take a close look at the upper part of the illustration. The action is quite similar to the action in Fig. 4-18, except that there are more conductors. In addition, the conductors in Fig. 4-19 are formed into loops. These loops have been sliced down the middle, so that you can see the inside of the conductor and study its electrical activity.

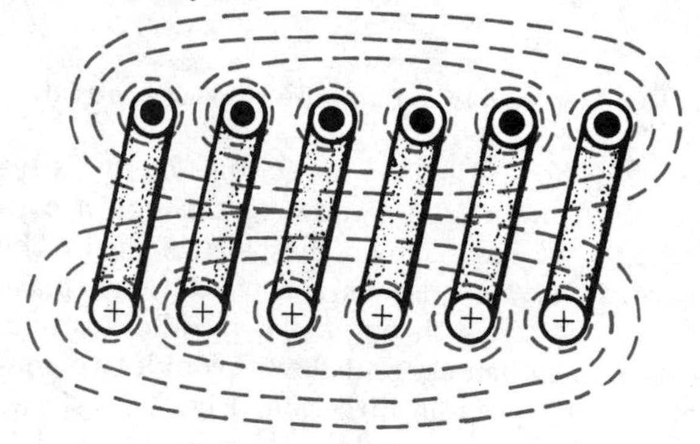

Fig. 4-19. Several parallel conductors are linked together with magnetic lines of force.

In Fig. 4-20 we see the entire loop or *coil.* This figure shows that all the parallel wires in Fig. 4-19 are merely separate parts of the same conductor. This conductor has been wound into a coil and sliced lengthwise. When a conductor is thus wound into a coil (as shown in Figs. 4-19 and 4-20), the magnetic lines of force around the individual loops will join together and *reinforce*

Fig. 4-20. A length of wire wound into the form of a coil (solenoid).

each other. Thus, it is possible to pack a long conductor into a small area, so that the magnetism is localized instead of being scattered over a wide area.

Ordinarily, the turns of a coil are closer together than the previous illustrations show. The usual practice is to wind the turns so tightly together that they almost touch each other (Fig. 4-21). Thus, there is less opportunity for the lines of force to *leak* between the turns—more lines of force are concentrated in the center and on the outside of the coil. One of the principal reasons for winding a conductor into a coil is to concentrate as many magnetic lines of force as possible in the smallest cross-sectional area.

When the lines of force enclose the loops of the coil (as shown in Fig. 4-21), a *north* magnetic pole is formed at one end of the coil, and a *south* magnetic pole is formed at the opposite end—just like the magnetic poles at the ends of a permanent magnet. The north or south magnetic pole of the

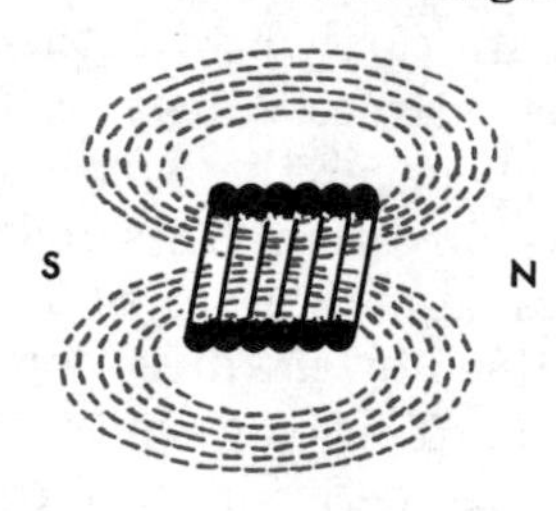

Fig. 4-21. Turns of a coil wound close together to prevent leakage of the lines of force between the conductors.

coil will react to the north or south magnetic pole of a permanent magnet—in fact, the coil will react in exactly the same manner as a permanent magnet reacts. If the coil of wire is suspended so that it can turn freely, it will be attracted or repelled by the poles of a permanent magnet, just as if the coil itself were a permanent magnet.

You will recall that the direction of the lines of force around a conductor can be determined by the *left-hand rule.* There is a similar rule for coils. This rule states: *Grasp the coil with the left hand so that the fingers point in the direction of the current flow and the thumb will then point in the direction of the north pole of the coil.* This rule is shown in Fig. 4-22.

A wire conductor wound into a coil (like the one shown in Fig. 4-21) is called a *solenoid.* Such a coil has many practical uses, for solenoids are employed in a great many electromechanical devices. If an iron rod is positioned so that one end

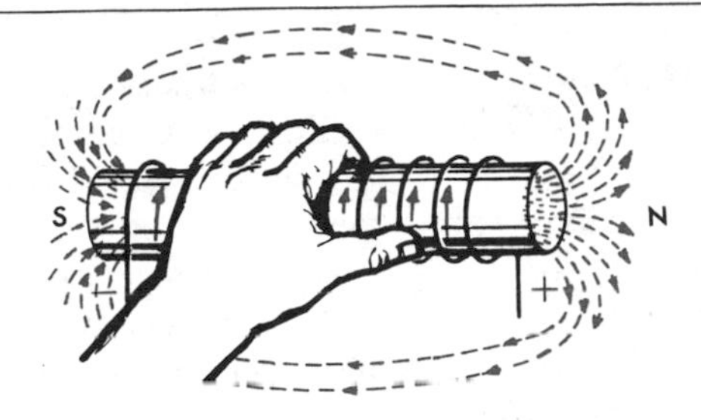

Fig. 4-22. The left-hand rule for a coil carrying a current.

is about to enter the coil, as shown in Fig. 4-23, an interesting reaction takes place. There is no reaction between the coil and the rod as long as no current flows through the coil. The instant the current begins to flow, however, a magnetic field begins to build up around the coil. The lines of force of the magnetic field assume a pattern similar to the one in Fig. 4-21.

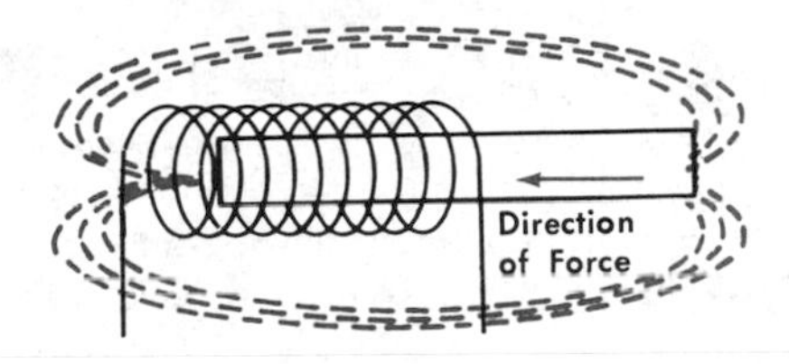

Fig. 4-23. A bar of iron in the field of a solenoid.

The iron core, being permeable, affords an easy path for the magnetic lines of force to follow. When the core is not centered in the coil, the electromagnetic field must extend further in order that it might flow through the core. It is almost as if the electromagnetic field had been "stretched" out of its normal position. The field exerts a force on the core (see Fig. 4-23) until the core is centered, as illustrated in Fig. 4-24, and the field is

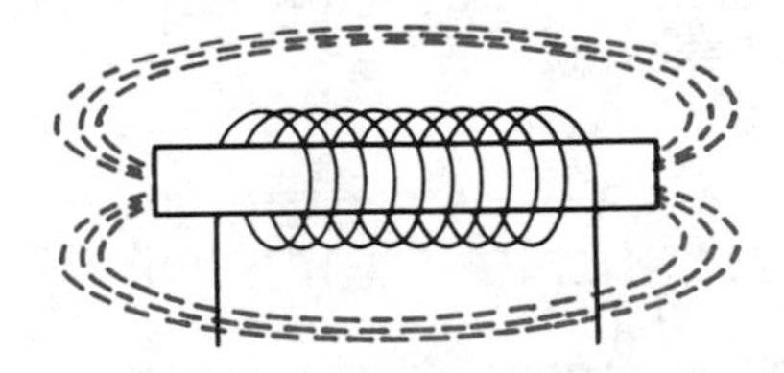

Fig. 4-24. The solenoid has drawn the bar of iron within itself.

again in its normal position. Solenoid-operated valves, plungers, locking devices, and many other electromechanical devices use this magnetic principle.

ELECTROMAGNETICS

A solenoid (coil) is relatively weak—magnetically speaking—when compared with the large amount of current required to energize it. The reason is that the air space within the coil (through which the magnetic lines of force must travel) is a poor conductor of magnetism. The magnetic properties of the coil are improved if a path with a lower resistance than that of air is provided for the lines of force. The resistance of a material to the passage of magnetic lines of force is called *reluctance.* In other words, the material is "reluctant" to let the lines of force pass through it.

A coil becomes magnetically stronger when a soft iron rod is placed in the coil. The reason is that iron has a lower reluctance than air. This lower reluctance encourages many more lines of force to appear than would appear if the coil had an air core. Furthermore, the iron rod concentrates the lines of force within the iron itself, rather than allowing them to spread over a wide area.

When a coil of wire is wound around a soft iron core, an electromagnet is created. The electromagnet is one of the most important electrical devices that man has discovered. Electromagnets provide a means by which electrical energy can be harnessed to perform certain mechanical work. Fig. 4-25 shows the basic principles of an electromag-

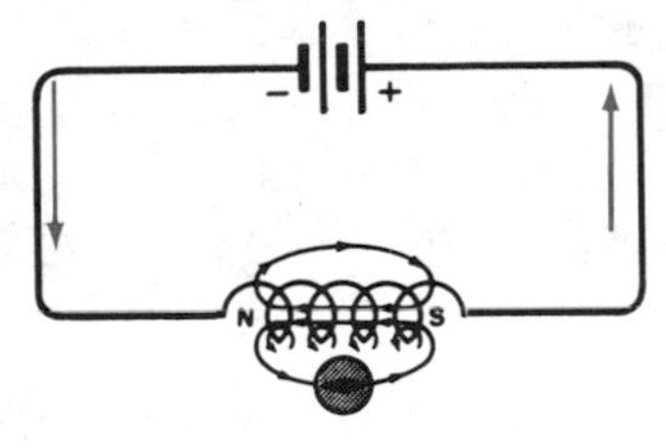

The Polarity of the Electromagnet Will Be Reversed if the Winding is Reversed As Shown Here, or if the Direction of the Current is Reversed.

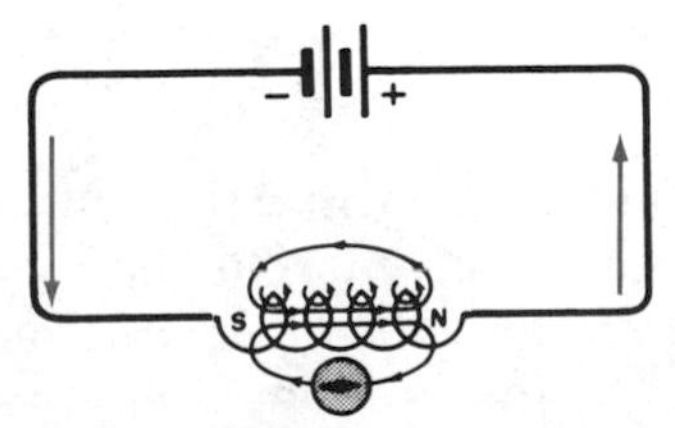

Fig. 4-25. Principles of an electromagnet.

net. The polarity of the electromagnet will be reversed if the winding is reversed, as shown here, or if the direction of the current is reversed.

The action of an electromagnet is quite simple. Until an electrical current passes through the coil, the device is powerless. When electrical current flows through the coil, the device comes to life. The magnetic field that always surrounds any wire carrying an electric current, is concentrated in the iron core. The iron core actually becomes a magnet as current flows through the coil. Like a magnet, the iron core has a north pole and a south pole. It reacts to the laws of magnetic attraction and repulsion just as a permanent magnet does. The iron core will continue to have the properties of a magnet as long as an electrical current flows through its coil.

Strength of an Electromagnet

The strength of an electromagnet depends largely upon (1) the construction of the coil and (2) the amount of current flowing through it.

For example, suppose a coil of wire has 100 turns. Let us send one ampere of current through the coil. One ampere of current flowing through 100 turns of wire produces a certain magnetic force. Instead of a coil of wire with 100 turns, let us substitute one with 200 turns. Now we have a coil with the same amount of current flowing through it, but with twice as many coils. Twice as many coils means twice as many magnetic lines of force. Therefore, exactly twice as much magnetomotive force (the force that drives the lines of force through the circuit) is produced with a 200-turn coil than with the 100-turn coil when the same amount of current flows through both coils. (See Fig. 4-26.)

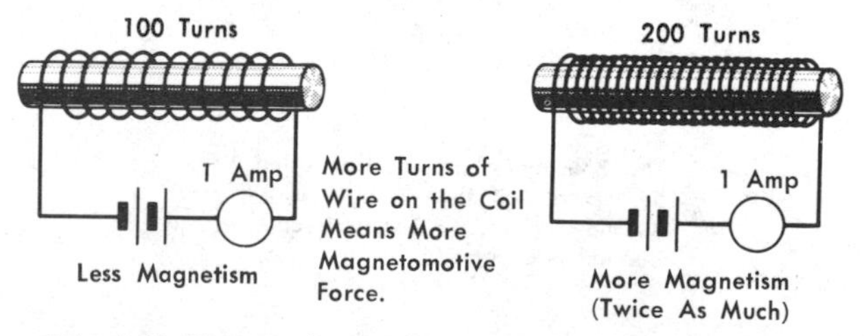

Fig. 4-26. The number of turns of wire in a coil affects the magnetic strength.

Let us return to the coil with 100 turns of wire. When one ampere of current passes through the coil, a certain amount of magnetomotive force is produced. Instead of doubling the number of turns of wire, now suppose we double the amount of current flow (from one ampere to two amperes)—see Fig. 4-27. Twice as much current means twice as many magnetic lines of force. The magnetomotive force generated in the coil has been doubled because the amount of current passing through the wire of the coil was doubled.

The magnetic force of an electromagnet can be increased (or decreased) by increasing (or decreasing) (1) the number of turns in the coil or (2) the amount of current flowing through the coil. These statements may be combined as follows: *The strength of an electromagnet depends on the number of amperes and the number of turns—that is, the number of ampere-turns.*

The substance through which the lines of force pass—including the iron core—form the *magnetic*

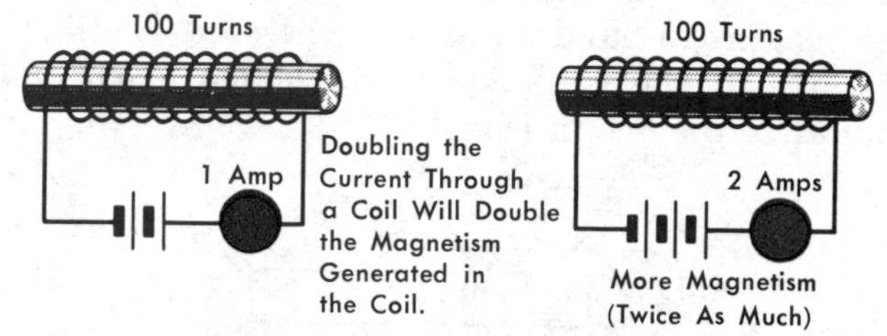

Fig. 4-27. The amount of current flowing through a coil affects the magnetic strength of the coil.

circuit. The expression, *number of lines of force per square centimeter,* is the measure of magnetic density. The total number of lines of force in a magnetic circuit is termed the *magnetic flux.* The relationship between *magnetic flux, magnetomotive force,* and *reluctance* in a magnetic circuit is similar to the relationship between *current, electromotive force,* and *resistance* in an electrical circuit. It can be expressed by the equation:

Magnetomotive Force =
Magnetic Flux times Reluctance

(Note the similarity to Ohm's Law: Emf = Current times Resistance)

TERMS APPLIED TO MAGNETIC PROPERTIES

The various properties of magnetism do have technical names, and it is wise to know their meaning, since these names are used in electrical men's everyday conversation and work.

The *maxwell* is a term we would like to mention. It is seldom used today; however, it will be found in many of the older electrical texts. The term maxwell is applied to a magnetic line of force and was named after James C. Maxwell. Maxwell was a brilliant Scottish physicist and scientist, as well as a physician, and many of his experiments were conducted in the field of magnetism. His research added to the understanding of magnetic phenomena, and the results of some of his discoveries have contributed much to our present knowledge of magnetism. This is quite remarkable, since Maxwell has been dead for over four hundred years, and his experiments occurred nearly two hundred and fifty years before those of some of his better known successors.

The maxwell is equal to one magnetic line of force. So 100,000 magnetic lines of force are equivalent to 100,000 maxwells. Generally speaking, electrical men are seldom interested in a single line of force. Even a weak magnetic field consists of hundreds, or even thousands, of lines of force. Furthermore, the term maxwell has almost disappeared from the terminology of magnetism. It has become the increasing practice to refer directly to the magnetic field as having so many "lines of flux" rather than to its consisting of so many "maxwells." It is becoming increasingly common to group the lines of force together and refer to them simply as the magnetic "flux," unless there is some occasion for determining with reasonable accuracy the number of lines of force present in a given magnetic field.

In describing the strength of a magnetic field, writers often refer to it as consisting of so many lines of force "per square inch." This method of expression is apparently very simple, yet all too often the newcomer to the study of electricity and magnetism fails to understand exactly what is meant by the term. In Fig. 4-28 is a drawing of an electromagnet made up of a bar of iron one inch square and long enough to be bent into the form of an exaggerated horseshoe. It has on it a coil of wire through which an electric current is passed. The electric current develops magnetism within the iron.

The cross section of the iron core is exactly one square inch. If there are 1000 magnetic lines of force within the iron, we have a magnet which has a magnetic strength of 1000 lines of force per square inch. If there are 50,000 lines of force within the iron, we have a magnet with a strength of 50,000 lines of force per square inch. If the current is strong enough and there are enough turns of wire to create 100,000 lines of force in the iron, there is a magnetic strength of 100,000

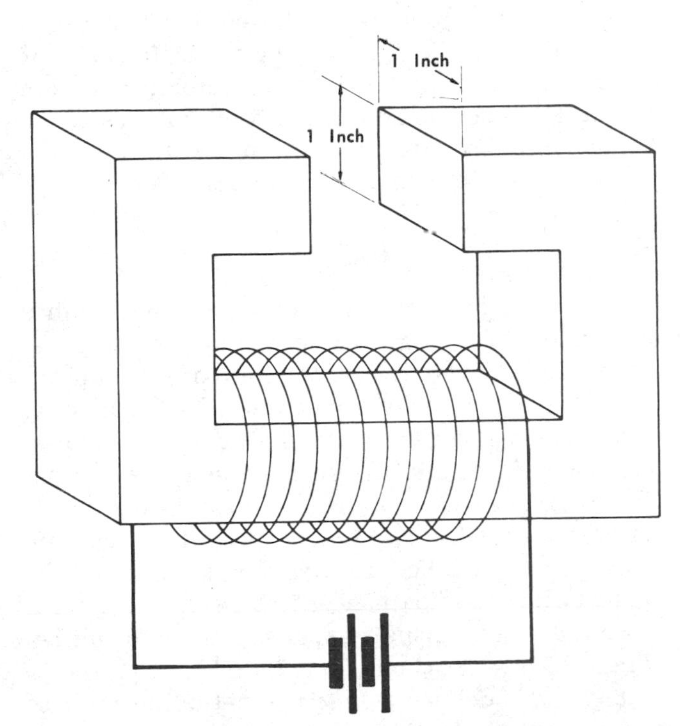

Fig. 4-28. How the cross-sectional area of magnetized iron is measured.

lines of force per square inch.

If the bar of iron were smaller, we would have a different magnetic strength. Suppose the size of the iron bar were reduced so it had a cross-sectional area of only one-fourth square inch. Then if there were 10,000 lines of force in the iron, there would be a concentration of magnetic strength equal to 40,000 lines per square inch. This is due to the fact that the 10,000 lines were concentrated in one-fourth square inch. By definition, regardless of the particular core's cross-sectional area, the magnetic strength is given in terms of how many lines of force there would be in the unit of area, one square inch. This was done for convenience, just as the unit of area, the square inch, was chosen for defining water pressure as pounds per square inch.

This matter is rather involved, and one can easily become confused over the terms. The number of lines of force within the iron is dependent upon the ampere-turns of the coil. The more ampere-turns, the stronger the magnet and the more lines of force present. The fewer the ampere-turns, the weaker the magnet and the fewer lines of force present. The size of the iron core will determine the concentration of the lines of force, the number of lines of force per square inch.

The metric system is used very extensively in electrical engineering, especially in the field of magnetics. Writers, instead of referring to the concentration of magnetic lines of force as being so many per square inch, often use the term so many lines of force per square centimeter.

GAUSS

Because a magnetic flux of any given number of lines of force might be spread over a comparatively large cross section of iron core or concentrated in a relatively small cross-sectional area, it is sometimes necessary to know both the number of lines of force and the area of the cross section. This problem has been met by formulating a unit for measuring the concentration of the magnetic flux. This unit, called a *gauss,* is a measure of the "density" of the flux concentration. It was named after an early German experimenter and mathematician, Karl F. Gauss (1777-1855).

A gauss is defined as one magnetic line of force per square centimeter. A piece of iron having a cross-sectional area of one square centimeter would have a flux density of one gauss if the iron had only one magnetic line of force. A density of 10,000 lines per square centimeter would be a density of 10,000 gausses, while a density of 100,000 lines per square centimeter would be 100,000 gausses.

An advantage of using the gauss as a unit of measurement is that it gives more information directly than do the other units of measurement. However, the beginner in the field of electricity and magnetism does not have to worry too much about these things. They are mentioned to acquaint you with the terms, since they occur frequently in discussions of magnetism.

PERMEABILITY AND RELUCTANCE

Reluctance refers to the opposition which any material presents to the passage of magnetic lines of force through it. It is thus comparable, though by no means identical, to resistance in an electrical circuit.

Often we are more concerned with the ease with which a substance or material will pass magnetic lines of force than with its opposition to such passage. This property is referred to as the permeability of the material and, in a sense, may be considered as magnetic conductivity. In other words, permeability is to magnetic lines of force what conductivity is to a flow of electricity. All materials, it should be remembered, will pass magnetic lines of force, because there is no insulator for them. However, not all materials permit the passage of the lines of force with equal ease. In fact, it is hundreds of times, even thousands of times, easier for the lines of force to pass through some materials than others. It is from 1800 to 2000 times easier for the lines of force to pass through some kinds of iron, for example, than through air. This can be said in a somewhat different manner: a piece of iron one inch square will permit the passage of from 1800 to 2000 times as many lines of force as the same space occupied only by air. Magnetic lines of force can certainly pass through the air; yet if the air is replaced with a piece of soft iron, the lines of force can pass many times more readily through it.

This ability of a material to permit the passage of magnetic lines of force is referred to as its permeability. Air has a permeability of 1. Other materials have permeabilities ranging from slightly less than unity to over 80,000. Iron has a permeability ranging from 1000 or so to more than 2000. Brass, however, has a permeability of less than 1. Each material has its own specific permeability, the exact value of which can be learned by referring to a table of permeability values found in handbooks of physical chemistry, etc.

WHERE ELECTROMAGNETS ARE USED

It is doubtful if any one could compile a complete list of applications in which electromagnets are used. They are used in so many devices and so many new uses are constantly being found that it would be impossible to name them all. For this reason we can cover only a few of their more important applications. The use of electromagnets for the purpose of handling scrap iron and steel has been used for years. This is a spectacular use and one with which most people are familiar.

One of the first uses of electromagnets was in telegraphy. Samuel F. B. Morse wound insulated wire onto an iron core to make his first electromagnetic telegraph instrument. Electromagnets have been used for that purpose ever since. Morse devised an electromagnet which would attract a movable iron bar or armature when the magnet was energized and would permit the armature to move away when, under the pull of a spring, the coil was de-energized. He energized the electromagnet by opening and closing a switch which is known today as a telegraph "key." This closing and opening of the switch, or key, caused electric current through the coil of the electromagnet used to attract the movable armature. As the armature moved up and down, it would strike adjustable "stops," producing the familiar sounds of the telegraph instrument.

The telephone, which is so much a part of our everyday lives, also uses electromagnets—uses many of them in fact. An electromagnet is used to reproduce the sound in the headphone we hold to our ear. Electromagnets are used in the tens of thousands of relays found in telephone "central" offices.

The great generators in power plants which produce our commercial electrical energy are able to generate electrical energy only because of the electromagnets that are rapidly rotating past other stationary electromagnets. Electric motors have magnetic fields which are produced by electromagnets. Here we find that the interaction of electromagnetic fields produces torque that turns the shaft of the motor, thus converting electrical energy into mechanical power.

Our way of life would be completely changed were it not for the common electrical relay. Many people have never heard of an electrical relay; yet it is so much a part of our everyday lives, it is difficult to imagine what living would be like without it. A relay is nothing more than an automatic electrical switch operated by one or more electromagnets. Relays are used in telephones, automobiles, automatic-door openers, elevators, traffic lights, oil burners, street cars, diesel-electric locomotives, and innumerable other devices.

Telegraph systems, telephones, relays, motors, radio sets, television sets, and thousands of other devices contain electromagnets. They find use anywhere that it is desirable to convert electrical energy into mechanical motion or force.

Summary

A magnetic field exists around any conductor that is carrying current.

The direction of the magnetic lines of force in a conductor is determined by the direction in which the current is flowing.

The *left-hand* rule for a wire states: "Grasp the wire with the left hand so that the thumb points in the direction of the current flow and the fingers will then point in the direction of the magnetic lines of force."

The *left-hand* rule for a coil states: "Grasp the coil with the left hand so that the fingers point in the direction of the current flow, and the thumb will then point in the direction of the north pole of the coil."

An electromagnet is a coil of wire (a solenoid) with a soft iron bar for a core.

The strength of an electromagnet depends on the number of turns of wire in the coil and the current flowing through them.

Questions and Problems

8. Is an electromagnet a solenoid? Explain.

9. State the left-hand rule for a wire that is carrying current.

10. Does a magnetic field exist outside of a solenoid? Explain.

11. If a solenoid and an electromagnet have an equal number of turns and equal currents, will their magnetic strength be equal? If not, which will be stronger, and why?

12. Does the magnetic field of an electromagnet differ from the magnetic field of a permanent magnet?

13. Why is an iron core used in an electromagnet?

14. What does the expression "magnetomotive force" mean?

INDUCED CURRENTS

The induction of current flow by magnetic fields was discovered by the English scientist Michael Faraday (1791-1867) around 1831. His studies of conversion from mechanical to electrical energy was the basis for the later discovery of the generator.

Magnetic Field Around a Wire

A current is induced in a wire when the wire is moved through a magnetic field (see Fig. 4-29). A current may also be induced in a wire when a magnet is moved as shown in Fig. 4-30. Notice that moving either the magnetic field or the wire will induce a current. The motion between a conductor and a magnetic field produces the current flow.

Fig. 4-31 depicts a current-carrying conductor A lying parallel to another conductor B. Note the permanent magnetic field around the current-carrying conductor. Since wires A and B and the magnetic field are stationary, no current is induced in conductor B.

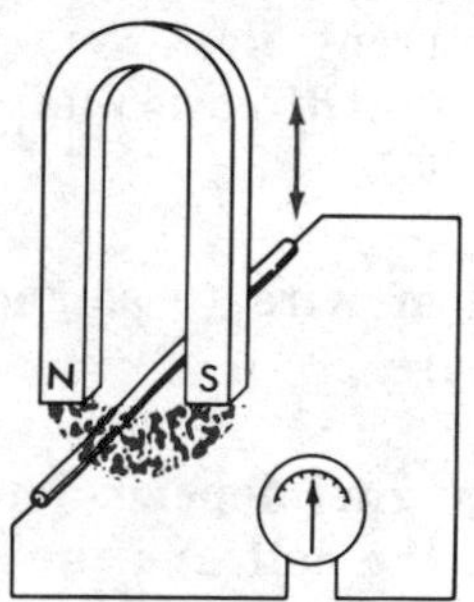

Fig. 4-29. Current induced in a wire moving through a magnetic field.

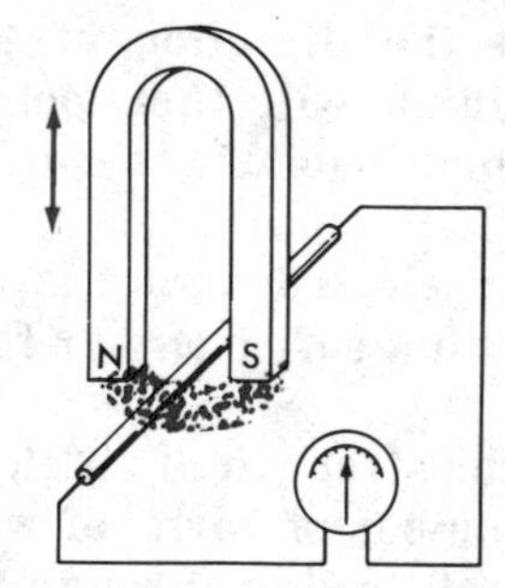

Fig. 4-30. Current induced in a wire by moving a magnetic field.

Moving conductor A or B will cause the magnetic lines to cut through conductor B and to induce a current. However, a more efficient method is available. For example, opening and closing a switch of the circuit in Fig. 4-31 will cause the magnetic field to collapse and to build up again. The magnetic lines of force will pass through conductor B in one direction during collapse and in the opposite direction during build-up. This is how a transformer, like the one in Fig. 4-32, operates. The field from coil A (the primary winding) induces a current in coil B (the secondary winding) each time the switch is opened and each time it is closed.

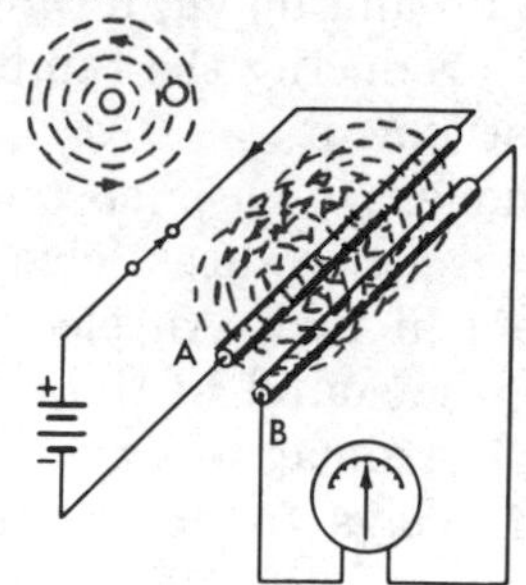

Fig. 4-31. Current-carrying conductor, parallel to another conductor.

The direction of the induced current depends on the directional motion of the magnetic field and the directional movement of the conductor through this field.

Think of the magnetic flux lines, from the north pole to the south pole, as being rubber bands that

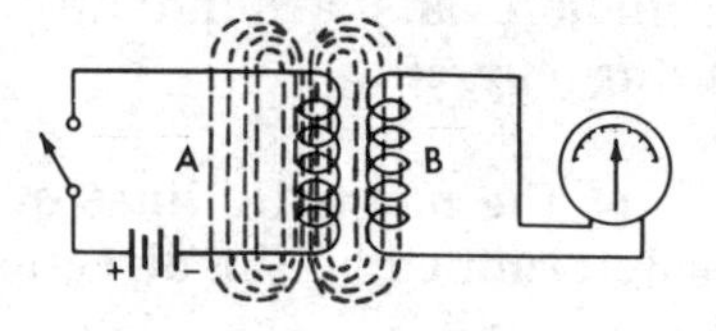

Fig. 4-32. Transformer effect.

stretch as the conductor moves against them. This is described by Fig. 4-33. The lines now flow counterclockwise around the conductor, and the induced current will flow away from you.

Generator Rule

The generator *left-hand rule,* also shown in Fig. 4-33, can be used to determine the direction of an induced current. The index finger points in the direction of the magnetic field (north pole to south

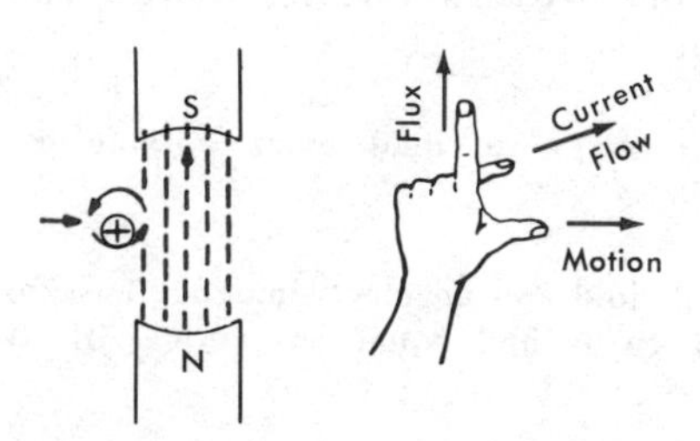

Fig. 4-33. Generator left-hand rule.

pole), and the thumb points in the direction the conductor travels. The middle finger points in the direction of the induced current flow.

Induced Emf

Thus, induced currents were produced when a conductor cut magnetic lines of force. Actually it was *not* current which was induced. What really happened was this—cutting the lines of force generated an *induced* emf. The induced emf forced electrons which were already in the wire to flow. It is correct to call it *induced* current so long as the emf is induced emf. The term "induced current" has become well accepted.

Let us replace the magnet producing an emf with a stronger magnet. Then, the moving conductor cuts a greater number of lines of force, and a *stronger* emf is produced.

Voltage Produced

The amount of voltage produced by a conductor cutting through a magnetic field is determined by (1) the speed of the conductor and (2) the number of magnetic lines of force cut. Maximum emf will be produced whenever the conductor crosses the field at right angles to the lines of force. On the other hand, no emf will be produced whenever the conductor is parallel to the lines of force.

One volt will be produced by a single conductor cutting through 100,000,000 lines of magnetic flux in one second. The voltage can be changed by adding more conductors, by forming coils as shown in Fig. 4-34, by changing the strength of the magnetic field, or by varying the speed of the conductors.

When a conductor is coiled, each turn is in series with the other turns. Therefore, voltages are added. If a conductor cutting a field which produces an induced emf of 10 volts is coiled into five turns, it will cut the same field and produce 50 volts.

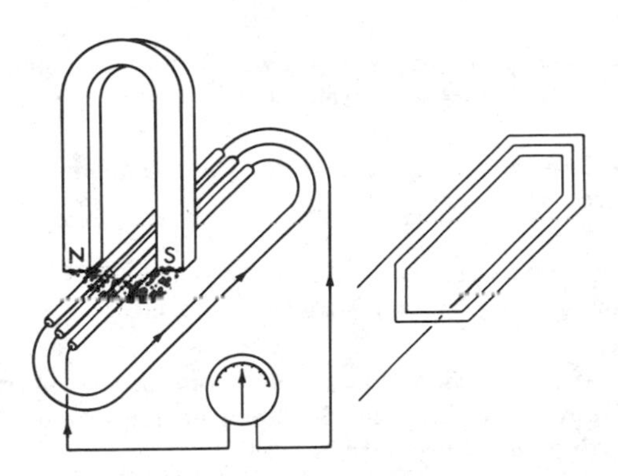

Fig. 4-34. Induced emf is increased by adding conductors.

Self-Induction

Inductive action in a coil is caused by the effect of its own magnetic field. The lines of force which expand and collapse about the coil—whenever the current is made, broken, or changed—cut the turns of the coil, developing an emf across the coil itself. This property of a coil is called self-induction.

Counter Emf

The emf developed as a result of self-induction has a polarity opposite to that of the electromotive force originally applied to the coil. This secondary emf *opposes* the original emf and is called a *counter emf.* Counter emf opposes a build-up of current through the coil. (It also opposes a decrease of the current.) As an object at rest is difficult to move and, once started, resists efforts to stop it, so counter emf slows the starting and stopping of current.

Summary

Mechanical energy can be converted to electrical energy by the movement of a conductor across a magnetic field.

A conductor cutting through magnetic lines of flux will induce a current in itself. Either the conductor or the magnetic field can be moved.

Induced voltage is determined by (1) the number of conductors cutting magnetic lines of flux, (2) the strength of the magnetic field, and (3) the speed at which the conductor cuts the magnetic field.

The direction of induced current flow is determined by the direction of the magnetic field and the direction of the conductor through the field.

Questions and Problems

15. Is it necessary to move the wire (or other conductor) in order to induce a current flow?

16. Explain why current flows in the secondary winding of a transformer.

17. If a conductor cuts 500,000,000 lines of force in one second, what voltage will be produced?

18. In what direction does current flow in a conductor that is moving parallel to the lines of force and from the north pole to the south pole?

19. Why does current flow in a conductor that is moving across a magnetic field?

20. In a transformer, what determines which winding is the primary and which is the secondary?

21. List several ways to increase an induced voltage.

22. Would an induced current be set up by a loop of wire revolving in a horizontal plane in the earth's magnetic field? Explain.

23. Why is the magnetic field around a coil stronger than the field around a single wire (conductor)?

24. What is *counter electromotive force?*

25. Define a *volt* in terms of time and lines of force cut.

5

ALTERNATING CURRENT

AC THEORY

An alternating current (ac) continually changes in potential—going from zero to maximum voltage and back to zero. In addition, ac periodically reverses its direction—from positive to negative. This is in contrast to a direct current (dc), which maintains a steady potential and flows in one direction only.

Fig. 5-1 shows how the voltage of an ac generator changes. At point A the voltage is zero. Immediately afterward it has a small, positive value. This value increases until it is maximum at B. A moment later the voltage, still positive, continues to drop steadily until it reaches zero again at C. Below C the voltage becomes negative. Therefore, the voltage is shown below the zero line from C to E.

Notice that the polarity does not reverse at B or D—these merely are the points of maximum positive or negative potential. The points of polarity reversal are at A, C, and E.

The complete series of voltage values, represented by the curve from A to E, represents one complete *cycle*. When the curve is continued, the cycle is repeated. The time necessary to complete one full cycle is called a *period*.

The number of cycles generated in one second determines the *frequency* of the ac voltage. The electricity supplied to most homes is 60 hertz, which means the voltage goes through 60 complete cycles—from zero to positive to zero to negative to zero—in one second.

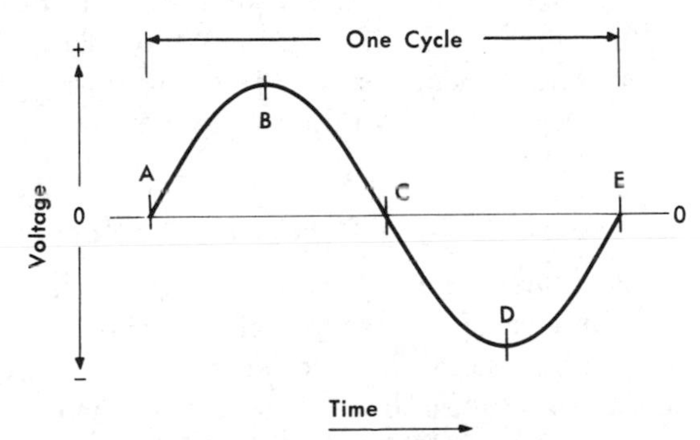

Fig. 5-1. Voltage changes in an ac circuit.

Advantages of AC

Alternating current can be transmitted more economically over long distances than dc. This is one of the great advantages of ac. It is easily transformed to higher or lower voltages. This is a very desirable characteristic for radio and television circuit applications. Many types of motors are designed for ac operation. Some ac motors operate without brushes, eliminating a common source of wear and motor maintenance problems.

Although ac differs from dc in many ways, practically all the basic principles of electricity that apply to dc also apply to ac.

GENERAL THEORY

When a conductor moves in a magnetic field, a voltage is induced in the conductor. This voltage

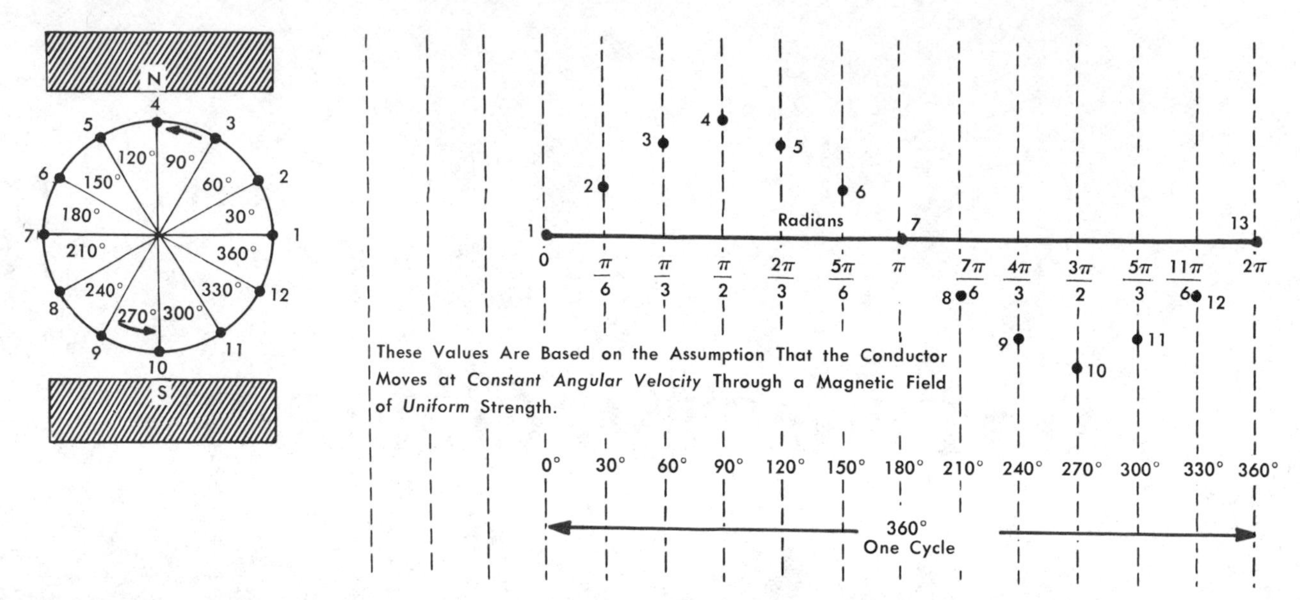

Fig. 5-2. Rotary path of a conductor in a magnetic field. Plotting the sine wave.

depends upon (1) the strength of the field (number of lines of force), (2) how fast or slow the conductor is moved across the lines of force, and (3) the direction the conductor moves with respect to the field.

Voltage Generated by a Revolving Conductor

Most electrical generators generate an emf by the rotary movement of a conductor within a magnetic field. This principle is illustrated in Fig. 5-2. The conductor moves through a circular path between the north and south poles of a magnet, cutting across magnetic lines of force as it moves. The position of the conductor is illustrated for each 30 degrees that the generator armature rotates. The magnetic lines of force are not shown. However, we may assume that they extend from one pole to the other and that they form a magnetic field of uniform strength between the poles.

The direction of rotation of the conductor is shown as counterclockwise, thus following the mathematical rule for the generation of angles. The same result occurs when the conductor is rotated in a clockwise direction.

At position 1 (see Fig. 5-2) the conductor is cutting the fewest number of lines of force per second; therefore, the induced voltage is very small. This voltage is plotted at point 1 on the graph at the right of Fig. 5-2. At position 2, more magnetic lines of force are cut each second. At position 3, still more lines of force are cut. The corresponding voltages are plotted at points 2 and 3 on the graph.

At position 4, the conductor is moving at right angles to the magnetic field and thus cuts the most lines of force per second. This corresponds to point 4 on the graph, the point where the greatest voltage is shown. The voltage decreases at points 5 and 6, until it reaches zero again at point 7. From points 7 through 12, the conductor is moving through the magnetic lines of force in the opposite direction. As a result, the polarity of the voltage is reversed. This reversed polarity is shown on the graph by the points below the center line of the graph.

Development of a Sine Curve

The manner in which an induced voltage varies from point to point, as a conductor rotates, is shown by the graph in Fig. 5-2. Points have been plotted for each 30 degrees of rotation. If other points are also plotted, they will lie on a curve known as a *sine wave*. (See Fig. 5-3.) This curve is called a sine wave because the *amplitude* (height) of any point on the curve is the sine of the angle through which the conductor has moved to that point. The graph is plotted horizontally in fractions of a second instead of in degrees of rotation. The frequency of rotation is 60 hertz.

Values of AC

A *peak* voltage is the maximum voltage during a complete cycle. It is represented in Figs. 5-2 and

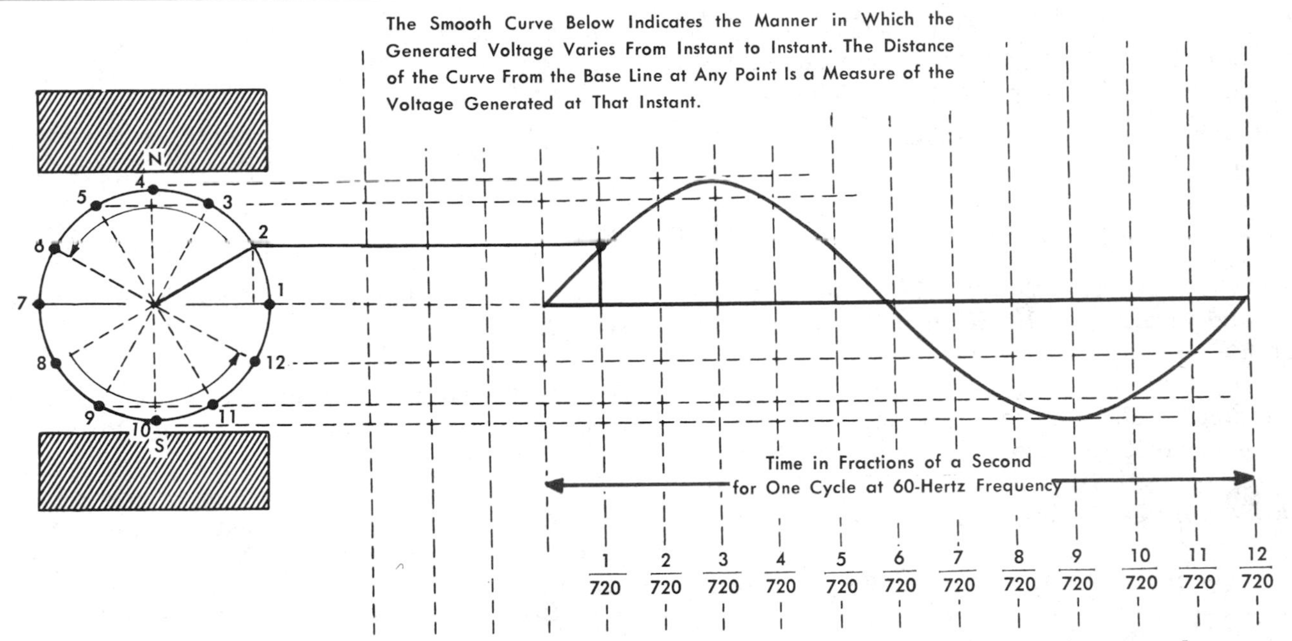

Fig. 5-3. The sine wave formed by the rotary path of a conductor through a magnetic field for one cycle.

5-3 by points 4 and 10. The highest value reached by the voltage may be positive or negative (E or −E).

The *effective* (rms) value of an ac voltage is that value which will produce the same effect as a numerically equal value of a dc voltage. The effective value of one ac ampere is an ac value that will produce the same heating effect as one dc ampere. If we take all possible values of *I* for one-half of a cycle and square them, then find the average of these squares and extract the square root, the result would be the *root-mean-square* (rms) value. If we figured the *rms* value, we would find that it is 0.707 *times* the peak value of the sine wave. Most electrical instruments and meters have their ac scales (if any) calibrated in *rms* units. This allows a direct comparison of ac and dc values.

The *average* voltage is merely the average value of the sine wave during one-half cycle. Geometrically, it is the height of a rectangle with the same area as the space between the sine wave and its base line for one-half cycle. (The average for a full cycle of sine wave would be zero.)

The relationships between the ac values we have been discussing are as follows:

$$\text{peak} = 1.414 \text{ times rms}$$
$$\text{rms} = 0.707 \text{ times peak}$$
$$\text{peak} = 1.57 \text{ times average}$$
$$\text{average} = 0.637 \text{ times peak}$$

Radian Measurement

It is quite common to measure generating angles and their corresponding portions of sine waves in *radians* instead of in degrees. Fig. 5-4 illustrates the meaning of the term *radian*. A *radian* is an angle which, if placed with its vertex at the center of a circle, will subtend an arc equal in length to the radius of the circle. Picture a clock. Let us say that the arc between the *one* and the *three* is as long as the minute hand. The angle formed by the minute hand and the hour hand at 1:15 is the radian. Or, assume that the circle in Fig. 5-4 is made of string. If we cut out the string between points A and B, and pull it until it is straight, and if the string is as long as the radius *OA*, the angle *AOB* will be *one radian*.

How many radians are in a circle? We know that the circumference of a circle equals $2\pi r$. Since an angle of one radian subtends an arc equal to the radius, there must be 2π radians in a circle.

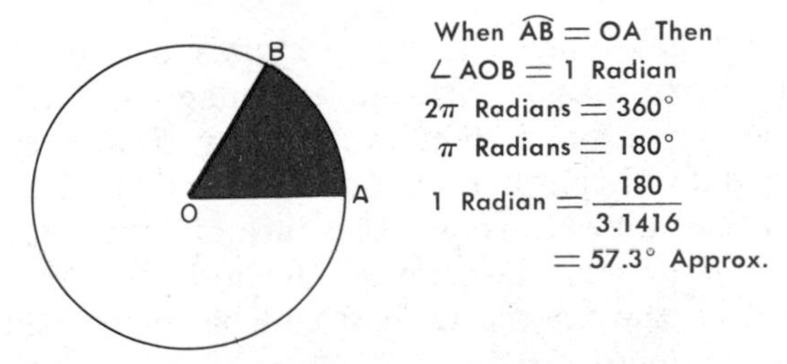

Fig. 5-4. Radian measurement.

The values on a sine waveform are often shown in radians instead of in degrees or seconds. For example, in Fig. 5-2, point *1* on the graph is "0 radians," point *4* is "$\frac{\pi}{2}$ radians," point *7* is "π radians," etc. The term *radians* is usually omitted, and the points are simply marked: "π," "$\frac{\pi}{2}$," "$\frac{3}{4}\pi$," etc. Whenever the symbol appears in this manner, the unit is understood to be the radian.

Summary

Alternating current changes in value continuously and reverses its direction of flow at regular intervals.

A complete cycle of alternating current consists of a complete trip from zero, through all positive voltage values, back to zero, through all negative voltage values, and back to zero.

The frequency of an ac voltage is the number of cycles through which the voltage passes in one second.

A common method for generating ac voltages consists of the rotation of conductors through a magnetic field.

The voltage induced in a rotating conductor depends upon the rate at which the lines of force are cut.

Questions and Problems

1. **What is the frequency of an ac voltage if one cycle is completed in 1/100 second?**
2. **List three ways of changing the voltage induced in a moving conductor in a magnetic field.**
3. **How many times does the current reverse its direction in one ac cycle?**
4. **State two advantages of alternating current.**
5. **What is the *maximum* voltage in an ac circuit when the rms voltage is 110 volts? What is the average voltage in this circuit?**
6. **Why is the *average* voltage in an ac circuit derived from one-half cycle?**
7. **Express the following degrees of rotation in radians: 45°, 90°, 180°, 225°, and 540°.**
8. **What is the effective voltage when the maximum voltage is 311 volts?**
9. **An ac ammeter reads 15 amperes (effective value). What is the maximum value of the current?**
10. **An ac voltage has an average value of 191 volts. What is the peak value?**

INDUCTANCE, REACTANCE, IMPEDANCE

INDUCTANCE

We have seen in the previous section how a conductor passing through a magnetic field will have a voltage *induced* in it. Any time a conductor cuts through magnetic lines of flux, voltage will be induced. To cut the lines of flux, either the *conductor* can be moved through the magnetic field or the magnetic field can be moved through the conductor.

Magnetic Field Around a Conductor

Let us consider a wire, as shown in Fig. 5-5, in which we suddenly cause a current to flow. The current flow will *build up* an electromagnetic field around conductor A. In building up or expanding, the flux lines of the magnetic field will be cut by the other conductor (shown at B), and a voltage will be induced in conductor B. As soon as the electromagnetic field, caused by the current flowing

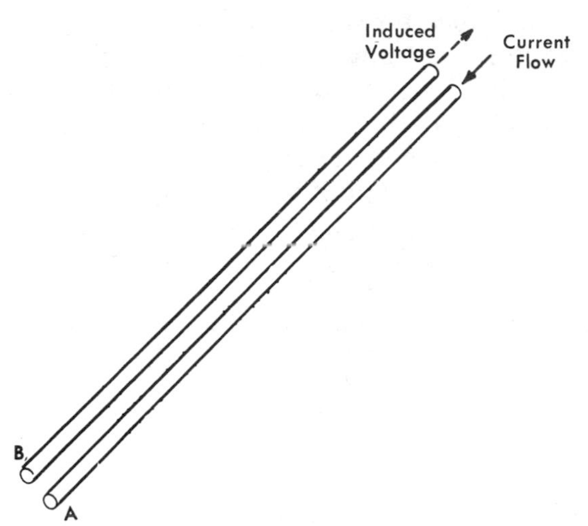

Fig. 5-5. Magnetic field around a conductor.

through A, builds up to its maximum, it will no longer be changing. Because there is no change, no flux lines will be cut by conductor B and no voltage will be induced.

To further illustrate this point, let us consider what happens when the current flow in A is stopped. We know that the electromagnetic field will collapse. In so doing it will again pass through the other conductor and will again induce a voltage in conductor B. Notice that, as long as the magnetic field is changing—either expanding or collapsing—a voltage will be induced in conductor B.

This induced voltage will cause in B a current flow which is opposite to the flow in A. By applying the left-hand rule, we can see that the two electromagnetic fields set up in conductors A and B will oppose each other. We also see that some energy is required before a conductor can be forced through a magnetic field.

These results are summarized by Lenz's law. It states that an induced current in a conductor will form a magnetic field which opposes the motion of the inducing field.

Inductive Circuits

Inductive circuits produce magnetic linkages within their own circuits or with other circuits. When a wire carries varying quantities of electrons, magnetic loops form links within the circuit. The number of linkages depends upon the circuit design. When the circuit wires are arranged as shown in Fig. 5-5, the cross-sectional area of the flux path is small, and the total linkages are almost zero.

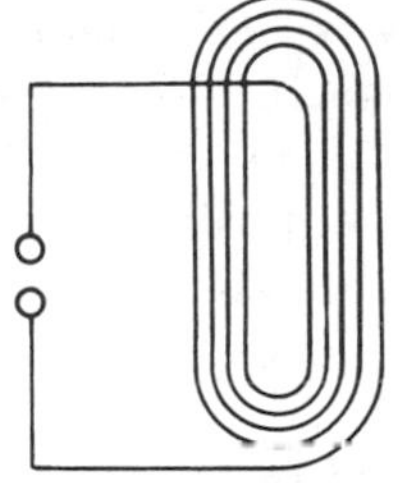
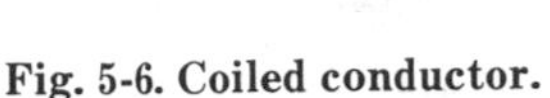

Fig. 5-6. Coiled conductor.

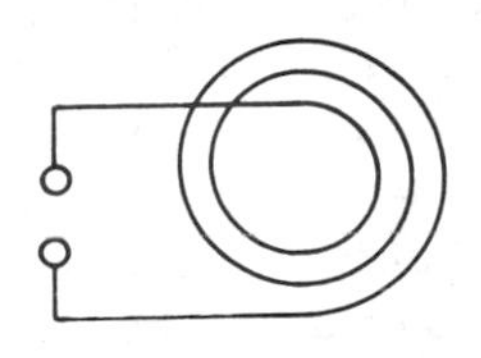

Fig. 5-7. Circular coiled conductor.

In Fig. 5-6 the conductor is shown wound into a coil. This circuit produces many more flux linkages per ampere than the one in Fig. 5-5, because the increased number of turns has increased the total flux. Furthermore, the increased flux is now linked with the circuit an increased number of times, since each line of flux tends to link with all turns of the coil. The coil in Fig. 5-7 is the same as the one in Fig. 5-6, except that its shape has been changed in order to provide the maximum cross section for the flux path.

In Fig. 5-8 an iron core has been added. This core provides a better path for the lines of flux and thus increases the strength of the magnetic field. By comparing the illustrations you will notice that the ability of an inductor to produce magnetic linkages depends upon:

a. The number of turns of wire in the inductor.
b. The area of the magnetic flux path.
c. The strength of the magnetic field.

When a circuit will produce 100,000,000 magnetic linkages for each ampere of electron flow, it has an inductance of one *henry.* In a circuit that has an inductance of one henry, electron flow changing at the rate of one ampere per second will produce a self-induced potential difference of one volt in the circuit.

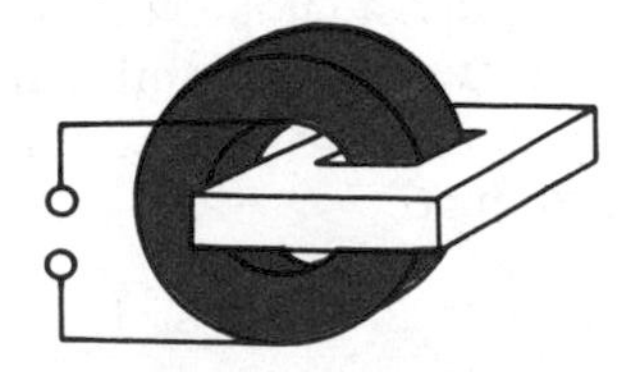

Fig. 5-8. Coiled conductor with an iron core.

REACTANCE

Inductive Reactance

In a dc circuit, the current does not change its rate of flow, but continues at a steady rate. The only opposition to current flow is the resistance of the conductor. In either a pulsating dc circuit or an ac circuit, however, other forms of opposition may appear, in the form of opposing voltages. These voltages, often referred to as *counter emf* or *back emf,* may be responsible for most of the opposition to the electron flow.

This counter emf is developed only in response to a *change* in the current rate, and it tends to oppose the change. That is, if the current tries to increase for any reason, it opposes the increase; if the current tries to decrease, it opposes the decrease. In an inductive circuit, this opposition to current change is called *inductive reactance.* Its symbol is X_L and its value is expressed in ohms.

The relation between inductance (L) and inductive reactance (X_L) is given by the equation:

$$X_L = 2\pi fL$$

where,

X_L is the inductive reactance in ohms,
f is the frequency in hertz,
L is the inductance in henrys.

Note that X_L is proportional to the frequency. Doubling the frequency (f) will double the reactance because the rate of change of electron flow (in amperes per second) is doubled. The larger the inductance or the higher the frequency, the higher the value of X_L.

An inductance value may be selected, at a given frequency, such that the inductive reactance is high enough—for all practical purposes—to stop the flow of ac through the inductor. However, a steady dc can still pass through the coil. Whenever a mixture of ac and dc is applied to such a coil, dc will pass through, but ac will be stopped. This principle is used often when ac and dc must be separated in a circuit. An inductor used for this purpose is called a *choke coil.* (See Fig. 5-9.) The circuit behavior of such a choke coil is the opposite of that of a capacitor. Where a mixture of ac and dc is applied to a capacitor, the ac flows through, while the dc is stopped.

Capacitive Reactance

When a capacitor is connected to a source of dc voltage, electrons are taken from one plate of the capacitor and added to the other plate. Thus, one plate is positively charged and the other is negatively charged. During the charging process, current flows through the voltage source and all parts of the electrical circuit between the capacitor plates. If the capacitor were a perfect one with no resistance (an unattainable ideal), the charging process would be instantaneous. Since there is some resistance in the capacitor and the circuit to which it connected, charging will require a definite time, even though it may be only a fraction of a second. When enough charge has been transferred to bring the voltage on the capacitor to the value of the source voltage, charging ceases. This occurs according to the equation mentioned earlier, $Q = CV$.

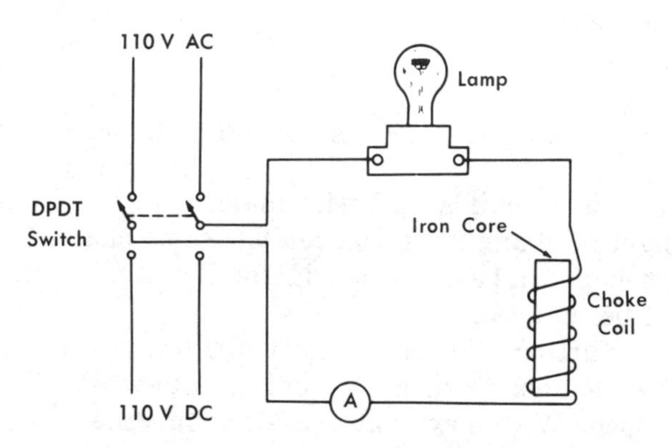

Fig. 5-9. Choke coil used as a dimmer.

At the instant when a voltage is first applied to an uncharged capacitor, the voltage on the capacitor is zero. Then, as just explained, it takes a little while for the capacitor to charge to the applied voltage. If the source voltage is increased or decreased, as would be the case with an alternating voltage, the capacitor charge increases or decreases, but it requires a little time for it to do so; that is, the capacitor "resists" any change in applied voltage.

The charging and discharging action allows a capacitor to pass ac, but it will not pass dc. The amount of ac passed by a capacitor increases as the frequency of the applied voltage increases, or as the size of the capacitor increases. But there is always some opposition to the passage of ac. This opposition is called *capacitive reactance.*

The symbol for capacitive reactance is X_c. Its value is given in ohms and can be computed with the following formula:

$$X_c = \frac{1}{2\pi fC}$$

where,

X_c is the capacitive reactance in ohms,
f is the frequency in hertz,
C is the capacitance in farads.

IMPEDANCE

Impedance is the *total* opposition to current flow. In all circuits this opposition includes resistance. In ac circuits and in pulsating dc circuits in which the current is constantly changing in value, this opposition also includes the inductive reactance and the capacitive reactance. The symbol for impedance is Z, and its value is given in ohms. The value of the impedance Z (in ohms) of a circuit may consist of the resistance R (in ohms) only, the inductive reactance X_L (in ohms) only, the capacitive reactance X_C (in ohms) only, or any combination of these opposition effects. The formulae for determining Z in terms of R, X_L and X_C are shown in Figs. 5-10, 5-11, and 5-12.

Ohm's law for ac:

$$I = \frac{E}{Z}$$

$$Z = \frac{E}{I}$$

$$E = I \times Z$$

where,

I is the current in amperes,
E is the voltage in volts,
Z is the impedance in ohms.

Note that the only difference between this formula and the one for current, voltage, and resistance is that impedance (the total opposition) has been substituted for resistance, which is only one of three forms of opposition that may be present in a circuit.

If the sum of E_R and E_L in Fig. 5-10 is compared with the applied line voltage, we will find that the sum $(E_R + E_L)$ is greater than the applied voltage. The reason is that voltage E_R and E_L do not reach their maximum values at the same instant. As shown by the curves and vector diagrams in Fig. 5-10, these voltages are actually "out of phase" with each other by one-quarter of a cycle or 90 electrical degrees. It is due to this phase difference that the total opposition in the ac circuit cannot be added, but must be combined by means of the equations shown previously. The conditions in Fig. 5-11 are similar to those in Fig. 5-10 except that the voltage E_C lags voltage E_R by the same amount that E_L leads E_R in Fig. 5-10. The voltage across a resistor is always "in phase" with the current flowing through it. The voltage (E_L) in an inductive circuit leads the current flow by 90 degrees. In a capacitive circuit, however, the *voltage* (E_C) *lags* the *current flow* by 90 degrees. The term "out of phase" indicates that the voltages of E_C and E_L do not pass through corresponding values at the same instant, as is true in a circuit with resistance only.

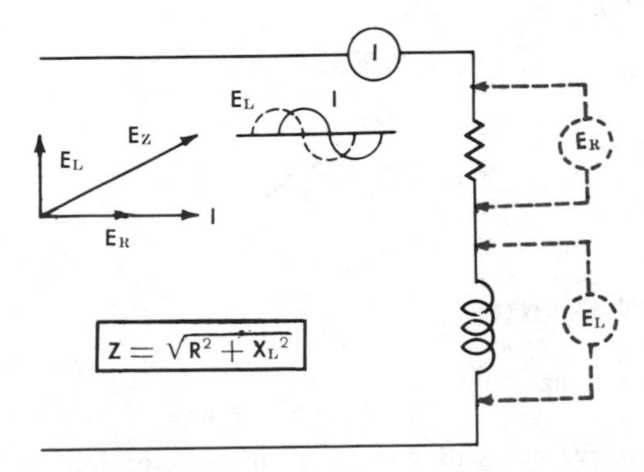

Fig. 5-10. Impedance—combination of inductance and resistance.

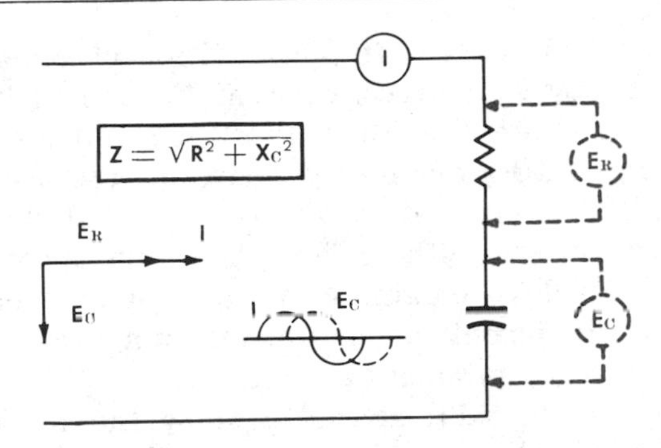

Fig. 5-11. Impedance—combination of capacitance and resistance.

In Fig. 5-12 the phase relationships are shown for R, L, and C in a series circuit. Voltage E_R and

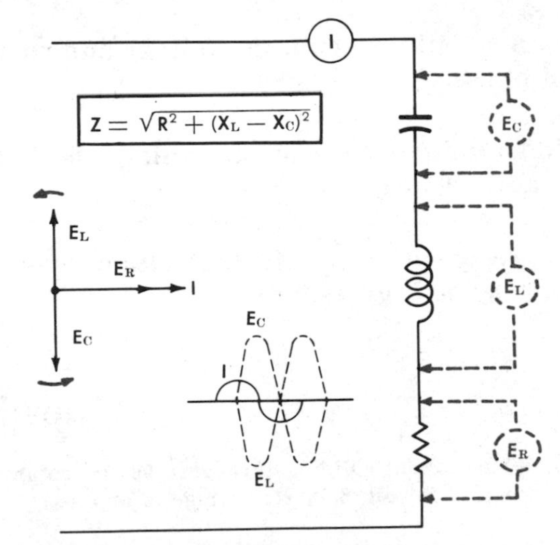

Fig. 5-12. Impedance—combination of capacitance, inductance, and resistance.

current I are always in phase with each other. E_L *leads* E_R by 90 degrees, whereas E_C *lags* E_R by 90 degrees. From the illustration you can also see that E_L and E_C are out of phase by 180 degrees. Also, if the values of E_L and E_C are equal, the two will cancel each other. Thus, the only opposition to current flow remaining in the circuit is from R. When R is the only opposition left in a circuit, it is called a *series resonant* circuit.

Figs. 5-10, 5-11, and 5-12 show the formulas for finding the total impedance (Z) of a circuit containing both resistance and inductive or capacitive reactance. Not shown is a circuit with capacitive and inductive reactance only. However, the formula in Fig. 5-12 is still used because there will always be some resistance in a circuit, even though there is no resistor. The value of R would probably be so small that it would not be used.

RESONANCE

In a circuit which contains inductance and capacitance, there is always one frequency at which the *inductive* reactance is equal to and is cancelled out by the *capacitive* reactance. This frequency is known as the *resonant frequency* of such a circuit at those particular values of inductance and capacitance.

A circuit which has resonance can exhibit either *series* resonance or *parallel* resonance, depending upon the circuit configuration. In a series-resonant circuit the maximum current flows at resonance and the voltages developed across the reactances are many times as great as the voltages applied to the circuit. A parallel-resonant circuit is characterized by high impedance and minimum current at resonance.

Summary

Lenz's law states that a change in current produces a magnetic field that induces a voltage which opposes the current change producing the field.

When a wire is wound into a coil, the flux linkages and the inductance are increased.

A current change of one ampere per second in a circuit with an inductance of one henry will generate one volt of induced potential.

In a resistive circuit, the voltage and current are in phase.

In an inductive circuit, the voltage leads the current by 90 degrees.

In a capacitive circuit, the voltage lags the current by 90 degrees.

Inductive reactance is determined by the magnetic linkages (100,000,000 linkages equal one henry) and the frequency of the circuit.

$$X_L = 2\pi fL$$

Capacitive reactance is determined by the value of the capacitor (in farads) and by the frequency of the circuit.

$$X_C = \frac{1}{2\pi fC}$$

The total resistance of a circuit with both resistance and inductive or capacitive reactance is called impedance. Its value is in ohms.

The formulas for computing the impedance of a circuit are:

$$Z = \sqrt{R^2 + X_L^2}$$
$$Z = \sqrt{R^2 + X_C^2}$$
$$Z = \sqrt{R^2 + (X_L - X_C)^2}$$

Questions and Problems

11. What current will flow through a .005-microfarad capacitor if a 30-volt, 60-hertz voltage is applied?

12. What current will flow through a 6-henry coil connected to a 110-volt, 60-hertz source?

13. State Lenz's law.

14. What current will flow in a circuit containing a 1-μF capacitor and a 2-henry coil connected in series to a 60-hertz, 110-volt source?

15. How many linkages must a circuit produce to have an inductance of 2 henrys?

16. In an inductive circuit, does the voltage lag or lead the current?

17. In a series circuit containing a 2.5-henry coil, a 100-ohm resistor, and a .02-μF capacitor, does the voltage lag or lead the current?

18. Explain how frequency enters into the computing of inductive or capacitive reactance.

19. Define impedance.

20. What is the inductive reactance of a 150-mH inductor operated at a frequency of 100 kHz?

AC POWER

Alternating current flows in a circuit when an alternating voltage is applied. If the circuit contains resistance *only,* the current (I) passes through zero and maximum at the same instant as the voltage (E). The current and the voltage then are in *step,* or in *phase.* The voltage and current are normally in phase in a resistive circuit only. For a resistive ac circuit, Ohm's law is the same as for dc: $E = I \times R$. In an ac circuit containing resistance only, the power (in watts) at any instant is equal to the resistance times the square of the instantaneous current ($P = R \times I_i^2$). The average power during an ac cycle is the average of the instantaneous power values. The square root of the average of all I_i^2 terms is I_{rms}; the *average* power equals $R \times I_{rms}^2$ ($P_{av} = R \times I_{rms}^2$).

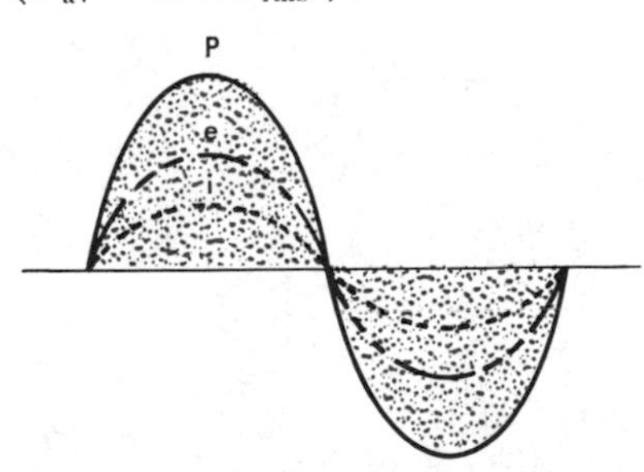

Fig. 5-13. The phase relationships between the voltage, current, and power in a purely resistive circuit.

Fig. 5-13 illustrates the phase relationships in a purely resistive circuit. Notice that all of the power is consumed by an ac circuit containing resistance only.

From the previous section in which reactance was discussed, we have seen how voltage and current are out of phase in an ac circuit containing capacitance or inductance, or both, in combination with resistance. The current may *lead* or *lag* the voltage. Counterelectromotive forces caused by the capacitance or inductance, or both, introduce voltage changes that must be considered when the power of an ac circuit is determined.

Inductive Circuit Phase Relationships

The relationships between voltage, current, and power in a purely reactive (inductive) circuit are shown in Fig. 5-14. The voltage and current are 90° out of phase. The power, illustrated by the shaded areas in Fig. 5-14, is both positive and negative. In a purely inductive circuit, the current resulting from an applied ac voltage goes through its instantaneous, peak, and zero values *behind* the corresponding voltage values. Therefore, the current is *lagging* the voltage. The energy from the power source establishes the magnetic field surrounding an inductance on the positive cycle, and power is stored. However, as the field of an inductor collapses on the negative half of the cycle, the power is returned to the power source. No power is being used up in the circuit.

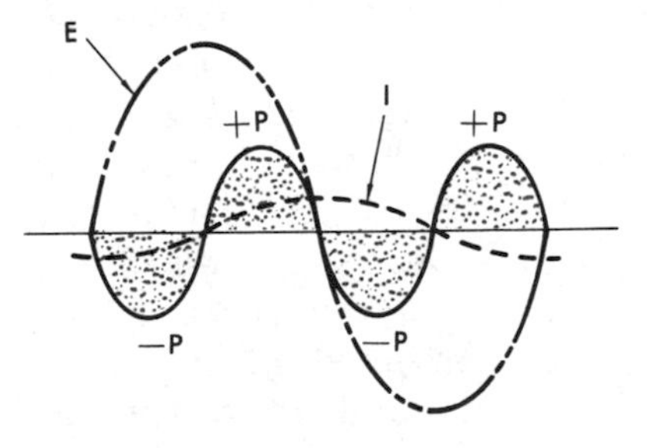

Fig. 5-14. The phase relationships between the voltage, current, and power in a purely reactive (inductive) circuit.

Capacitive Circuit Phase Relationships

The relationships between voltage, current, and power of another purely reactive circuit—a capacitive circuit—are the opposite from those in an

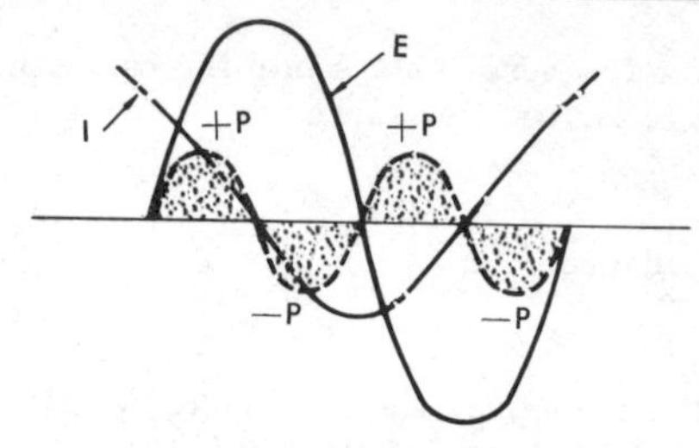

Fig. 5-15. The phase relationships between the voltage, current, and power in a purely reactive (capacitive) circuit.

inductive circuit. The power, shown by the shaded areas in Fig. 5-15, again is both positive and negative. As the energy from the generator or other power source charges a capacitor on the positive cycle, power is stored. As the capacitor discharges on the negative half of the cycle, however, the power is returned to its source. No power is being used up in the circuit. Current flows into a capacitor until the emf across the capacitor is the same as the applied voltage. The current then stops, and the voltage is at maximum. Thus, the current *precedes* the voltage in this reactive circuit.

In a resistive circuit we have shown that all the power is consumed in the circuit and that no power is returned to the voltage source. In a purely reactive circuit we have shown that no power is consumed and that all the power is returned to the voltage source.

Circuit Impedance

In a circuit that contains both resistance and reactance, some power will be used up in the resistance and some power will be returned to the voltage source by the reactance. Even the best coil has some resistance, in addition to its inductance. Even the best capacitor has resistance associated with its capacitance. Resistors, inductors, and capacitors often are connected in series. Thus, an alternating current flowing through an electrical circuit may meet opposition from a resistance (R), an inductive reactance (X_L), and a capacitive reactance (X_c). (Capacitive reactance is considered negative.) The combined effect of resistance, inductive reactance, and capacitive reactance is known as *impedance*. Impedance, resistance, and reactance are measured in *ohms*. The relationships between voltage, current, and power in a resistive and a reactive (capacitive) circuit are shown in Fig. 5-16.

We explained that, in a circuit containing resistance only, the voltage and current are in phase. We also explained that, in a circuit containing inductance only or capacitance only, the voltage and current are 90° out of phase. The circuit in Fig. 5-17 is a combination of resistance and inductance in series. In the vector diagram of Fig. 5-18, we see that the voltage across the resistance is in phase with the current, but the voltage across the inductor leads the current by

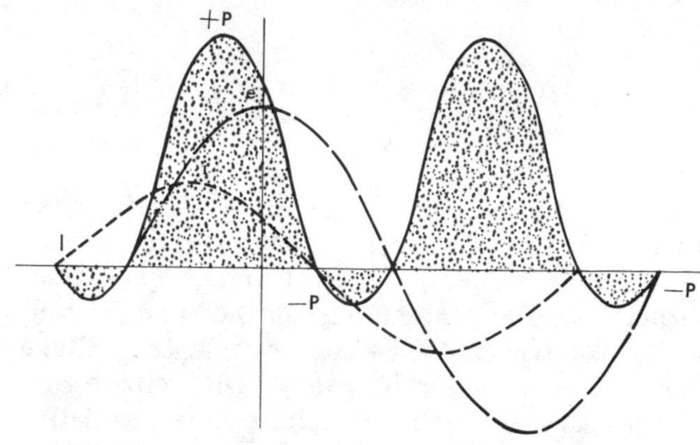

Fig. 5-16. The phase relationships between the voltage, current, and power in a resistive and reactive circuit.

90°. Therefore, we cannot add or subtract by Ohm's law as was done with a dc circuit. Using the vectors in Fig. 5-18, however, we can plot the voltage across the resistance and the voltage across the inductive reactance. We can then determine a value of applied voltage in a circuit containing both resistance and reactance.

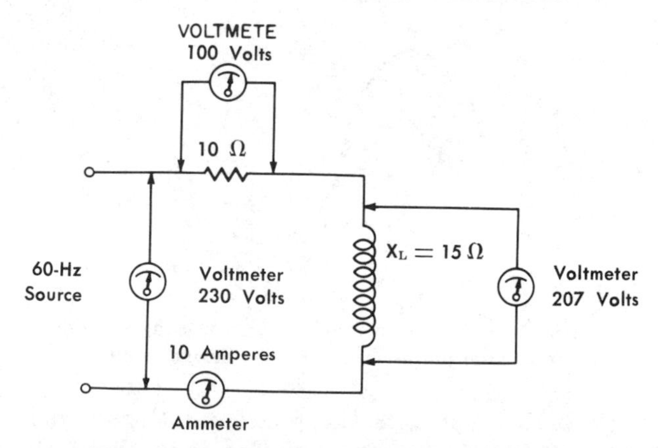

Fig. 5-17. Resistance and inductance in series.

From geometry, we know that the length of the hypotenuse of a right triangle is equal to the square root of the sum of the squares of the other two sides $c = \sqrt{a^2 + b^2}$. Therefore, we can determine the applied (overall) voltage of the circuit as follows:

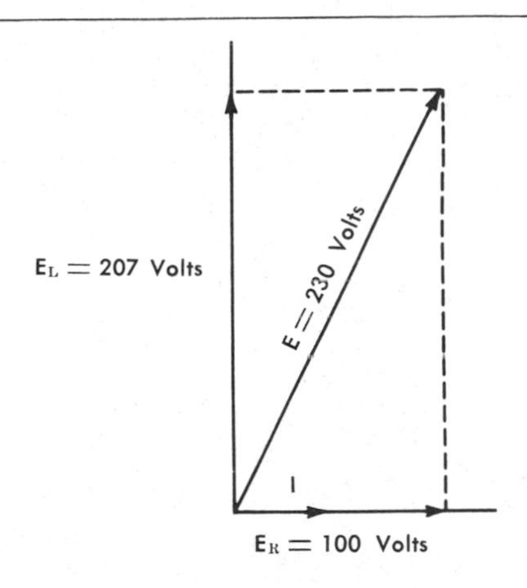

Fig. 5-18. Vector diagram for circuit containing resistance and inductance in series.

$$E = \sqrt{(E_R)^2 + (E_L)^2}$$
$$E = \sqrt{(100^2 + 207^2)}$$
$$E = 230 \text{ volts}$$

where,

E_R is the voltage difference, in volts, across the resistance,

E_L is the voltage difference, in volts, across the inductive reactance.

When there are ac voltages in the circuit, Ohm's law for ac is similar to Ohm's law for dc, except that *impedance* is substituted for resistance:

$$I = \frac{E}{Z}$$
$$Z = \frac{E}{I}$$
$$E = I \times Z$$

Summary

In a purely resistive circuit, all of the power is used up by the circuit, and the voltage and current are in phase.

***Phase* refers to the relationship between ac voltage (emf) and current at a given instant.**

In a purely reactive circuit, the power is returned to the voltage source, and the voltage and current are 90° out of phase.

Current *lags* voltage by 90° in a purely *inductive* circuit, and *leads* voltage by 90° in a purely *capacitive* circuit.

Impedance (Z) is the combined opposition to current flow offered by reactance and resistance and is expressed in ohms.

When a circuit has both resistance and reactance, some of the power is used up in the resistance, and the remainder is returned to the voltage source by the reactance.

Questions and Problems

21. What is the phase angle of a purely resistive circuit? Of a purely reactive circuit?

22. Can the voltages in a circuit with both resistance and inductance be added so that the sum of the voltage drops across the individual units equals the applied voltage? Explain.

23. Does the voltage across a capacitor lead or lag the current in a circuit?

24. Does resistance in a circuit consume any power?

25. Find the impedance of a circuit in which a resistor having a resistance of 12 ohms is connected in series with a coil having a 60-hertz reactance of 7.54 ohms.

26. A resistor having a resistance of 150 ohms is connected in series with a capacitor having a 60-hertz reactance of 106 ohms. What will the impedance of the circuit be?

27. What is the value of the current through a coil with a resistance of 7 ohms and a 60-hertz reactance of 24 ohms, when connected to a source of 110-volt, 60-hertz alternating current?

28. What would the current value be through the coil described in Question 27 if it were connected across a 110-volt dc supply circuit?

6

ELECTROMAGNETIC INDUCTION

GENERATORS AND ALTERNATORS

Direct current is a *one-way* flow of electrons. A two-way flow of electrons—a current which flows in one direction and then reverses and flows in the opposite direction—is an alternating current.

An electric *generator* is a machine that converts mechanical power into electrical power. Commercial generators consist of an armature coil which turns around magnetic poles.

Alternating current voltage cannot be obtained directly from cells or batteries, but usually originates in a special kind of generator called an *alternator*.

The mechanical energy that turns the armature coil comes from several sources, such as a steam engine, a turbine driven by water or wind, an internal combustion engine, or an electric motor. This source is called a prime mover. The mechanical energy required depends upon the electric power desired. In other words, the larger the electric power required, the larger the mechanical energy needed to generate it.

GENERATORS

Generator Types

There are two types of generators—ac and dc. One of the main differences between them is the way they take the current from the generator. A pair of collector rings or slip rings is used in the ac generator, and a commutator is used in the dc generator. The other parts of a generator are the field and the armature discussed earlier. The field provides the magnetic lines of force. These lines are cut by the armature winding as it turns within the field. The *dynamo* is a machine for transforming mechanical energy into electrical energy. It is called a *generator* or an *alternator*. A generator can produce either ac or dc, depending on its construction, but an alternator produces ac only.

AC Generator Action

The principle of the ac generator is shown in Figs. 6-1A through 6-1E. The figures also show the progressive development of an ac sine wave during one revolution. The external circuit is connected to the slip rings by the brushes. Each brush always contacts its own slip ring.

One side of the circuit is always connected to side A of the armature loop and the other side to B. The current through the loop flows in one direction during one-half of the revolution and in the opposite direction during the other half. This action gives the sine-wave voltage shown in Fig. 6-1E.

DC Generator Action

The operating principle of a dc generator is shown in Figs. 6-2A through 6-2E. The voltage developed in the armature loop is ac, like the voltage in the ac generator. However, the commutator reverses the connections during every half cycle,

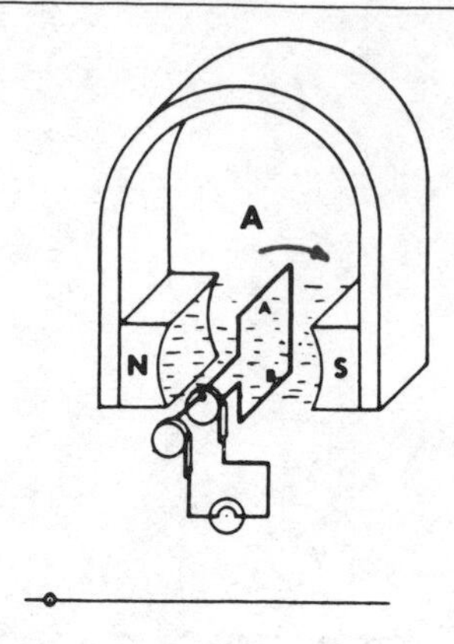

(A) Coil sides A and B are moving parallel to the lines of force at this instant, and no emf is generated.

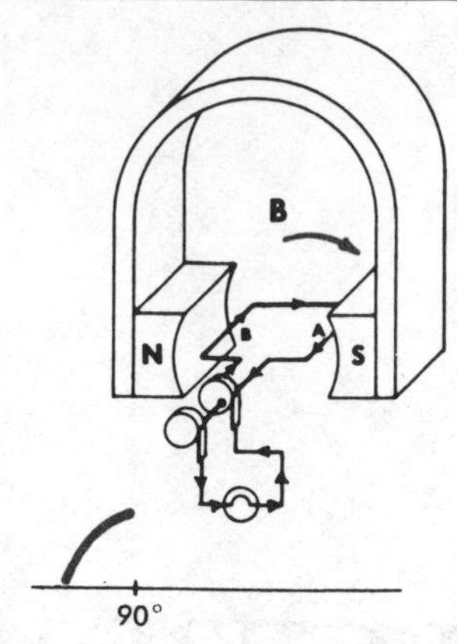

(B) Coil sides A and B are moving directly across the lines of force. At this position the generated emf is maximum.

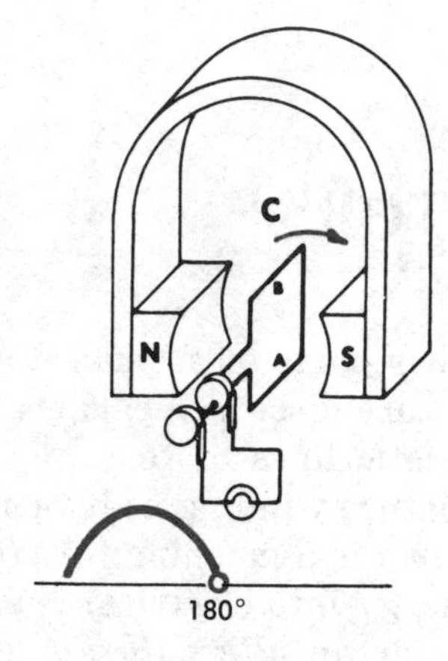

(C) A and B are again moving parallel to the field flux, and the voltage is zero. See voltage curve above.

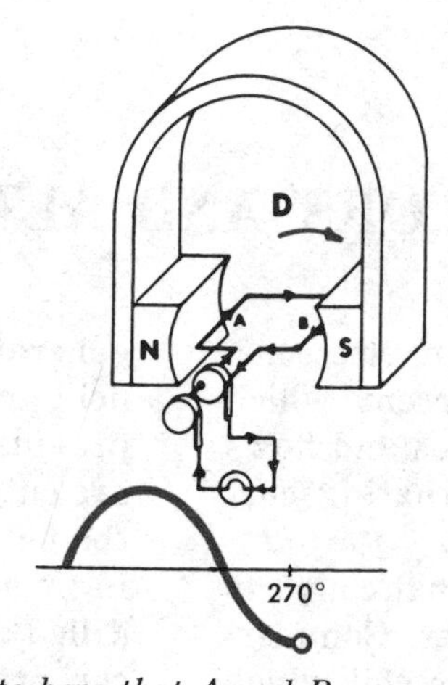

(D) Note here that A and B are now moving under opposite poles, and the coil voltage is reversed.

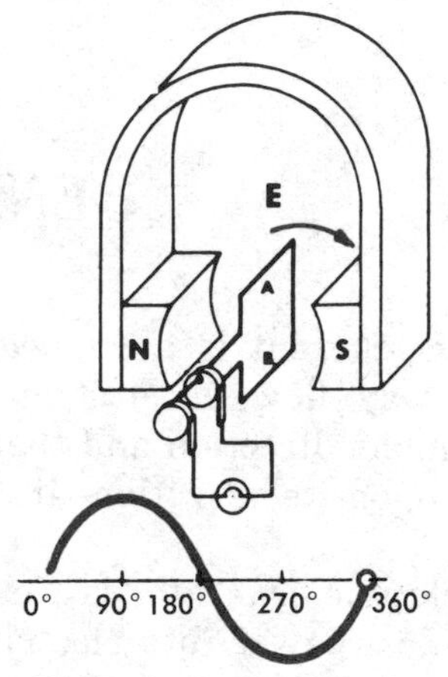

(E) Coil sides are back in the original position and the cycle is completed.

Fig. 6-1. Ac generator.

so that pulsating dc (rather than ac) is delivered to the circuit.

The Magnetic Field

Figs. 6-1 and 6-2 show a permanent magnet as the field supply. This type of field is used in magneto-generators, which are small machines designed for telephone-line signaling or gas-engine ignition, among others. Electromagnetic fields supplied by field coils, either separately or self-excited, are used in most generators and motors.

Series-, Shunt-, and Compound-Wound Generators

The field coils of generators (and motors, too) can be wound in any of the ways just named. However, series-wound generators will not be explained here because they are rarely used.

A shunt-wound dc generator is shown in Fig. 6-3. It has a separately excited field at A and a self-excited field at B. Fig. 6-4 shows a compound-wound generator. It combines shunt and series field windings into one unit.

Residual Magnetism

When a generator remains idle for some time, current is obviously not flowing through the coils. Consequently there will be no magnetism in the field, except for a small amount left over from the last time the generator was operated. This retained magnetism is called *residual* magnetism. It is usually enough to gradually build up the field current when the generator is restarted. Sometimes the residual magnetism drops to zero or even reverses its polarity. The generator can be easily restored to normal by lifting the brushes from

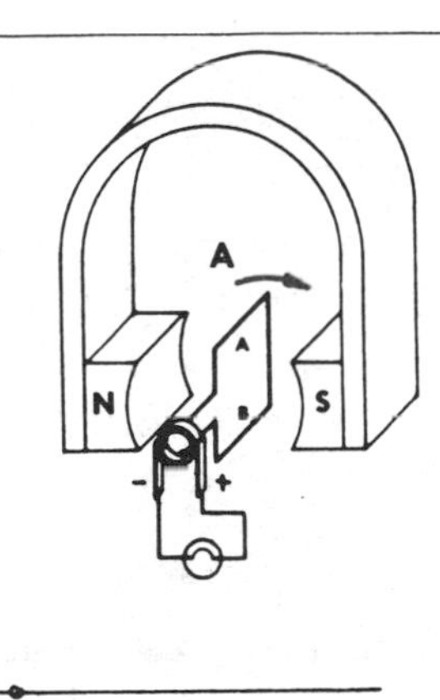

(*A*) *Coil sides are generating no voltage, and brushes are short-circuiting the coil.*

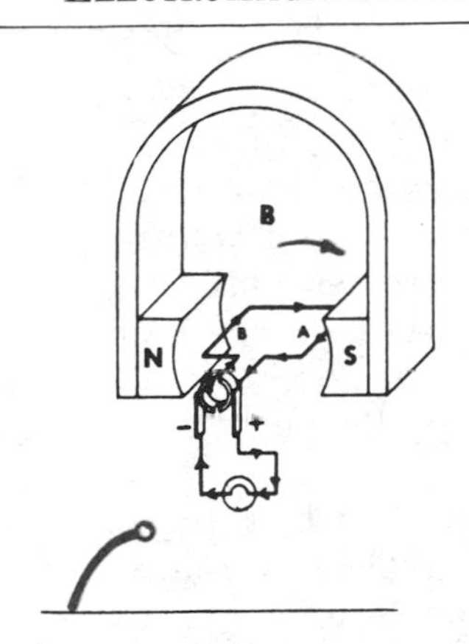

(*B*) *Maximum voltage is generated at this point. Brushes collect current from segments.*

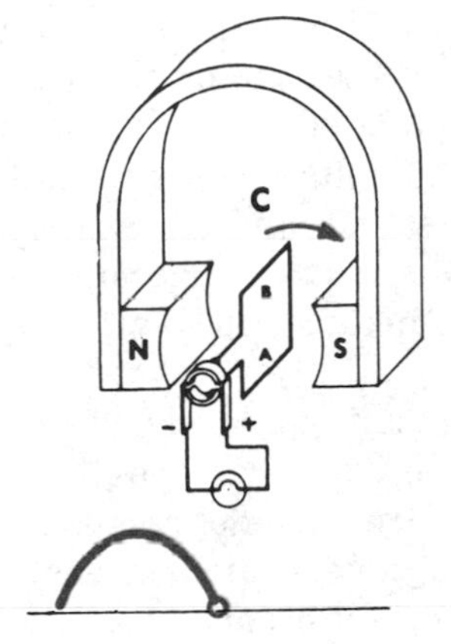

(*C*) *Generated voltage falls to zero as conductors A and B again move parallel to lines of force.*

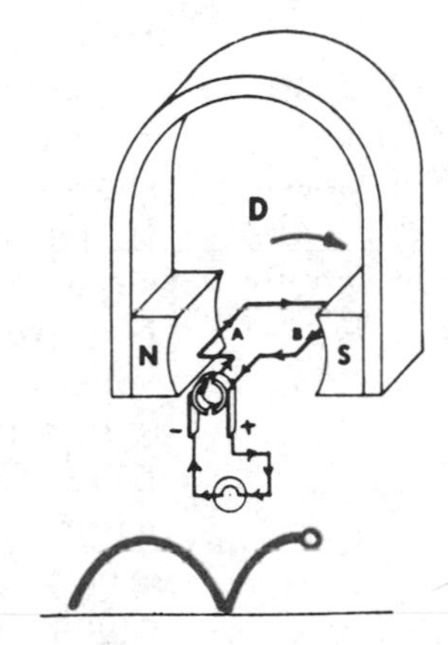

(*D*) *Coil voltage is reversed. Note that segments under brushes reversed at the same time.*

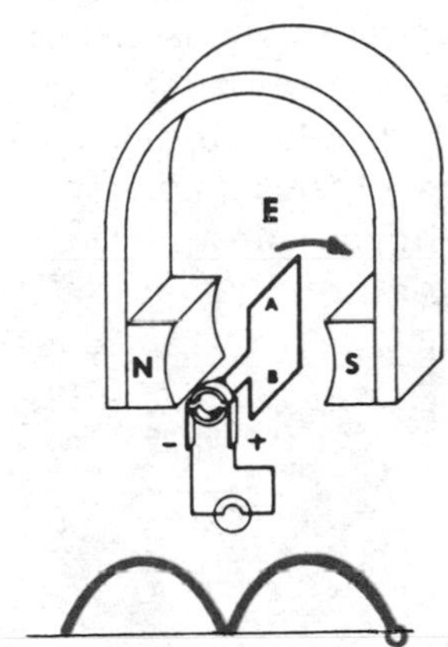

(*E*) *One revolution completed. Note that in the external circuit the current flow is in the same direction at all times.*

Fig. 6-2. Dc generator.

the commutator and sending a surge of current through the field coils in the proper direction.

The field polarity can be reversed in a dc generator, of course. However, in such dc applications as battery charging and electroplating, the correct polarity must always be maintained.

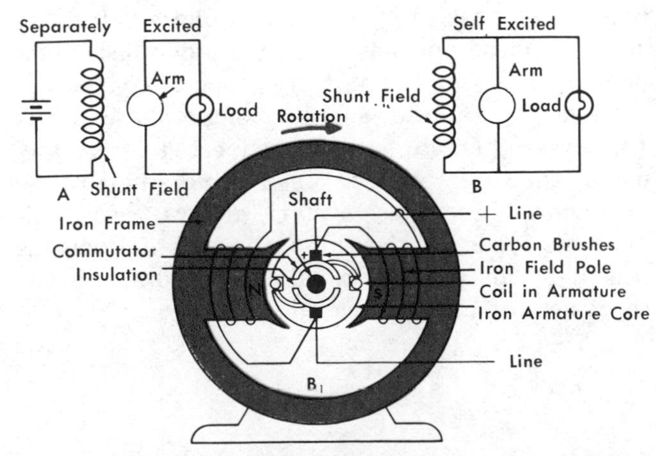

Fig. 6-3. Shunt-wound dc generator.

Number of Poles

In the previous illustrations we have shown one pair of pole pieces only, in order to keep the discussion simple. Actually, most machines contain several pairs for greater efficiency and reduced heating effects.

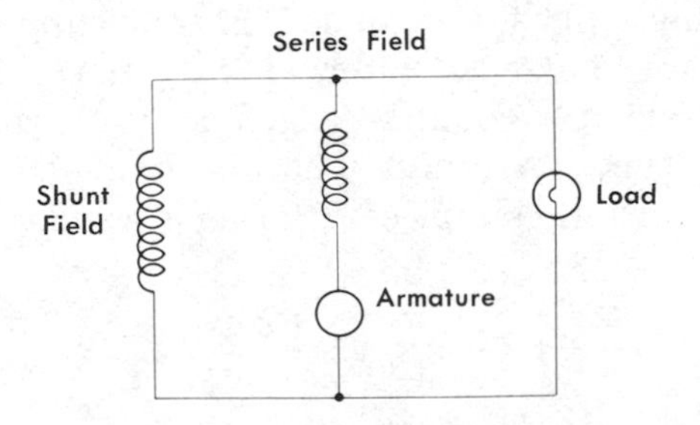

Fig. 6-4. A compound-wound generator.

ALTERNATORS

An alternator starts out with a zero voltage. It then builds up a voltage in one polarity. This voltage increases until the maximum is reached and then decreases again to a zero value. The voltage then builds up again to a maximum value (but in the opposite polarity) and then decreases to zero.

Current must be supplied from some source to energize the field magnet of an alternator. Sometimes the field is magnetized by current from an outside source, such as a battery of storage cells. (See Fig. 6-5.) A small direct-current dynamo, called an exciter, is generally used to energize the field coils of an alternator. By this means the current through the field is kept constant, and the voltage of the alternator does not vary if its speed is uniform. Most direct-current generators are *self-exciting* dynamos. They supply current to energize their own field coils.

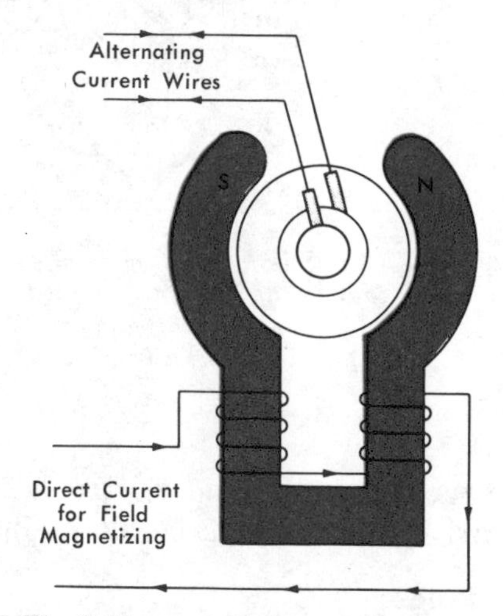

Fig. 6-5. The field magnets of an alternator are magnetized by a small dynamo or by a storage battery.

SINGLE-PHASE AND THREE-PHASE CIRCUITS

A single-phase circuit is one that supplies voltage from one set of alternator (generator) coils. The alternator could be either a single- or three-phase type. A single-phase component, when broken into its simplest form, is always served by two lines. Modern single-phase house service is generally three lines—one from each side of a transfomer secondary coil and the third a ground or neutral line from a center tap on the secondary.

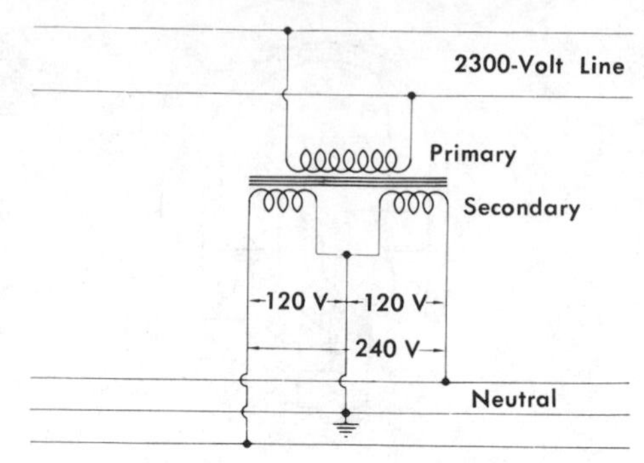

Fig. 6-6. A three-wire ac circuit supply.

Two voltages are available from the service shown in Fig. 6-6: 120 volts when a circuit is between L_1 *or* L_2 and neutral, and 240 volts when a circuit is between L_1 *and* L_2. Older house service had two lines, L_1 and neutral, supplying only 120 volts.

In a three-phase circuit the voltage is supplied from three sets of coils from a three-phase alternator. The coils are spaced 120 degrees apart, and, even though the voltage generated in each coil is equal, there is a timing difference ($\frac{1}{180}$ second in 60-hertz, 3-phase) between the voltage peaks of each phase. When only three-phase components are served, the system requires only three lines; however, four and sometimes six lines are used.

Be sure to check the phase voltage before connecting a single-phase load. Voltages vary in the different supply systems. When three lines are used, the return is through the other two in each phase. When four lines are used, the fourth is a ground or neutral return. When six lines are used to a three-phase component, each phase is used as a single phase and has its own single-phase components built into the unit. By properly connecting into a three-phase supply, single-phase loads (motors, lights, etc.) can be connected across any one of the three phases. When using single-phase loads on a three-phase system, always try to balance the loads on the three phases as closely as possible. This presents a balanced load across the three sets of coils in the alternator.

Summary

An electric generator is a machine that converts mechanical power into electrical power. An alternator is a special kind of generator or dynamo which produces alternating current voltages.

The three main parts of a generator are the magnetic field, the armature, and the current-collecting system.

Generator action is based upon the rotation of conductors through a field of magnetic flux.

One of the principal differences between ac and dc generators is the system of collecting current. The terms series-, shunt-, and compound-wound refer to the way the field windings are connected to the armature.

Questions and Problems

1. Define "electric generator."
2. Define "alternator."
3. What is the purpose of the field in a generator?
4. What is the purpose of the armature in a generator?
5. Is the voltage developed in the armature of an ac generator different from that of a dc generator? Explain.
6. Draw the diagram for a compound-wound generator and label it.
7. How would you restore a field that had lost its residual magnetism?
8. Which type of generator uses a commutator as the current-collecting device?
9. What is the source of power for the field magnets of a self-excited generator? For the field magnet of an alternator?

MOTORS

An electric motor is a machine that converts electrical energy into mechanical energy. Notice that this definition is just the opposite from that for electric generators. In fact, most generators will operate as motors if current is applied to them. However, it is impractical to use them interchangeably.

There are many more types of motors than generators because motors are more widely used and more easily controlled. Motors range from those developing fractional horsepower to ones developing thousands of horsepower.

Fundamental Motor Action

Whenever current is sent through a conductor in a magnetic field, the conductor will move. Notice that this action is just the opposite from that for generators.

Fig. 6-7 shows how the field around a conductor reacts with a magnetic field to produce motion. Fig. 6-7A shows a cross section of the wire. The dot within the circle indicates that current is flowing toward the viewer. The arrows show the direction of the circular field produced by the current.

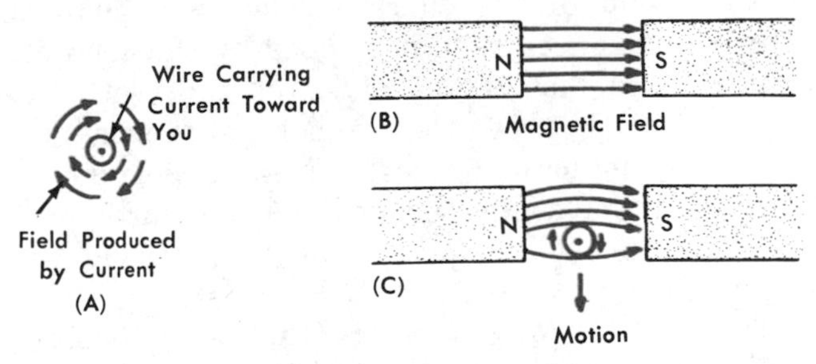

Fig. 6-7. Motor action.

At Fig. 6-7B the magnetic lines of force are going from the north pole of a magnet to a south pole, in the conventional manner. When A and B are combined, the result is Fig. 6-7C. On one side of the wire, the lines of force around it add to the lines produced by the magnetic field. On the other side, the lines of force around the wire oppose the lines produced by the magnetic field. As a result, the conductor moves toward the weaker field (downward, as shown by the arrow).

Fig. 6-8 shows the right-hand rule as applied to motors. The rule states that, when the forefinger points toward the magnetic lines of force and the middle finger points toward the current in the conductor, the thumb will point in the direction the conductor is moving.

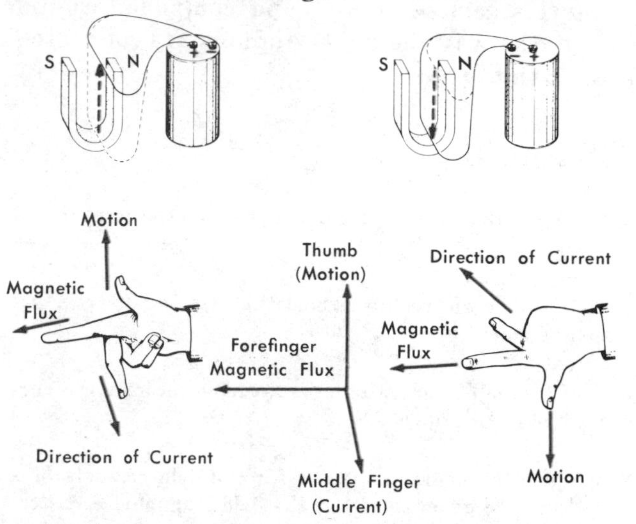

Fig. 6-8. The right-hand rule for motors. The forefinger always points in the direction of the magnetic flux—from N to S—the middle finger always points in the direction of the current—from negative to positive—and the thumb always points in the direction in which the conductor moves.

Fig. 6-9 shows how a conducting loop acts in a magnetic field. The two small circles represent the cross section of the loop; the cross and dot, the direction of the current. The small arrows show the magnetic field developed by the current. You can see that these fields, reacting with the lines of force between the two poles of the magnet, force the conductor nearest the south pole downward and the conductor nearest the north pole upward. This action pivots the loop around the center point, P.

Another way of explaining this is to consider the loop as a solenoid. When the current flows in the direction indicated, the electromagnet formed will have a north pole at the top and a south pole at the bottom. These poles will be attracted by the unlike poles of the field magnet. Hence, the loop will turn in the direction shown in Fig. 6-9.

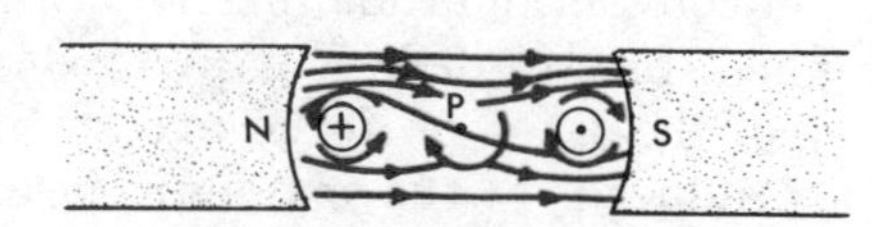

Fig. 6-9. Motion of a current-carrying loop in a magnetic field.

Practical motors have many turns of wire for the armature coil, instead of just the one shown here. Also, more than one set of poles will normally be used.

DIRECT-CURRENT MOTORS

Types of DC Motors

Like the electric generator, dc motors come in three types—the shunt-wound (Fig. 6-10), the series-wound (Fig. 6-11), and the compound-wound (Fig. 6-12).

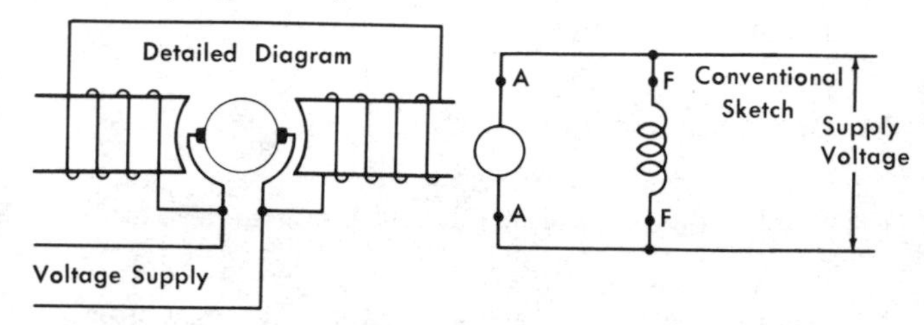

Fig. 6-10. Shunt-wound dc motor.

The shunt motor gives a more constant speed than the other types do. However, its limited starting torque makes it useless where constant starting and stopping are required. It will not "run away" under no-load conditions.

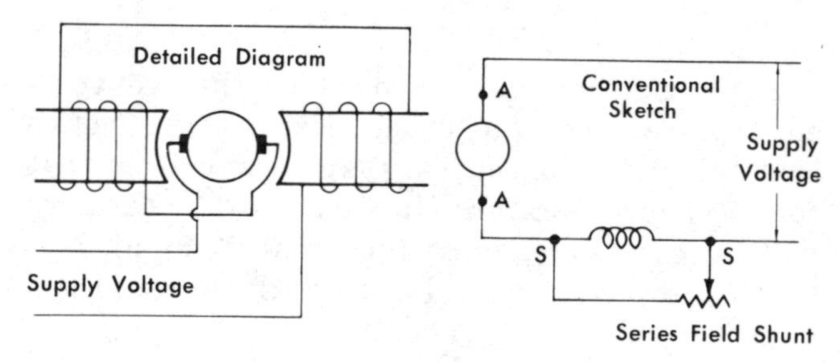

Fig. 6-11. Series-wound dc motor.

The series motor has excellent torque at slow speeds and will operate at high speeds under light loads. However, it will "run away" under no-load conditions. Therefore, some sort of regulating device is normally used.

The compound motor combines the best features of the two previous types. It has good starting torque and will not "run away" under no-load conditions.

Construction

The motor frame is made of iron because it completes the magnetic circuit for the field poles (Fig.

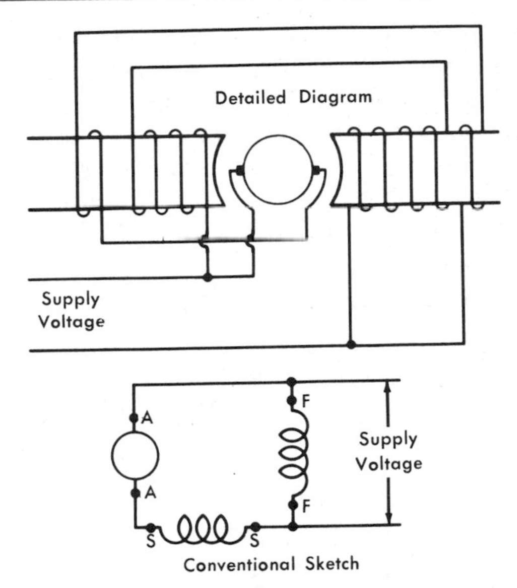

Fig. 6-12. Compound-wound dc motor.

6-13). The end construction of the frame varies. If the motor is to be used in dirt- or dust-free locations, the end plates (bells) of the motor frame can be left open so that air can circulate through the machine.

If the motor will be exposed to dirt or dust, the end plates are covered to keep dirt or dust out of the machine.

A closed construction is used when the motor must be operated under severe conditions, such as in cement plants, flour mills, and saw mills.

The field poles are made either of solid iron or thin strips of iron called *laminations*. The laminated construction helps reduce eddy-current losses.

The three types of field windings have already been mentioned. The shunt type is characterized by many turns of wire, the series type by few turns, and the compound type by a combination of the two.

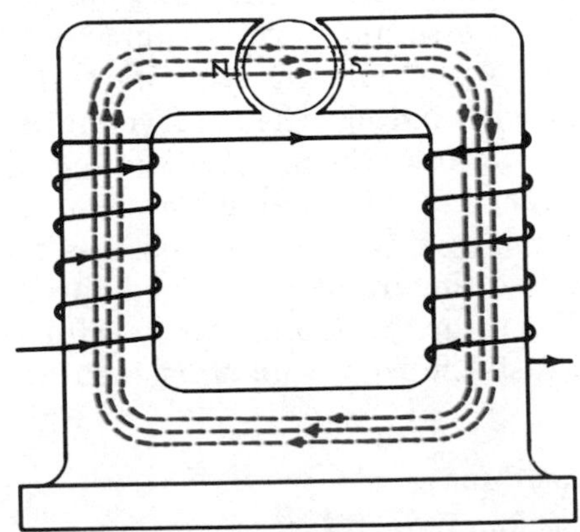

Fig. 6-13. Magnetic circuit for the field poles of a motor.

The Armature

The armature core is made of thin sheets of iron pressed tightly together. The core which becomes part of the magnetic circuit for the field, has a number of slots around its surface. The armature coils are wound in the slots. The ends of the armature coils are connected to the commutator bars. These bars are insulated from each other and from the armature shaft.

Brushes

Brushes apply electrical power to drive the motor by sweeping the commutator. They are made of carbon, graphite, copper, or a mixture of these materials. Small springs hold the brushes against the commutator.

ALTERNATING-CURRENT MOTORS

Most of today's motors are operated by ac. In the average home, the record player, the refrigerator, the washing machine, the food mixer, and many other appliances are powered by small single-phase, 120-volt, 60-hertz motors. In industrial plants, power is supplied by 120-, 240-, or 480-volt, 60-hertz single-, or three-phase motors.

Many of the construction principles are the same in both the ac and dc motors. Magnetic attraction and repulsion applies to both. The ac motor has an alternating supply voltage (usually 60 hertz). Therefore, its magnetic field will alternate. The alternating field provides the rotary motion.

Types of Single-Phase AC Motors

Series (universal) motors can be operated from either an ac or a dc source. Fig. 6-11 shows that the same current flows through the field and armature windings. Therefore, when the polarity is reversed, the magnetic fields of both the armature and the field coils reverse their polarities together. Thus, the reaction between the armature and the field winding is the same for both ac and dc motors.

Induction motors are used in nearly all fans, refrigerators, and similar appliances. These motors are inexpensive to build and maintain. They require no brush assemblies or commutators, since the armature fields are "induced" by the alternating fields of the field coils (*stator* windings)

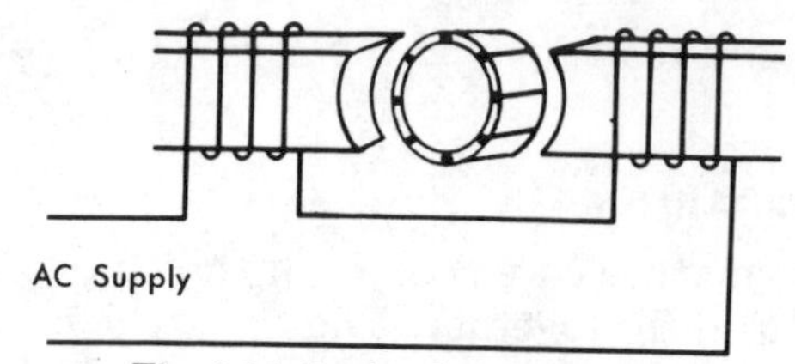

Fig. 6-14. Squirrel-cage motor.

connected to the supply voltage. The rotating assembly (*rotor*) is made up of laminated iron and heavy copper conductors. The conductors are arranged in a cylindrical form that looks like a "squirrel cage." Hence, these motors are sometimes called squirrel-cage motors (see Fig. 6-14). The motor shown here would not start by itself. However, it would run in either direction, once started by an impulse from an outside source.

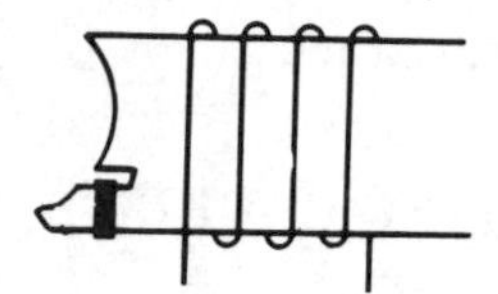
Fig. 6-15. Phase-shifting device for starting small motors.

This initial impulse may be produced by many methods. Most of them depend upon a phase shift between the rotor and stator fields. The feature shown in Fig. 6-15 is used in many record-player motors. A heavy copper strap, placed around part of each pole of the stator, produces the phase shift. The rotation is further smoothed out by angling the "squirrel cage" so that each bar overlaps the stator fields.

Another common starting method is used by "split-phase" motors. Some devices, such as fans start under almost no load. Fig. 6-16 shows the simplest type of starting method for them. Note that there are two windings. One, the running winding, is connected across the supply voltage. The other, the starting winding, is connected through a switch. This switch is closed when the motor is not running. As soon as the motor reaches a certain speed, a centrifugal device opens the switch. The motor then operates as an induction motor.

A variation of the split-phase motor is obtained by adding a capacitor in series with the starting winding. This motor has more starting torque and may therefore be used on such devices as air compressors and refrigerators which start under a load.

Repulsion motors contain a wound armature and commutator bars. The brush assembly is not connected to the supply voltage. The brush and bar mechanism provide a number of low-resistance windings. These windings impart a transformer action between the rotor and stator when voltage is applied to the stator, producing a large starting torque.

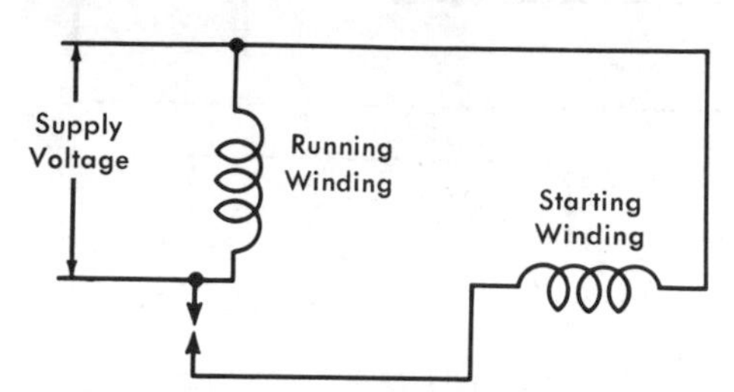

Fig. 6-16. Split-phase motor.

Repulsion-induction motors are a combination of repulsion-start, induction-run characteristics. The armature and brush assembly provides repulsion action for starting. After a certain speed is reached, the armature switches automatically to induction-type operation. These motors are used quite often in factories and in other industrial applications.

Three-Phase Motor

A special three-phase service entrance is necessary when three-phase motors are to be used. Three-phase current is furnished by the power utility company and is usually required when the motor load is in excess of 5 horsepower.

A three-phase motor does not require any special electrical components. It has strong starting torque, high operating efficiency, and costs less than a comparable size single-phase motor. The three-phase motor has three individual windings, each with its own phase voltage impressed across it. Therefore, the currents will start at different times in each winding when the voltages are first impressed across them. The timing sequence causes a very strong starting torque.

Each winding might be visualized as the thumb and first two fingers of your hand spread evenly around the rotor, providing a powerful magnetic hold to grasp the rotor and start turning it (Fig. 6-17).

Two-Phase Motor

You probably will never see a two-phase motor used. However, a discussion of the two-phase

Fig. 6-17. Three-phase motor.

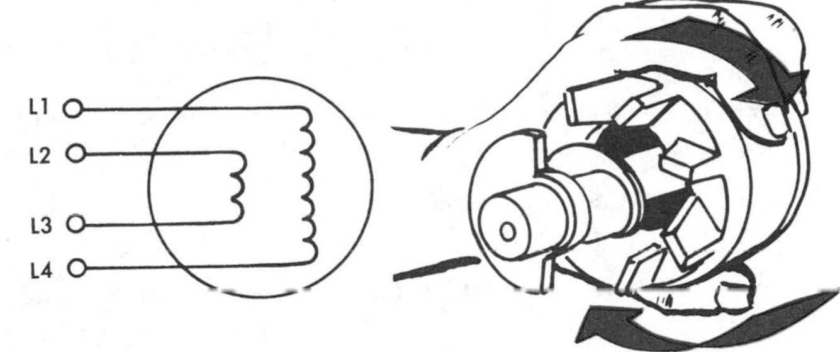

Fig. 6-18. Two-phase motor.

motor can be helpful in understanding single-phase motors.

The two-phase motor has two separate windings, each with its own phase voltage. On starting, the currents will start at different times through the two windings, resulting in a strong starting torque. When running, the motor operates quite efficiently. No special starting components are required for this type of motor.

The two coils might be visualized as the thumb and first finger of your hand grasping the rotor in a powerful magnetic grip and starting it turning, as shown in Fig. 6-18.

It is possible to design a motor so that once it is up to speed, it will run on one winding. It *is not* possible to design a motor so it will start electrically with only one winding. There must be at least two windings with different timing of the current through them in order to provide the necessary magnetic action to twist the rotor on start, causing it to revolve.

Summary

An electric motor is a machine that converts electrical energy into mechanical energy.

Motors come in a wide range of sizes, from fractional horsepower to thousands of horsepower.

Motor action is the result of a current-carrying conductor reacting in a magnetic field.

Motion of a current-carrying conductor in a magnetic field is determined by the right-hand rule which states, "With the forefinger pointing toward the magnetic lines of force (north to south) and the middle finger pointing toward the direction of current flow in the conductor, the thumb will point in the direction the conductor is moving."

The series-wound (universal) motor can operate on both ac and dc.

The alternating current in the field of ac motors rotates the armature.

Both ac and dc motors operate on the principle of magnetic attraction and repulsion.

Many power companies require a three-phase installation when motor loads exceed 5 horsepower.

Each of the three windings in a three-phase motor is energized by its own current. The three currents are out of phase with each other. The timing sequence of the phases causes a very strong torque.

A three-phase motor has a stronger torque, higher efficiency, and lower cost than a comparable single-phase motor.

Two-phase motors are rare. Their operation is similar to that of a three-phase motor.

No starting components are needed for either multiphase motor.

Questions and Problems

10. What are electric motors used for? How are they rated?

11. Why are frames of motors and generators made of iron?

12. Why are field poles and armature cores made of iron?

13. Why are the iron cores and poles laminated?

14. Name three types of ac motors.

15. What type of motor would be most suitable for high-speed, low-power applications?

16. What types of motors would be most suitable for starting under load?

17. In which types of ac motors are commutators used?

18. How do other types of ac motors manage to operate without commutators?

19. What are the main parts of a simple motor?

20. What energy change takes place in a motor?

21. Can a series-wound dc motor be made to run on ac? Explain your answer.

TRANSFORMERS

A transformer is an electrical device that operates by the effects of mutual induction between windings in a changing magnetic field to produce changes in voltage and current. It can also provide electrical isolation and impedance matching between circuits.

There are many types of transformers, some large and some small. Some of the common types are the doorbell transformer, the power transformer which hangs on a power pole and furnishes homes with electricity, and the transformers that power our television sets, radios, and children's toys such as electric trains. Various transformers are shown in Fig. 6-19.

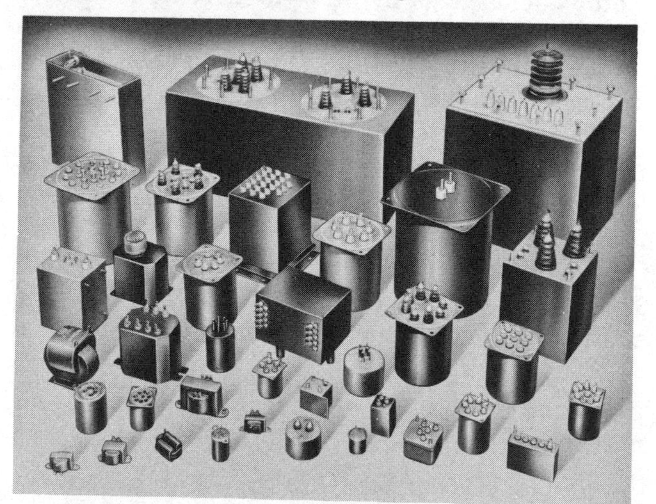

Courtesy Eisler Engineering Co.

Fig. 6-19. Various styles and sizes of small transformers.

A common type of transformer construction consists of two separate windings of insulated wire wound around an iron core. One winding is known as the *primary* winding and the other as the *secondary* winding. The primary winding of a transformer receives energy from an ac voltage source. Changing magnetic lines of flux are set up by the changing current flowing through the primary. The changing current in the primary winding causes an electromagnetic field to build up and collapse. This field cuts through the secondary coil winding and induces a voltage in the secondary. In this way the electrical energy is transferred from the primary to the secondary by *induction*. Notice that the frequency of the electromagnetic waves does not change as the power is transferred from the primary to the secondary windings.

Cores

This fluctuating magnetic energy produced by the primary current is efficiently coupled to the secondary coil by means of a laminated-steel core. The efficiency with which the magnetic energy is coupled is primarily due to the good permeability (magnetic conductivity) of the steel core. The efficiency of a transformer core can be increased by using a core material that has a greater permeability.

Shown in Fig. 6-20 and Fig. 6-21 are two types of laminated cores which are used in transform-

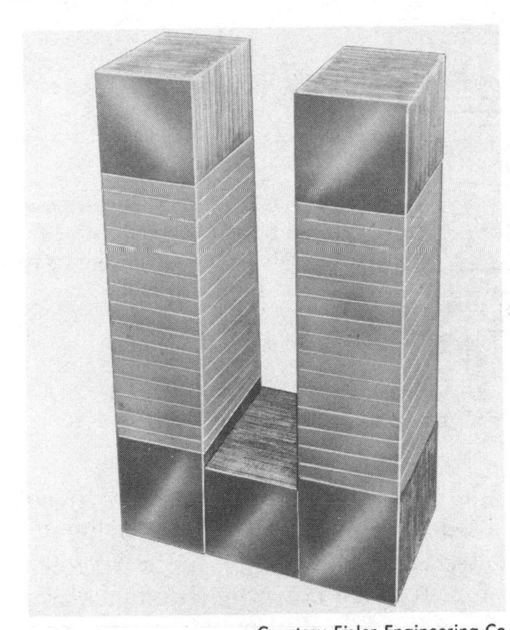

Courtesy Eisler Engineering Co.

Fig. 6-20. The laminated core of a core-type transformer.

ers. Other types and shapes are made for special purposes.

Power Losses

Transformers are designed to keep power loss at a minimum. One of the principal unavoidable losses is the *iron loss*. Iron losses occur in the core and result from two factors, *hysteresis* and *eddy currents*. Hysteresis loss is caused by the resistance that the molecules of the iron offer to being

Courtesy United Transformer Co.

Fig. 6-21. The laminated core of a shell-type transformer.

shifted each time the alternating current is reversed. (The molecules of the core are shifted 120 times each second in a 60-hertz transformer.) The resistance to this shifting is the result of friction, and the friction produces heat. Losses from hysteresis cannot be eliminated; however, they may be reduced considerably by the use of soft steel or a special transformer steel containing silicon. The molecules of these metals shift more easily and produce less friction and heat.

Another loss is due to *eddy currents*. If transformer cores were made of solid steel, as shown in Fig. 6-22, the alternating magnetic flux produced by the transformer primary winding would induce currents in the transformer core. Such currents are known as *eddy currents*. They are called eddy currents because they *eddy* or circulate entirely within the iron core and are really short-circuited currents flowing within the core material. They produce heat for the same reason that any short-circuited current produces heat. A current of electricity will be produced in a conductor, such as in the solid cross section of steel core shown in Fig. 6-22, whenever an *alternating* magnetic flux is passing through it. This, of course, would be the case if the conductor were a part of the core structure of a transformer. Eddy currents may be broken up by slicing the core into thin sections and insulating them from each other, as shown in Fig. 6-23. Eddy currents cannot be entirely eliminated, but they can be reduced to a point where the loss is negligible if the laminations are very thin.

USES OF TRANSFORMERS

Many transformers are used by power companies. There is probably one right in your neigh-

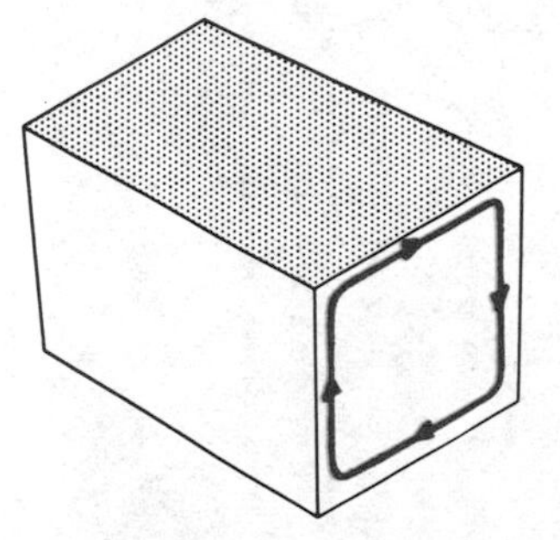

Fig. 6-22. Eddy currents are set up in a solid steel transformer core through which alternating magnetic lines of force are passing.

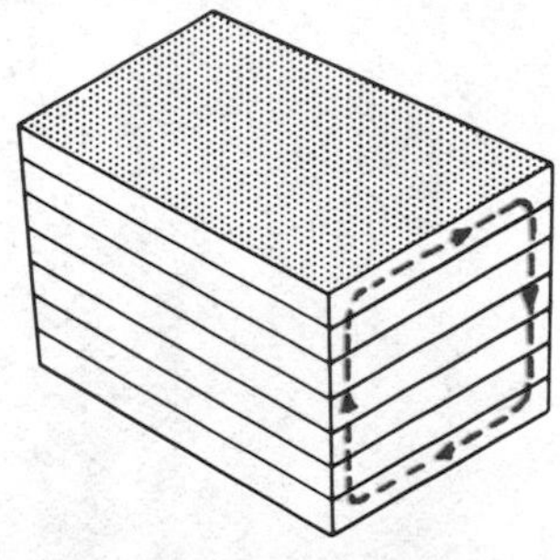

Fig. 6-23. Slicing (laminating) a steel transformer core into thin sheets reduces eddy currents considerably.

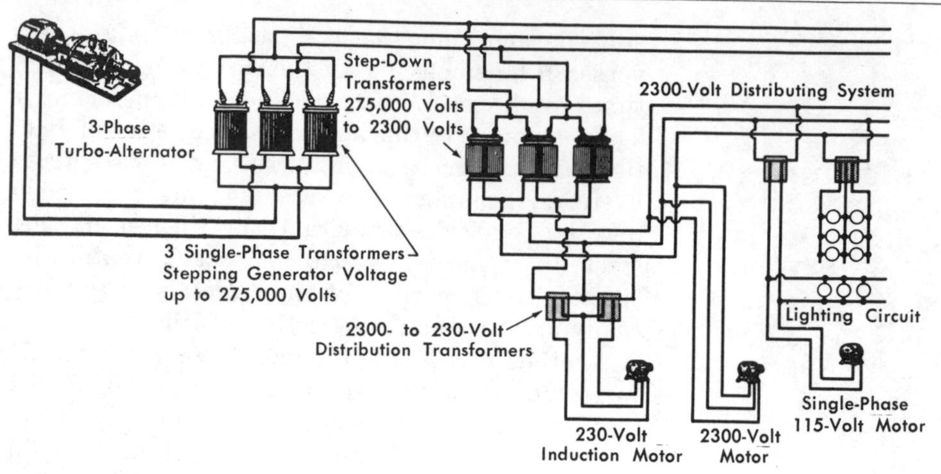

Fig. 6-24. Transformers are used for many purposes in a power transmission and distribution system.

borhood. Fig. 6-24 gives some idea of how transformers are used in stepping up the voltages for distribution or transmission over long distances and then stepping these voltages down at the location where they are to be used.

Voltage Step-Up

Stepping up the voltage to very high levels may seem rather foolish; however, there is a very good reason for doing so. It is done to avoid the large losses of power which would otherwise occur in the transmission lines. If the voltage is stepped up to relatively high levels, it is possible to transmit the power over long distances with small loss. It is dangerous to attempt to use the power at such high voltages. For that reason, the voltage is always stepped down again to a safer value before the power is put to use.

We see in Fig. 6-24, three large transformers used to step up the voltage to 275,000 volts for transmission over long-distance transmission lines. At a distant location, a bank of three transformers is used to step down that 275,000 volts to 2300 volts. Shown in Fig. 6-25 is a typical substation for stepping down long-distance transmission line voltage. Additional transformers step down the 2300 volts to lower voltages for use in homes, in factories, and for street lighting. These distribution transformers are usually mounted on the utility poles near the homes or business places of the customers, as shown in Fig. 6-26.

Theory of Operation

Since a transformer operates on the principle of *induction,* caused by changing currents flowing through the windings, it will not operate if pure dc is applied to the primary coil. When dc of a fluctuating nature is applied, the ac component causes a transfer of energy from the primary to the secondary. The current in the primary coil of a transformer must be *interrupted* or must be varied. That is the reason transformers are generally used with alternating current.

Fig. 6-25. A typical substation where several power transformers are used.

Fig. 6-26. A distribution transformer mounted on a utility pole.

Since the current in the primary coil in a transformer is constantly changing value, it follows that the magnetic flux in the core is also constantly changing. The magnetic flux and its polarity are dependent upon the current flowing in the primary coil. As the direction of the current in the primary coil changes, the direction of the magnetic flux also changes.

It should be remembered that the polarity of the voltage induced in the secondary coil of the transformer depends upon the direction the magnetic field is moving. The magnetic field builds up to a maximum and falls to zero; then it reverses and builds up to a maximum in the opposite direction and again falls to zero. The voltage in the secondary coil will build up, then fall off and reverse, just as the voltage in the primary coil does. The polarity of the voltage induced in the secondary coil will be opposite to the direction of the current in the primary coil if the primary coil and the secondary coil are wound in the same direction.

The strength of a magnetic field produced in a coil will be directly proportional to its *ampere turns*. One ampere of current will produce a certain number of lines of force in one turn of a coil. Either two amperes in one turn of a coil or one ampere in two turns of a coil will produce twice as many lines of force. You can see that the voltage induced in the secondary coil will depend upon the number of ampere turns of the primary coil and upon the number of turns in the secondary winding. This can be expressed by the equation:

$$\frac{E_P \text{ (Primary Volts)}}{E_S \text{ (Secondary Volts)}} = \frac{N_P \text{ (Primary Turns)}}{N_S \text{ (Secondary Turns)}}.$$

A given number of ampere turns for the primary winding will cause a certain voltage to be induced in each turn of the secondary coil. Two turns of the secondary winding will induce twice that amount; three turns, three times that amount, etc. This results from the fact that all the turns of the secondary coil are in series with each other. The total voltage is equal to the sum of all the individual voltages. A transformer having the same number of turns on the primary and secondary windings will have a voltage induced in the secondary winding equal to the voltage applied to the primary winding.

Another method that is sometimes used for calculating primary and secondary voltage relations is the "turns per volt" concept. For example, suppose that a certain transformer is to have 200 turns in the primary and that 100 volts ac will be applied to it. If we divide the number of volts into the number of turns, we get a figure of 2 turns per volt. Now, every 2 turns in the secondary, being subject to the same magnetic field as the primary, will have 1 volt induced. Therefore, if we desire an output voltage of, say, 50 volts, we merely multiply volts desired times turns per volt and get, in this case, 100 turns. Notice that this is the same figure that you would have gotten if you had used the preceding equation.

A transformer is shown in Fig. 6-27 which has 100 turns on the primary coil and 100 turns on the secondary coil. When 100 volts ac is applied across the primary coil of the transformer, 100 volts ac will appear across the terminals of the secondary coil.

Voltage Step-Up and Step-Down

If, instead of having 100 turns in the secondary coil, as shown in Fig. 6-27, the transformer had 1000 turns, as in Fig. 6-28, we would have 1000 volts across the secondary coil. A comparison of the number of turns in the primary coil with the number of turns in the secondary coil is called the *turns ratio*.

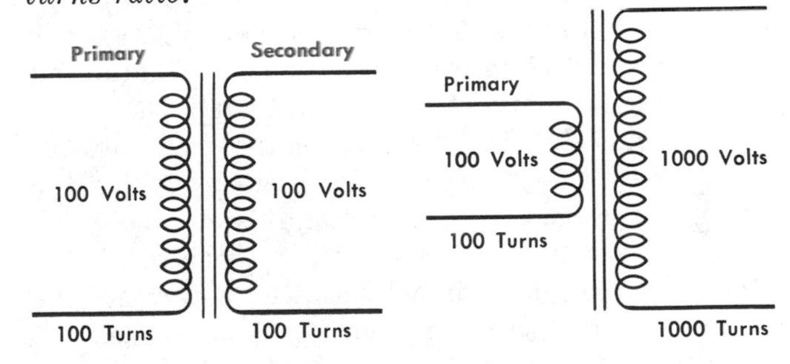

Fig. 6-27. When the secondary has the same number of turns as the primary, it also has the same voltage.

Fig. 6-28. Step-up principle. (When there are ten times as many turns on the secondary as on the primary, the voltage across the secondary is ten times that across the primary.)

We can obtain lower voltages than those applied to the primary coil of a transformer by using fewer turns on the secondary coil than on the primary coil. Fig. 6-29 shows how we can transform 100 volts in the primary winding to 50 volts in the secondary winding by using 50 turns on the secondary winding and 100 turns on the primary winding.

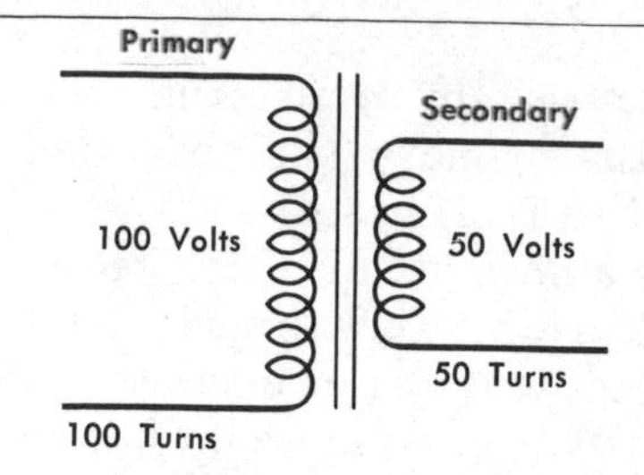

Fig. 6-29. Step-down principle. (When there are only one-half as many turns on the secondary as on the primary, there is only one-half as much voltage developed.)

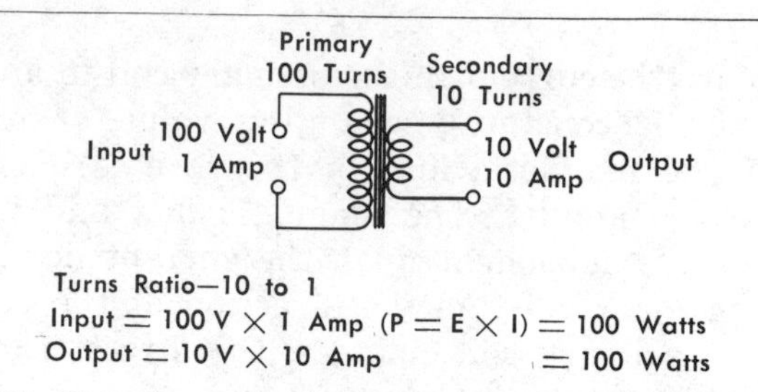

Fig. 6-30. Voltage, current, and power relationships in a step-down transformer.

Theory of Operation

Since it is so easy to increase or decrease voltage merely by altering the turns ratio of a transformer, one might assume that power might be increased or decreased. This assumption, of course, is not valid, since it violates the law of conservation of energy. It is impossible to get as much power out of a transformer as is put into it, because no device can be made to operate at 100% efficiency. There is always some loss. However, if we assume a transformer to be operating at an efficiency of 100%, the amount of transformed energy from the transformer is neither increased nor decreased. Only the ratio of the values of voltages and currents is changed.

Fig. 6-30 illustrates the relationship between the power input and the power output, together with the relationship between the turns ratio and the voltage-to-current ratio of a transformer.

The current in the secondary coil always changes by the inverse of the ratio by which the voltage changes. If the voltage is doubled, the current is halved. If the voltage is raised to 10 times its original value by the transformer, the current in the secondary coil will be reduced to one-tenth the value of the current in the primary coil.

This means that the *voltage* in the primary winding is multiplied by the turns ratio to find the voltage in the secondary winding. To find the current in the secondary coil, *divide* the current in the primary coil by the turns ratio.

The method of transferring electrical energy by the transformer is *indirect*. It first converts the electrical energy into magnetic energy; then it reconverts magnetic energy into electrical energy. Because of this conversion process, the transformer can perform duties which have made it invaluable in the field of electrical science.

THE COOLING SYSTEM

The transformer core loss and the copper loss in the transformer are converted to heat. Unless there is some means provided for continuously removing the heat from the case and windings, they would get progressively hotter and eventually result in failure due to overheating.

In the oil-filled transformer, oil is used not only for its electrical insulating properties, but as a medium for transmitting heat from the core and coil to the outer walls of the transformer tank. These walls then radiate heat to the surrounding air, thus helping to keep the transformer temperature from rising to dangerous levels. The oil bath that surrounds the core and coil does not quite fill the containing tank; enough space is left above the oil surface for expansion as the oil heats. This also helps establish a circulation of oil through heated parts of the transformer.

As the oil is heated, it expands so that its weight per gallon is reduced; i.e., one gallon of heated oil weighs less than one gallon of cold oil. This action causes the heated or expanded oil to rise vertically through the cooling ducts. This rising oil results in a downward column of oil coming in contact with the inside of the tank wall. The tank wall then dissipates the heat carried by the oil to the surrounding air. The oil is cooled, becomes heavier, and assists in setting up the circulating circuit.

Compared with the amount of water needed to carry out the same amount of heat, there would be required 2.46 gallons of oil or 2.5 gallons of Chlorextol liquid for each gallon of water.

Summary

A varying voltage applied to the primary winding produces a varying primary current, which in turn develops a varying flux through the iron core. This varying flux cuts all windings, thereby inducing in each of them a voltage proportional to the number of turns.

The ordinary transformer is a device used either to step up or step down ac voltage, or in some cases, a pulsating dc voltage.

A laminated-iron core is a stack or bundle of thin sheets or strips of iron which are insulated from each other by an oxide film. This arrangement of thin sheets or strips tends to limit or confine the *eddy currents* induced in the iron and thus reduces undesirable heating of the iron.

Hysteresis losses refer to the energy lost in reversing the molecules of the transformer core with each alternation.

Since a transformer operates on the principle of induction, it will not operate if pure dc is applied to the primary coil. Pure dc may burn out the winding.

Questions and Problems

22. Why is it better to use a laminated-iron core than a solid core in a transformer?

23. Is it true that hysteresis can be eliminated in a transformer?

24. If the primary coil of a transformer had 100 turns with 100 volts applied and the secondary coil had 500 turns, what would the voltage across the secondary coil be?

25. How much current would flow in the secondary coil described in Question 24 if 2 amperes of current were flowing in the primary coil?

26. Should a doorbell with a voltage rating of 24 volts be operated from a step-up or step-down transformer if the primary coil of the transformer is connected to a 120-volt source?

27. What is the turns ratio of the transformer described in Question 26?

28. A transformer with a 120-volt primary coil and a 12-volt secondary coil has a power output of 6 watts. What is the current in the secondary coil? What is the current in the primary coil?

7

MEASUREMENT AND CONTROL

METERS

DIRECT-CURRENT METERS

Galvanometer

The permanent-magnet, moving-coil galvanometer (called the D'Arsonval galvanometer) is the basis of most dc meters. (See Fig. 7-1.) It contains a coil of wire wrapped around a movable aluminum form and placed between the poles of a permanent magnet. This unit is similar to the armature of a motor. However, instead of rotating 360 degrees, the unit is held by two coil springs so that it cannot rotate too far on its jeweled bearings. A stationary soft-iron cylinder completes the magnetic path. The coil rotates around this cylinder.

When current flows in the coil, the coil acquires a magnetic polarity and turns against the spring, trying to line up its poles with the opposite poles of the permanent magnet. The greater the current, the greater the deflection will be. If the current is reversed, a meter with a zero-center scale will indicate its amount and direction. Eddy currents induced in the aluminum form will act as a brake and will "dampen" the meter movement without affecting its accuracy.

Voltmeter

A voltmeter is used to measure voltages. It must always be connected in *parallel* with the circuit being measured. The polarity of the circuit must also be noted, so that the current is applied in only one direction. The pointer is usually made to swing in only one direction. The springs act against the rotation and return the moving coil and pointer to zero. The dial is marked in volts, the range being secured by means of a *multiplier* (a current-limiting resistor connected in series with the moving coil). (See Fig. 7-2.)

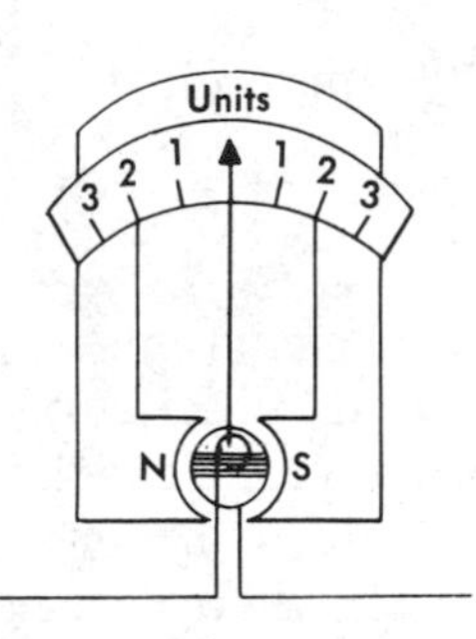

Fig. 7-1. Moving-coil D'Arsonval galvanometer.

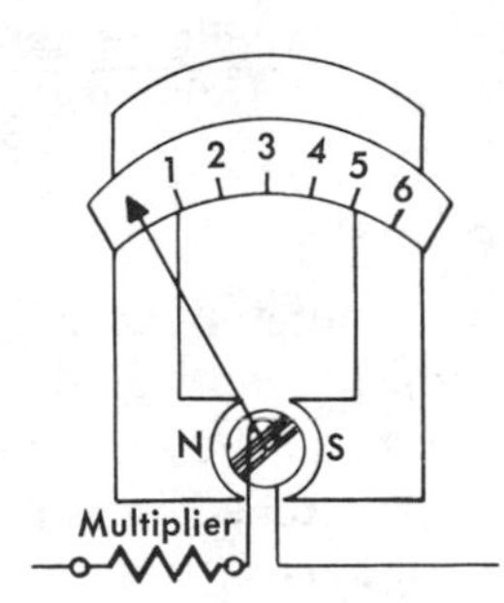

Fig. 7-2. Dc voltmeter.

For example, the moving coil of a typical dc voltmeter has a resistance of 5 ohms and requires a current of 0.01 ampere (10 milliamperes) for full-scale deflection. What must the value of the series resistor be if the meter is to measure 150 volts? For this meter to register 150 volts at the top of its scale, a multiplier resistance must be added which will limit the meter current to 0.01 ampere at 150 volts.

Total resistance (R_T) of the circuit is found from Ohm's law:

$$R_T = \frac{E}{I} = \frac{150}{.01} = 15{,}000 \text{ ohms}$$

where,
R_T is the total resistance in ohms,
E is the applied voltage in volts,
I is the current in amperes.

Since the resistance of the meter movement (R_M) is five ohms, this value is subtracted from R_T to obtain the multiplier resistance (R_{mu}).

$$R_{mu} = R_T - R_M = 15{,}000 - 5 = 14{,}995 \text{ ohms}$$

where,
R_{mu} is the resistance of the multiplier in ohms,
R_T is the total resistance in ohms,
R_M is the resistance of the meter movement in ohms.

This instrument would be known as a:

a. 10-milliampere movement

b. .05-volt (50-millivolt) sensitivity movement

$$E_M = I \times R_M$$

where,
E_M is the voltage of the meter in volts,
I is the current in amperes,
R_M is the resistance of the meter movement in ohms.

c. 100-ohms-per-volt movement

$$\frac{15{,}000 \text{ ohms}}{150 \text{ volts}} = 100 \text{ ohms/volt}$$

Ammeter

An ammeter is used to measure current. It must always be connected in *series* in the circuit. The ammeter is the same as a voltmeter except that its dial is marked in amperes. A drop in voltage is shown across a low resistance when current is passing through the resistance (see Fig. 7-3). This resistance is called a shunt, and the range of an ammeter can be changed by changing the value of the shunt. For example, in the galvanometer-voltmeter movement previously described (the one showing a drop of .05 volt on full-scale deflection), the value of the shunt needed to extend the range to read 1000 amperes can be found as follows:

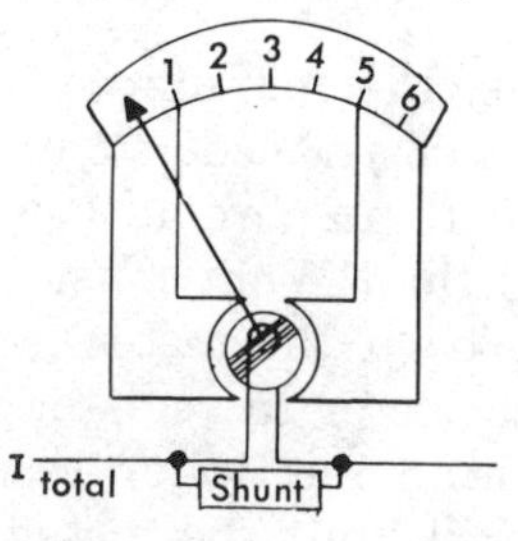

Fig. 7-3. Dc ammeter.

$$R_{shunt} = \frac{E_{shunt}}{I_{shunt}} = \frac{.05}{1000 - .01} = .00005000005 \text{ ohms}$$

where,
R_{shunt} is the resistance of the shunt in ohms,
E_{shunt} is the voltage of the shunt in volts,
I_{shunt} is the current in the shunt in amperes.

The practical value would be .00005 ohms. Ohm's law is applied only to the shunt, since it carries almost all the current.

Voltammeter

Some meters have a switch which will connect either shunts in parallel with the moving coil to measure amperes, or resistors in series to measure volts. They are known as voltammeters and can measure volts or amperes over wide ranges.

Ohmmeter

A direct-reading ohmmeter is used to measure resistance. It should always be connected *directly across* the resistance being measured, with all power disconnected.

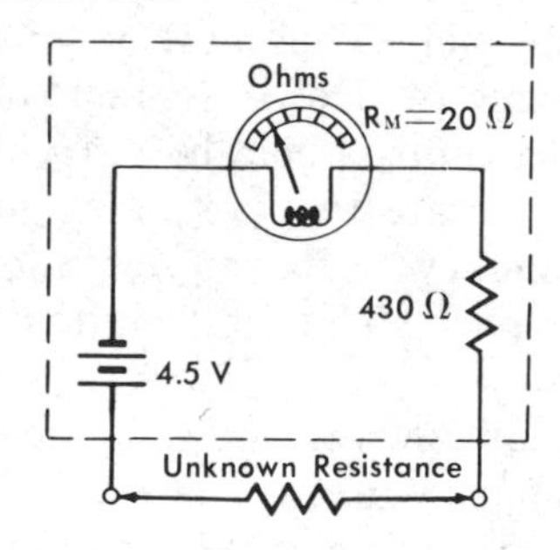

Fig. 7-4. Basic ohmmeter circuit.

The ohmmeter is a milliammeter with a low-voltage dc power supply, usually a battery of known voltage. The unknown resistance is connected across the terminals. The battery is usually contained in the meter case (shown by the dotted lines in Fig. 7-4). The dial is marked in ohms.

The higher the resistance under test, the lower is the current that will flow through it, and the lower the meter reading will be. A low resistance, on the other hand, will allow a high current to pass and will result in a high meter reading. "Zero ohms" is commonly found at the extreme right of the meter scale.

Watt-hour Meter

A watt-hour meter is used to measure electrical *energy*. It operates like a small motor. The armature is connected across the line in series with a high resistance (like a voltmeter) to measure voltage. The field coils are connected in series with the line (like an ammeter) to measure the current. The voltage across the armature and the current in the field coils together give the same effect as volts times amperes, which gives watts. The higher the voltage and/or current, the faster the armature rotates, and the higher the wattage reading will be. An aluminum disc is the armature. This disc rotates between the poles of a permanent magnet. The eddy currents induced in the disc increase, resulting in a "dragging" effect. The rotating shaft of the disc is connected by gears to the registering mechanism, consisting of several dials and pointers that indicate the amount of energy used, usually in kilowatt-hours.

ALTERNATING-CURRENT METERS

Dynamometer Voltmeter

A dc galvanometer cannot directly measure ac because the voltage fluctuates faster than the needle can move, resulting in a zero reading. However, if the permanent magnets are replaced with two electromagnets, the meter can be used on ac

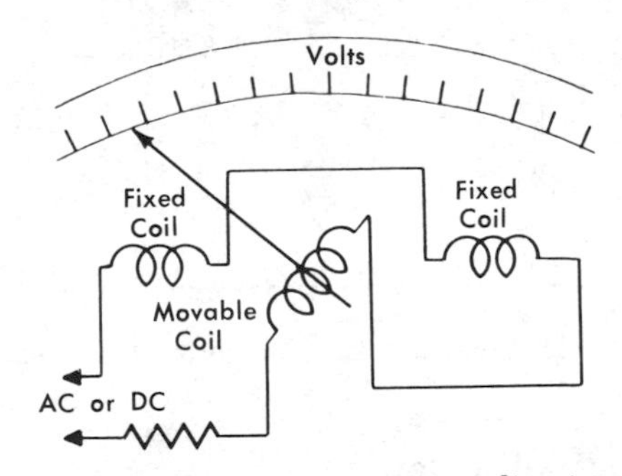

Fig. 7-5. Dynamometer-type voltmeter.

because the current in the fixed coils and in the moving coil would change at the same time. Such an instrument is known as a dynamometer voltmeter. Its coil windings consist of many turns of fine wire, which keeps the current low. This type of instrument is most efficient in the higher ranges. (See Fig. 7-5.)

Dynamometer Wattmeter

If the dynamometer field coils are connected in series with the line—like an ammeter—they will turn the moving coil and pointer and will measure the current. (See Fig. 7-6.) Then if the moving coil is connected across the line in series with a resistor—like a voltmeter—they will measure the voltage. Together, the field current and the armature voltage will turn the moving coil and pointer to indicate instantaneous power (calibrated in watts).

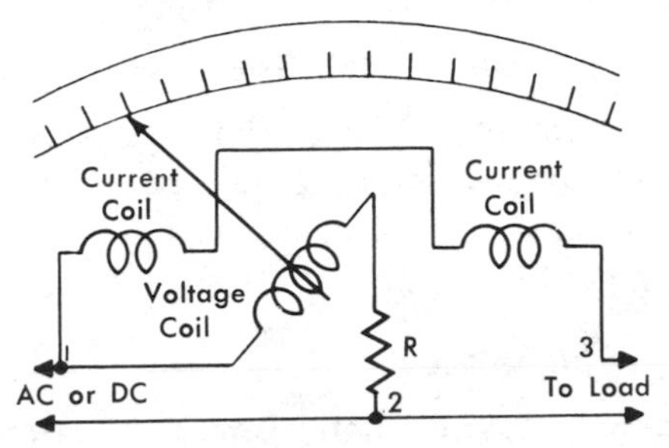

Fig. 7-6. Dynamometer-type wattmeter.

Movable-Iron Voltmeter

Two iron bars or "vanes" are used in this meter. (See Fig. 7-7.) One is stationary, and the other is attached to the meter shaft, which is connected to a pointer. A magnetic field is furnished by a coil consisting of many turns of fine wire. Because the bars are magnetized alike by the magnetic field, they repel each other. The repelling action causes the movable piece to rotate and to move the pointer. A small spiral spring holds the pointer at zero when no current is flowing.

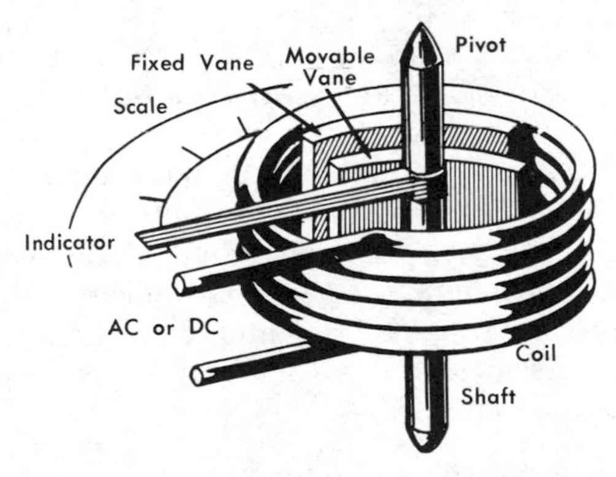

Fig. 7-7. Movable-iron vane for a voltmeter.

Movable-Iron Ammeter

The movable-iron ammeter is the same as the ac voltmeter except that its winding is a single loop made of a copper bar.

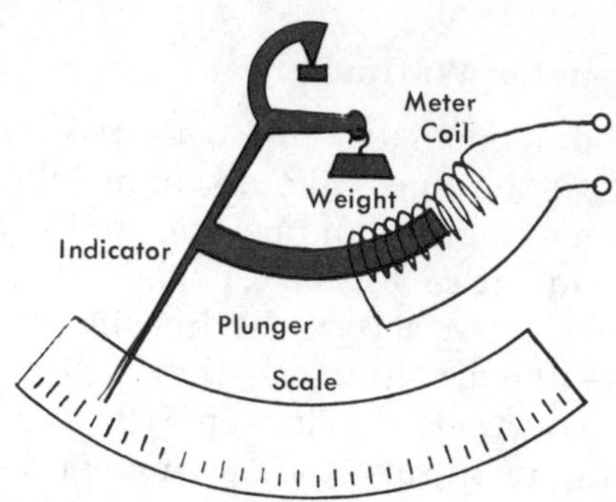

Fig. 7-8. Plunger-type iron vane.

Plunger-Type Iron Vane Meter

This instrument is based upon the attraction of an iron core by a solenoid. It can be used on either ac or dc and as a voltmeter or ammeter, depending upon the solenoid winding. (See Fig. 7-8.)

Rectifier-Type Meter

The rectifier-type meter permits alternating current or voltage to be measured with a dc meter. A small copper-oxide, full-wave rectifier changes the ac voltage to dc voltage. (See Fig. 7-9.)

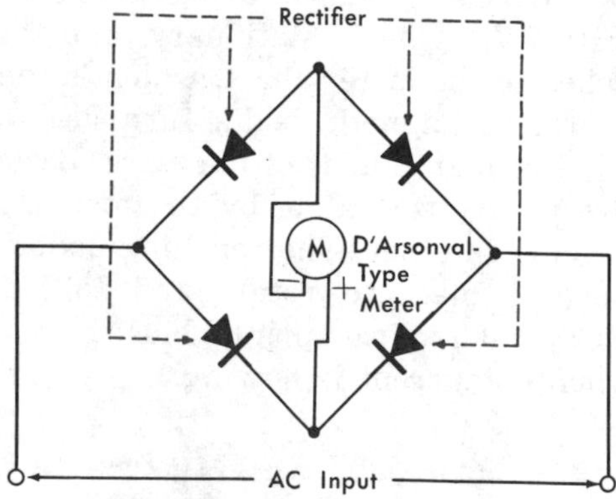

Fig. 7-9. Rectifier-type meter.

This feature is employed in most multimeters (i.e., meters which will make multiple measurements). With the proper switching of the rectifier and the multipliers, the meter movement will indicate a wide range of ac and dc voltages. The rectifier circuit ensures that only dc will be applied to the movement.

Electrothermal-Type Meter

Another way to change ac to dc so that a dc meter can be used is by *electrothermal* action. The ac or dc current flowing in a resistance wire heats the junction of two unlike metals. This junction is known as a *thermocouple*. (See Fig. 7-10.)

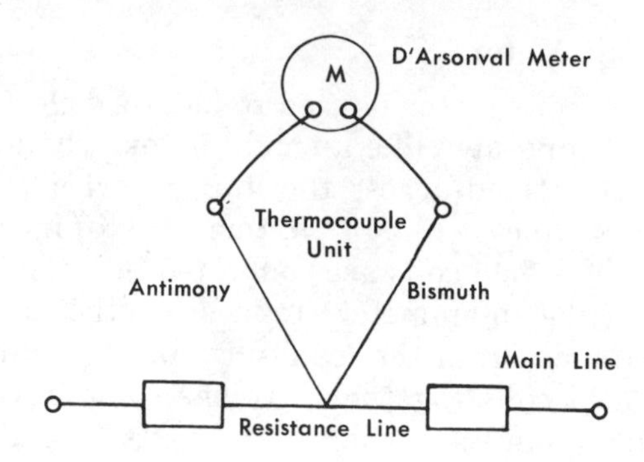

Fig. 7-10. Electrothermal-type meter.

Electrostatic Voltmeter

This type of voltmeter draws no current. Instead, it depends upon the force between two charged bodies. When two plates are connected across a source of supply, one plate becomes positively charged, and the other plate becomes negatively charged. The attracting force between these charged bodies pulls them toward each other, the deflection being opposed by the tension of a spring. The higher the voltage, the greater the deflection. (See Fig. 7-11.)

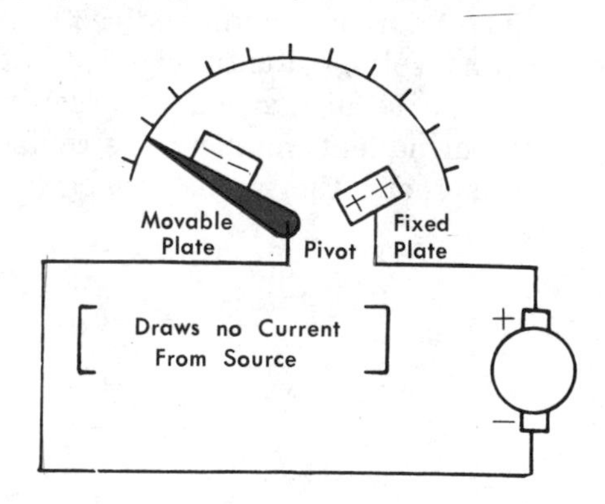

Fig. 7-11. Electrostatic-type voltmeter circuit.

Electronic Voltmeter

An instrument which draws almost no current is the electronic voltmeter. It is known also as a *vacuum-tube voltmeter* (vtvm). The D'Arsonval movement is used, and a vacuum-tube amplifier is added to make the meter more sensitive to extremely small currents.

Summary

Meters are divided into two types—ac and dc.

The basic dc meter is the D'Arsonval galvanometer.

A dc voltmeter consists usually of a galvanometer movement and a current-limiting resistor in series.

Voltmeters are always connected across a circuit and ammeters are always connected in series with a circuit.

An electrostatic voltmeter draws no current from the circuit being measured.

Questions and Problems

1. Why does an electrostatic voltmeter draw no current?
2. What must the size of the shunt be to make a 28-ohm, 1-milliampere meter movement read 0 to 25 amperes?
3. What series resistance is needed to make the meter described in Question 2 read 0 to 150 volts?
4. Is the resistance of an electrothermal-type meter the same for ac as it is for dc?
5. What is the ohms-per-volt rating of the meter described in Question 3?
6. Why must the power in the circuit be turned off when an ohmmeter is placed in the circuit?
7. What type of meter will make both ac and dc voltage readings?
8. What are some advantages of a vacuum-tube voltmeter?
9. With the meter described in Question 2 (28-ohm internal resistance, 0- to 1-milliampere movement), what resistance would be needed for a 0- to 30-volt scale? For a 0- to 300-volt scale?

ELECTRICAL SWITCHES

A switch is a device for connecting or disconnecting, or for changing the connections in an electrical circuit. A switch is like a drawbridge. When the switch is open, the electron traffic patiently lines up, unable to cross the gap. The minute the switch is closed, the electrons—like the evening rush-hour traffic—rush across and continue on their way. Fig. 7-12 shows some of the various switches used.

Fig. 7-12. Typical electrical switches.

General Types

The most common switch is the *single-pole* switch. It provides the on-off function in most electrical equipment. The symbol (—o/o—) for this type of switch illustrates its operation. There is one stationary *contact* (terminal). The moving part (pole) is attached to the other terminal. To show that the switch can be open or closed, a broken line (—o/o—) is sometimes added.

Another type of switch is the *double-pole* switch, which has four terminals and two poles. These poles are insulated from each other.

—o/o— —o/o— The double-pole switch opens *both* wires to a light or other device, and thus breaks all connections between it and the line.

The *three-way* switches have four terminals and, usually, one pole. *Flush-type* three-way switches have only three terminals. Two of the terminals are permanently connected together with a shunt wire. These terminals can usually be detected by the strip of sealing wax in the groove between them on the base of the switch. This wax covers the shunt wire. Sometimes the shunt is indicated by the word "connected" stamped on the porcelain base. The shunted terminal of a three-way switch is known as the "marked" ("common") terminal. This terminal usually has a screw of a different color from the others, often of a dark or oxidized finish.

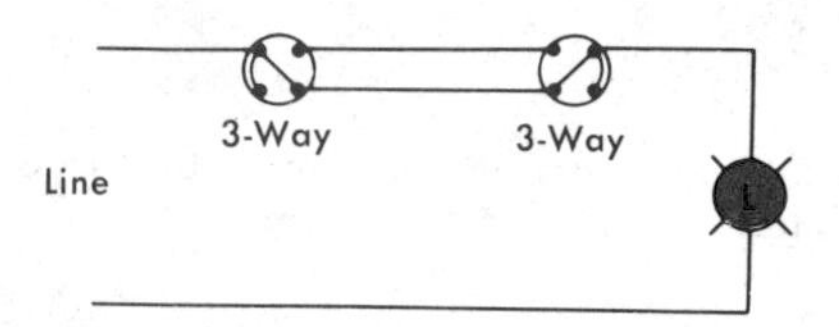

Fig. 7-13. Three-way switches used to control a lamp.

Two three-way switches can be installed in different sides of a room, for example, so that the lights in the room can be turned on or off at either switch. Figs. 7-13 and 7-14 show the connections for three-way switches.

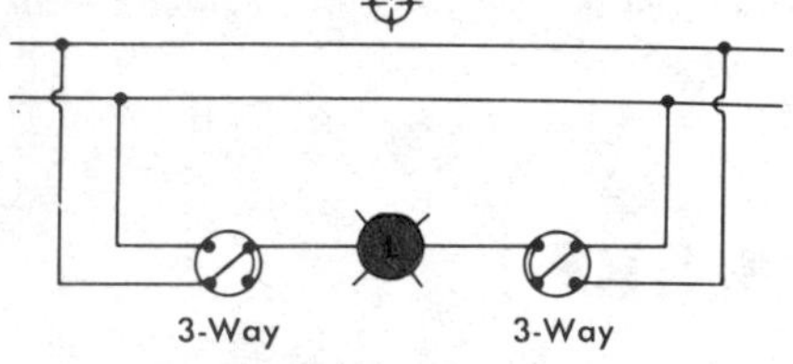

Fig. 7-14. Alternate wiring of three-way switches to control a lamp.

The *four-way* switch has four terminals and two poles. It is easily distinguishable from the other switches because its poles are always connected to adjacent terminals. Four-way switches are normally used where a light or a group of lights are to be controlled from more than two places. They are used in combination with three-way switches also, to control a light from as many places as desired (Fig. 7-15).

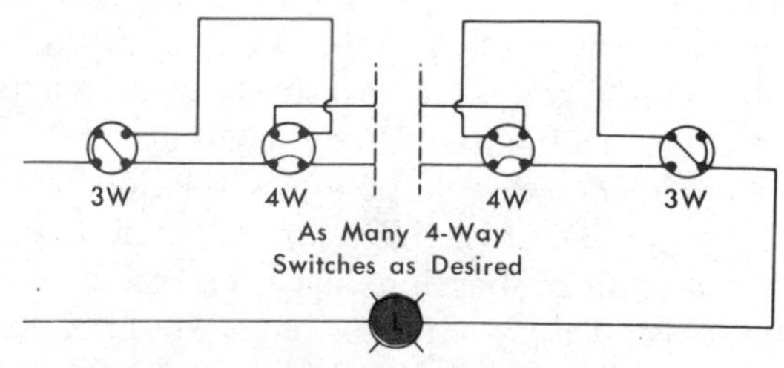

Fig. 7-15. Four-way switches used to control lamps from more than two locations.

Electrolier switches have one line or *main* terminal and two or more *circuit* terminals, with two or more poles electrically connected together but mounted on different levels, so that a variety of controls can be obtained.

Electrolier switches are used to control one or more circuits—such as several sections of a heater element in an electric range, two filaments in a three-way lamp, or several lights on a chandelier.

Fig. 7-16 shows how an electrolier switch can be used to turn on one or more lights at a time.

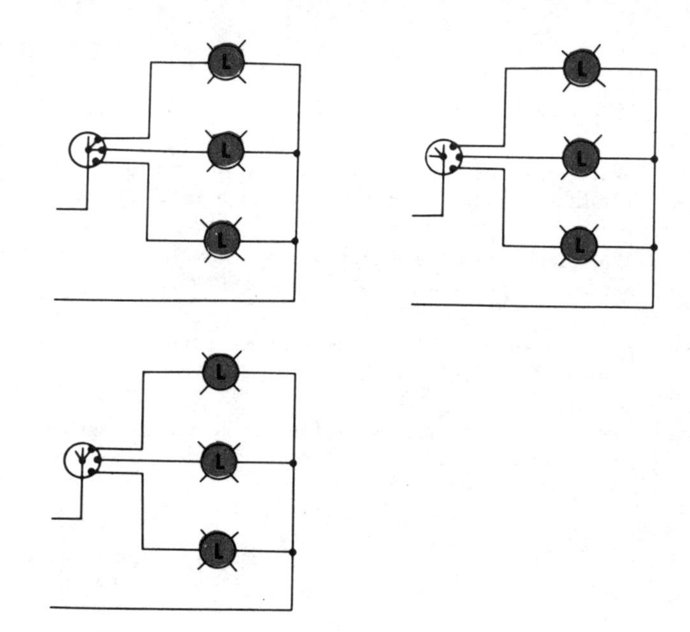

Fig. 7-16. Electrolier switches can turn on one or more lights at a time.

Fig. 7-17 shows the electrical symbols of the various switches we have discussed. Each type is shown in both the open and the closed positions.

Switches without an "off" position are known as nonindicating. Single-pole, double-pole, and electrolier switches are indicating switches; three-way and four-way switches are nonindicating switches.

Knife switches (Fig. 7-18) are a form of switch in which the moving part, usually a hinged blade, enters or clips into a set of contacts. Unless of the double-throw type, they must be so connected that there will be no voltage on the blades when the switch is open. Also, single-throw knife switches must be so mounted that they will not be closed by gravity. Double-throw switches should be mounted horizontally.

Fig. 7-18 shows various knife switches used for heavy-duty switching.

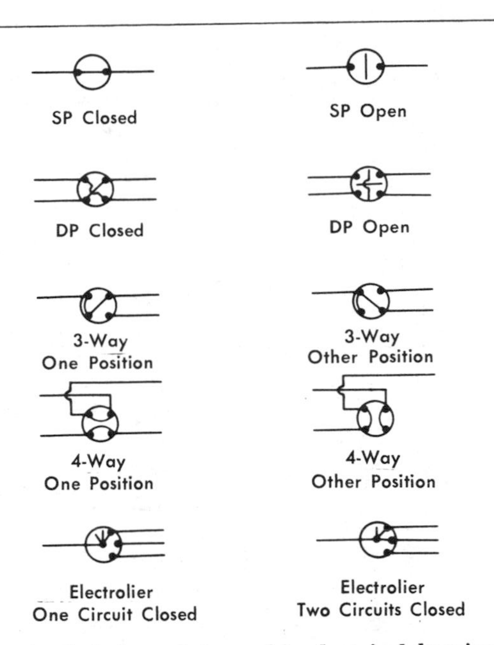

Fig. 7-17. Switch symbols used in electrical drawings.

As a general rule, all knife switches should be enclosed in metal boxes, as shown in Fig. 7-19.

Fig. 7-20 is a switchboard-type of disconnect switch rated at 20,000 amperes. It is used in power plants, for circuits of 600 volts or less. The switch contacts are designed to carry heavy currents without overheating.

Another type of switch is the motor-starting, push-button switch shown in Fig. 7-21.

For safety, a switch that disconnects a grounded conductor should also disconnect the ungrounded conductors at the same time.

APPLIANCE SWITCHES

Switches are used to control most appliance circuits. The action of a switch is to provide electrical continuity or flow in a circuit when it is closed and to stop electrical flow when it is open. Switches are activated in several ways: manually, mechanically, by level, by pressure, and by heat; also there are the newer electronic switches, which we will not cover in this discussion.

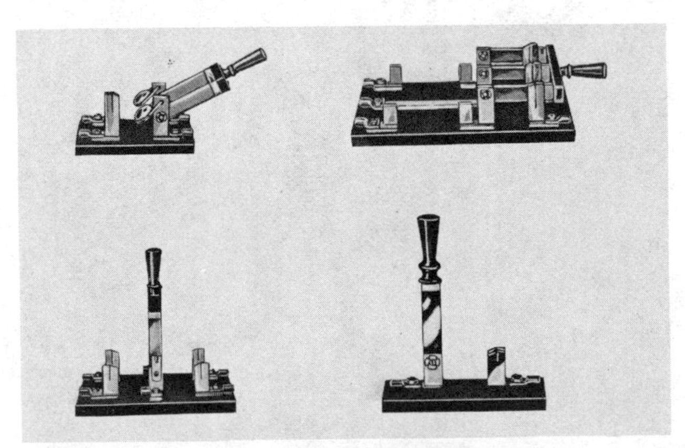

Courtesy Metropolitan Electric Co.
Fig. 7-18. Knife switches.

Courtesy Metropolitan Electric Co.
Fig. 7-19. Knife switches should be enclosed in metal boxes.

Manual Switches

These switches are operated by hand, usually by a flip-flop motion (toggle switch), a turning motion (rotary switch), or a push-pull motion

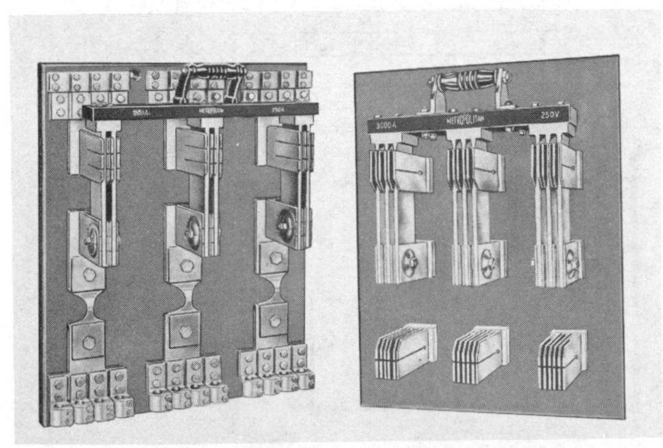

Courtesy Metropolitan Electric Co.
Fig. 7-20. Heavy-duty switches.

Fig. 7-21. The motor-starting, push-button switch.

(push-button switch). Depending on their design, they can control one or more switch contacts (multiple switch). The switching mechanism or switch arms can be single-pole single-throw, simple "on-off" switches, or more complicated single-pole double-throw and double-pole double-throw switches (Fig. 7-22).

Some manual switches, such as bell switches, are closed only while they are being pushed (momentary contact). Others, such as many fluorescent-lamp switches, make a set of momentary contacts while being pushed or turned and another set of contacts when released.

Mechanically Operated Appliance Switches

These switches generally fall in the class of timer switches such as those used in washers, dryers, ranges, and refrigerators. With the exception of washer timers, these switches are activated by an electric clock motor that slowly turns a cam (Fig. 7-23). Spring-loaded switch arms

Agitate	Switch Function
High	BU-O to BU
Medium	BU-O to OR
Low	BU-O to GY-P
Spin	
High	R-W to BU
Low	R-W to OR

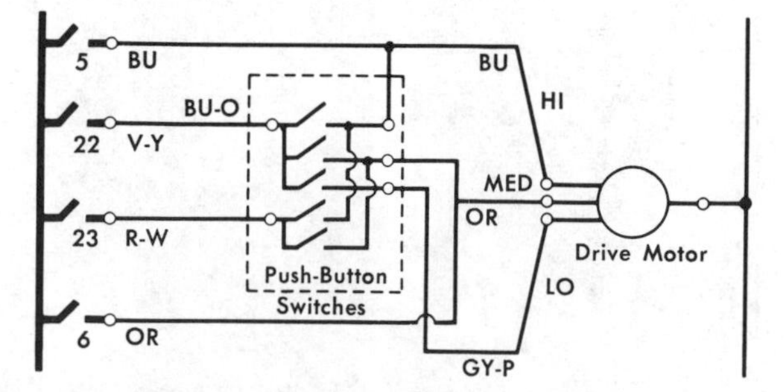

Fig. 7-22. Complex speed-control switch.

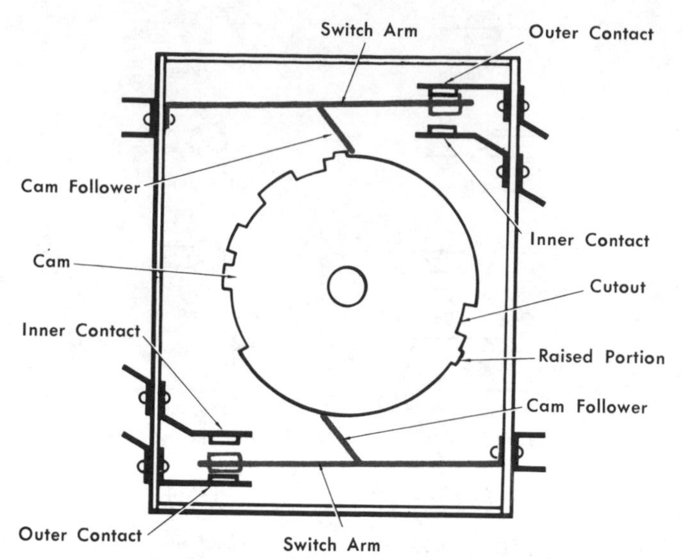

Fig. 7-23. Multiple-function cam.

close as the cutout in the cam allows them to drop and open as the turning cam forces them open.

Washer timers, because of the greater number of switch arms and cycles controlled, are more complex and use an intermediate spring-wound escapement (Fig. 7-24) to move the switch cams a designed number of degrees rapidly, as the escapement spring unloads. The clock motor slowly winds the escapement, storing energy in the stretched spring.

Magnetic Switches

Magnetic switches operate by the action of a magnetic field. This magnetic field can be from a permanent magnet or electromagnet. A permanent

Fig 7-24. Timer with direct-drive escapement.

magnet switch is sometimes used to control the lights on some refrigerators. A permanent magnet attached to the door of the refrigerator attracts a soft-iron switch arm, opening the switch contacts and the circuit to the lights. As the door is opened, the permanent magnet swings away from the switch arm, causing it to lose its magnetic effect on the arm and closing the switch to the light circuit.

Mercury Switches

Some switches are activated by the *switch level* (mercury switch). The switch in the 24-volt room thermostat just discussed is a prime example. A small sealed glass tube (Fig. 7-25) with a drop of mercury inside and a set of wires extending into the tube at each end make up the mercury switch.

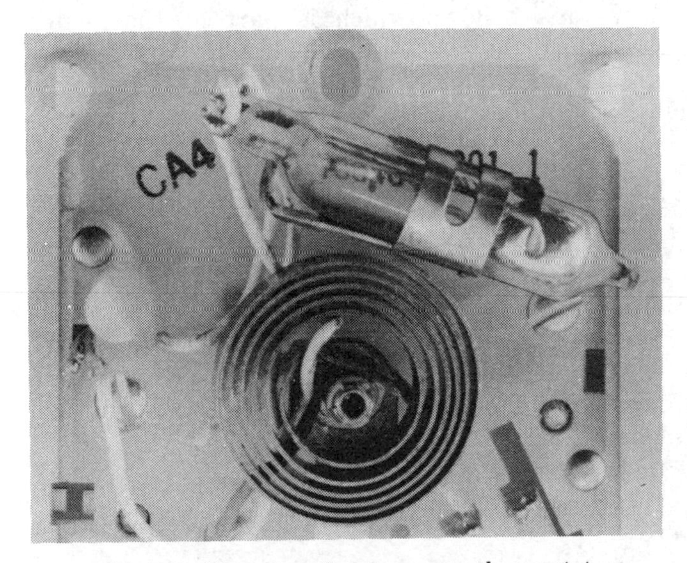

Fig. 7-25. Mercury switch in a room thermostat.

As the tube tilts, the mercury runs to the lower end, covering the ends of the exposed wires. The electrical circuit containing these wires is completed through the mercury. A similar switch is used to stop the spin of many automatic washers when the top lid is raised.

Pressure-Actuated Switches

Many appliance switches are activated by *pressure*. Dishwasher overflow switches are pressure activated. In some, the pressure or weight of the water on a rubber-like diaphragm activates a switch and opens the circuit to the water fill valve. On others, the water level causes a float to rise that reduces the pressure on a spring-loaded arm and causes the switch to open the circuit. Washer water level is controlled in a similar manner. Air is trapped in a pressure dome and tube (Fig. 7-26). As the water in the tub rises, the pressure becomes greater in the tube. This air presses on a rubberlike diaphragm that activates a switch. Other washers have used a float arrangement to activate the water-level switch.

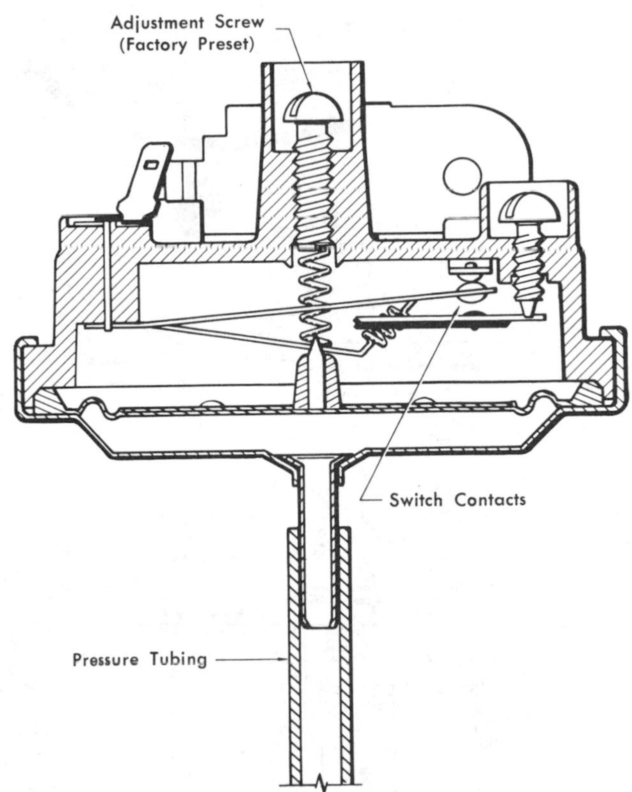

Fig. 7-26. Water-level switch.

Heat-Actuated Switches

Many switches are *activated by heat* and are generally called thermostats. There are two general principles of control: bimetal and hydraulic.

Different metals expand and contract at different rates. When two differing metals are bonded together, they form a bimetal strip. As heat is applied, the two strips expand or grow at different rates, causing the bimetal strip to warp or bend toward the side that has the smaller expansion or growth rate (Fig. 7-27). This movement is used to operate switches. The room thermostat (Fig. 7-25) or fan limit switches on furnaces are good examples of adjustable bimetal thermostats. Most fixed-temperature thermostats used on appliances such as dryers, refrigerators, air conditioners, motor temperature controls, etc., use this principle.

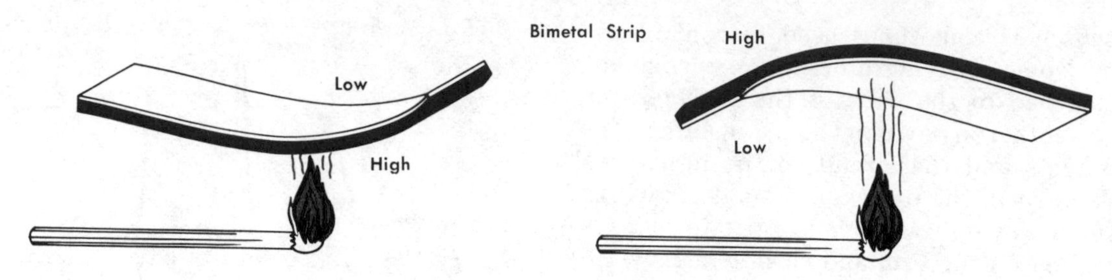

Fig. 7-27. Principle of the bimetal strip.

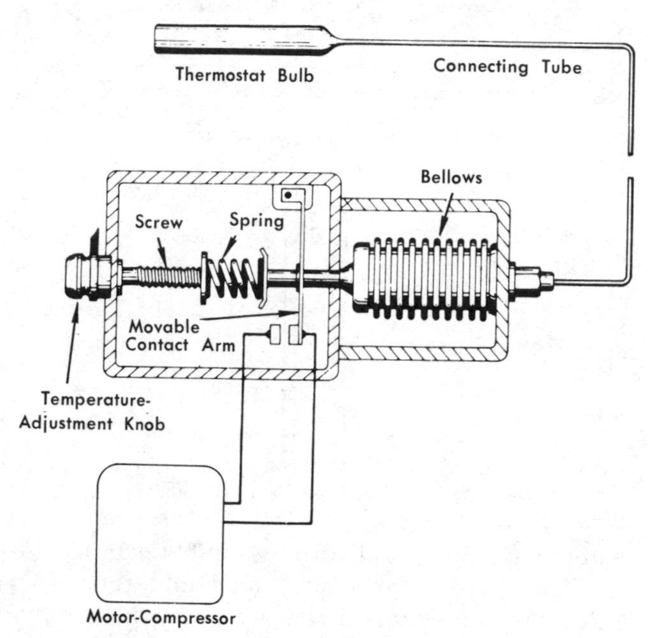

Fig. 7-28. Refrigerator thermostat assembly.

Heat-actuated hydraulic switches (Fig. 7-28) use a sealed hydraulic system to operate a switch. Liquids, like metal, expand as they heat. The sealed hydraulic system is generally filled with a liquid oil, mercury, freon, etc.; however, it can be filled with a gas—the principle of operation remains the same. A sensing area of the system "feels" the temperatures that require control. This sensing area is connected by a small tube or capillary to a bellows which is part of the switch assembly. The sensing area, capillary, and bellows are filled with liquid and form the sealed hydraulic system. As the sensor is heated or cooled, the expanding or contracting liquid flowing through the capillary tube expands or contracts the bellows. This movement of the bellows actuates the switch in the thermostat. This type of control is often designed so that it is adjustable over a temperature range. For example, a refrigerator thermostat may be adjustable from 32° F to 45° F, or a range oven thermostat may be adjustable from 200° F to 550° F, etc. This is generally accomplished by spacing the thermostat switch contacts by turning the thermostat dial.

Summary

A switch is a device for connecting or disconnecting, or for changing, the connections in an electric circuit.

A pole of a switch consists of the parts necessary to contact one conductor of a circuit. A switch may be single-pole or multiple, depending upon the number of single poles that are operated simultaneously.

The blade of a knife switch is the movable part which clips into a set of contacts.

All switches must be connected into the ungrounded wire of the line.

Manual switches may be toggle, rotary, or push-pull. A doorbell switch is a momentary type.

Mechanically operated switches are timers for washers, dryers, and refrigerators. They are operated by a clock motor.

Mercury switches use a rolling drop of mercury to make and break circuits.

Pressure-operated switches are used to control water level.

Switches operated by heat can be either bimetallic or hydraulic. The bimetallic switch is the fixed

type composed of a bimetal strip and contact points. It is usually found on overload switches, fixed thermostats, and limit switches.

Hydraulic thermostatic switches have sensing tubes or bulbs charged with a liquid or gas. As the temperature of the sensor changes, the liquid or gas will expand or contract. The expanding material will flex a diaphragm or bellows, causing switch contacts to open or close.

Questions and Problems

10. What type of switch has a built-in shunt?
11. Is an electrolier switch an indicating or a nonindicating switch?
12. How is a double-pole switch used in a circuit?
13. If a circuit has two three-way switches and you wish to install some four-way switches, how many can you put in the circuit?
14. What is a device for making, breaking, or changing the connections in an electric circuit called?
15. How should a single-pole knife switch be mounted?
16. What switch makes contact by means of a moving liquid?
17. What natural force operates a thermostat?

RELAYS AND MOTOR CONTROLS

RELAYS

A *relay* is a switch operated by an electromagnet. From a previous discussion of the solenoid, you will remember that adding an iron core to a coil of wire will concentrate and strengthen the magnetic field. Fig. 7-29 shows an electromagnet and a relay switch. The iron *armature* is attracted to the core of the electromagnet. Notice that the armature is also the pole of a switch, which is held open by the spring. When the armature is attracted by the electromagnet, the circuit controlled by the relay will close. Although this is a simple relay, it is typical of all relays.

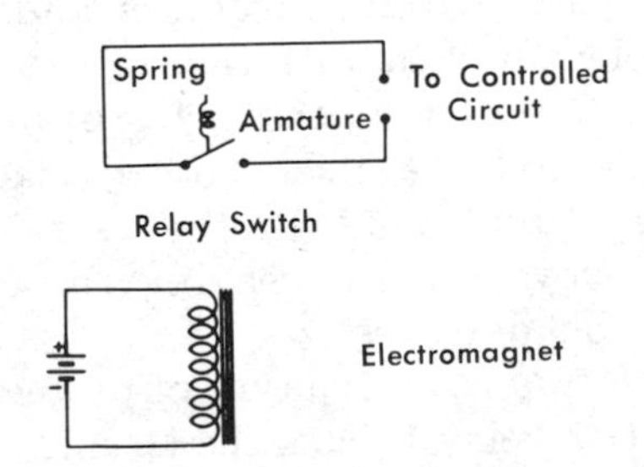

Fig. 7-29. An electromagnet and a relay switch.

Relays operate on either ac or dc power. Those used with ac have laminated cores and a *shading coil*. A shading coil is a single, short-circuiting loop of copper which covers half the surface of the core or pole piece. Such a coil is shown in Fig. 7-30. This shading coil, acting as a secondary winding of the relay coil, prevents the armature from releasing momentarily as the ac changes polarity.

Fig. 7-30. The shading coil of an ac relay.

Uses of Relays

Relays are used for switching circuits located some distance from the operating point (Fig. 7-31). The low-voltage, low-power relay circuits

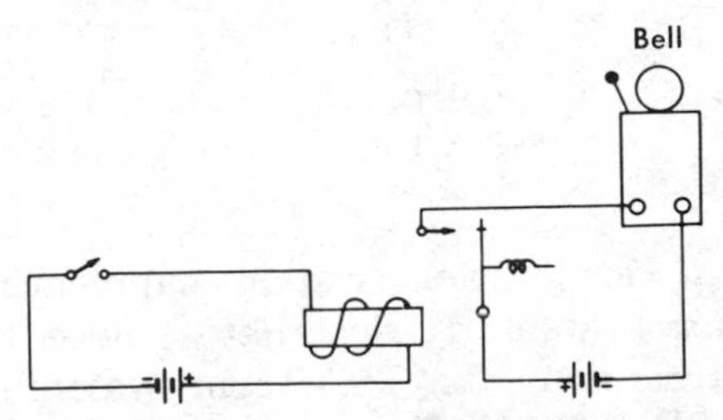

Fig. 7-31. Remote-control relay circuit.

can safely switch high-voltage or high-amperage circuits, as illustrated in Fig. 7-32. There are overload relays, which will open a circuit to protect it from excessive loads. Relays are also needed where manual switches would be too slow. The dial telephone is a good example. When a person picks up the receiver and dials a number, between 45 and 60 relays have operated by the time the call has been answered.

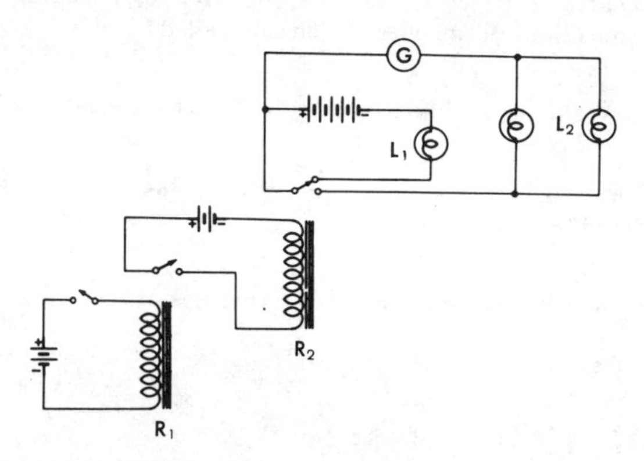

Fig. 7-32. A low-power relay circuit used to switch a high-voltage or a high-amperage circuit.

Because relays have a minimum of moving parts in the switch, they are quite dependable. Switching is quickly and accurately done by a relay. There are many ways a relay can be used in a circuit.

Contact Designation

The various switch actions are called *contact arrangements.* They are referred to by standard designations (abbreviations) established by relay manufacturers. Each of the 17 types are numbered. For example, "1" designates a single-pole, single-throw, normally open relay (SPSTNO).

1. SPSTNO
2. SPSTNC
3. SPSTNODB
4. SPSTNCDB
5. SPDT
6. SPDTDB
7. DPSTNO
8. DPSTNC
9. DPSTNODB
10. DPSTNCDB
11. DPDT
12. 3PSTNO
13. 3PSTNC
14. 3PDT
15. 4PSTNO
16. 4PSTNC
17. 4PDT
18. 5PSTNO
19. 5PSTNC
20. 5PDT
21. 6PSTNO
22. 6PSTNC
23. 6PDT

S = Single
O = Open
D = Double
C = Closed
P = Pole
B = Break
T = Throw
M = Make
N = Normally

A form designation system supplements the numbering system. It designates more complex contact arrangements. This form system is illustrated in Fig. 7-33. These contact arrangements

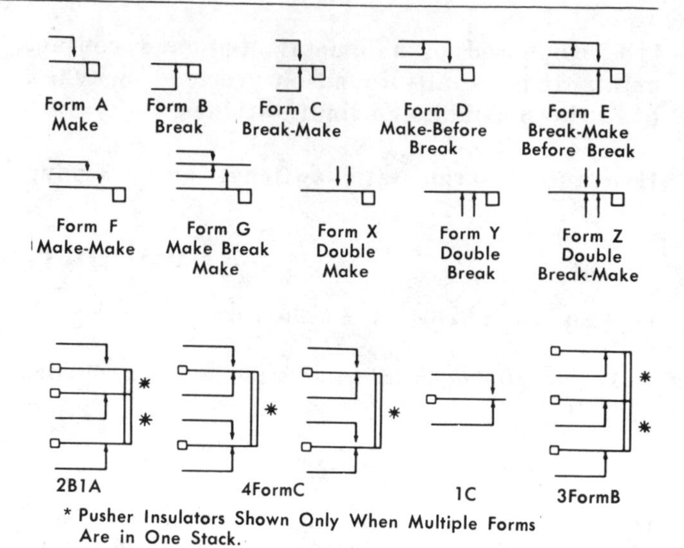

Fig. 7-33. The form system of relay contact designation.

may be used singly or in combinations, up to the maximum number of contact spring assemblies a specific relay can handle.

Classes of Relays

All relays, no matter how complex, are grouped into four classes. These classes are *pilot, control, protective,* and *regulating.*

The *pilot* relay (Fig. 7-32) is used to control another relay. Such a relay can be used as a remote control. For example, relay R_1 in Fig. 7-32 could be located at a distance from relay R_2. Closing or opening the switch in the R_1 relay circuit would energize or de-energize relay R_2. This would switch R_2 from circuit L_1 to circuit L_2.

Because of the switching action of relay R_2, it is classed as a *control* relay. Control relays switch one or more power circuits.

Many *protective* relays are used in complex industrial circuits. Often, machines and other devices must be supplied with the correct voltage or current. Failure of a machine, for example, could damage the machine itself, other machines, or the product being made. Protective relays guard against this by quickly breaking the circuit the instant trouble develops.

The *regulating* relay is a type of protective relay. The protective relay breaks the circuit so that the difficulty can be corrected manually. On the other hand, the regulating relay forestalls or corrects the difficulty automatically. Such a relay either may adjust a device to avoid a malfunction, or may correct the difficulty after a protective relay has already broken the circuit.

MOTOR CONTROLS

The armature resistance of a dc motor is very small. At the instant of starting it is zero. Therefore, if a dc motor is switched directly to its *rated* voltage, too much current will flow to the armature. For example, the armature resistance of a 5-hp, 220-volt motor is approximately 0.4 ohm. The instant the motor is connected to the line, the current through the armature becomes 220 volts ÷ 0.4 ohm or 550 amperes. The full load current (I_{FL}) of a motor of this size is only 20 amperes. Hence, the starting current is 27½ times the full load current!

Starting Devices

Available resistance must be connected in series with the armature to limit the starting current to approximately 1½ times the full-load current. A generated, or counter, emf is produced when the armature begins to rotate. This counter emf increases as the motor speeds up. At the same time, the armature current decreases. The starting resistance can then be cut out gradually, as shown in Fig. 7-34.

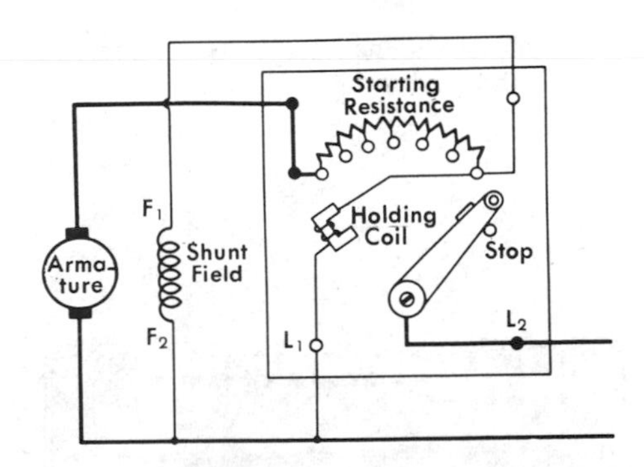

Fig. 7-34. A starting resistance for a large dc motor.

This starting device or *starter* gradually cuts the starting resistance out of the armature circuit as the arm is slowly rotated. All the resistance is cut out when the arm contacts the holding coil, which is merely an electromagnet. The arm is held by the electromagnet as long as the motor is running and the voltage is maintained. This starter may be used with both series and compound-wound motors.

Fig. 7-35. Motor controllers.

The low-voltage holding coil may be connected in series with the shunt field of the starter. Such a starter is used only for starting shunt-wound or compound-wound motors.

Controllers

A *controller* for electric motors is a starter which provides a safe, convenient method of starting and stopping, accelerating and decelerating, reversing, and dynamic braking.

There are two types of controllers—the face-plate type and the drum type. (See Fig. 7-35.) Each type is operated by a single handle.

A magnetic *contactor,* a device to obtain current-limit acceleration, is usually placed in the controller circuit. The contactor is controlled by a push-button station placed in the cover of the controller or placed at a remote location.

Automatic controllers are widely used because of their numerous advantages over the manual types. With a manual-type controller, the motor may take too much current, opening the circuit breakers or blowing the fuses, if the motor is accelerated too quickly. On the other hand, the magnetically operated switches of an automatic controller will safely cut out the starting resistance before this can happen.

Summary

A relay is an electrical switch operated by an electromagnetic field.

An ac relay has a shading coil to prevent the armature from releasing each time the ac cycle changes polarity.

The armature of a relay is operated by or attached to the contacts of a switch.

Relays are used in many circuits to perform many jobs.

All relays can be classified, according to their basic functions, as pilot, control, protective, or regulating relays.

Questions and Problems

18. Explain how the construction of an ac relay differs from that of a dc relay.

19. Why are relays better than manual switches?

20. Explain the purpose of a pilot relay.

21. Explain the purpose of a control relay.

22. Explain the purpose of a protective relay.

23. Explain the purpose of a regulating relay.

24. What is the switching action of a relay with a number-4 contact arrangement?

25. List three ways relays are commonly used, either on an automobile or in your home.

26. How do dc motor starters control resistance during acceleration?

27. What harm can result when a high starting current is applied to a large motor?

28. Does the counter emf developed by a motor increase or decrease as the motor speeds up?

FUSES

A fuse is a safety device. When excessive current flows through it, a soft metal strip inside the fuse will melt and will open the circuit. Every wiring system must be properly fused.

All fuses are made to blow at once when a short circuit develops. They will also blow when the circuit is overloaded, but not at once. Fuses will blow in one to fifteen minutes on a 50% overload, although most fuses will withstand a 10% overload indefinitely.

There are two types of fuses—renewable and nonrenewable. In the *renewable-type* fuse, the metal strip can be replaced after the fuse has blown. The different styles of renewable fuses are shown in Fig. 7-36. The length of this style of fuse ranges from two to seven inches, depending upon the type of circuit.

The *nonrenewable-type* fuse consists of a cylinder of hard fiber, plastic, or glass, with a chamber

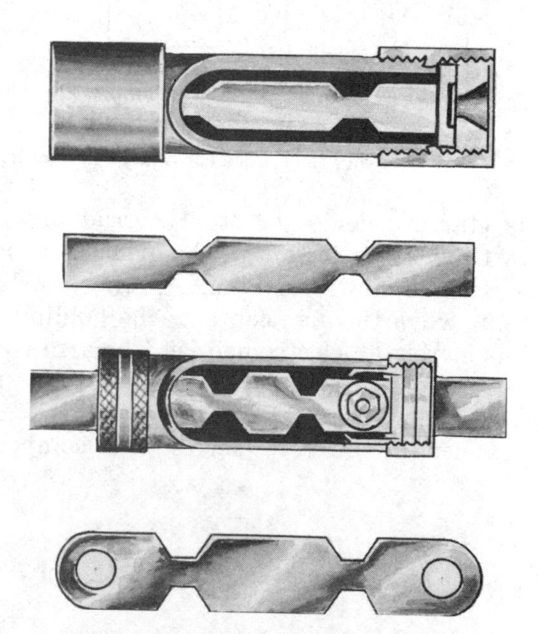

Fig. 7-36. Typical renewable fuses.

through which the fuse filament passes. This filament is soldered at each end to a ferrule. One of the most popular types is the screw-in socket fuse. Some have a plastic top with a mica window, and some an all-glass top through which the filament can be seen. The ratings of this type of fuse do not exceed 30 amperes. The screw-in socket fuses are used for fusing branch circuits and feeder circuits in house wiring systems. The nonrenewable-type fuses are shown in Fig. 7-37.

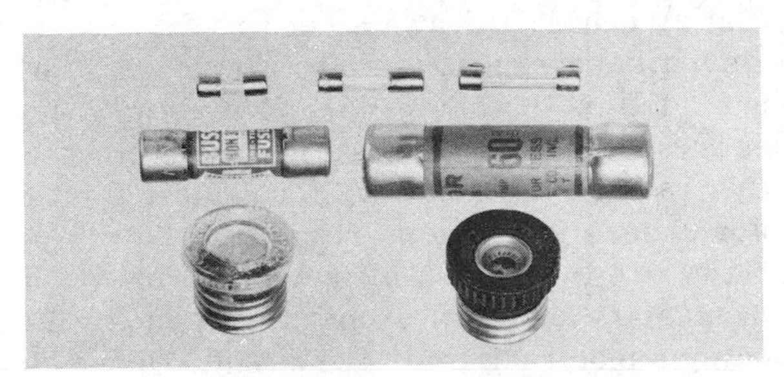

Fig. 7-37. Typical nonrenewable fuses.

Sometimes fuses blow needlessly when motors on washers, oil burners, refrigerators, and other household appliances are started. Such motors consume moderate current while running normally. But, from the instant a motor is first started until it has accelerated to almost normal operating speed, it draws current far in excess of normal. During this short starting time, a fuse often blows, unless it is a time-lag fuse.

The *time-lag* fuse looks like any other fuse externally, but is so constructed that it blows instantly on a direct short circuit. It blows eventually on a small, continuous overload and will withstand a big overload for a fraction of a minute. The cross section of a time-lag fuse is shown in Fig. 7-38.

Another style of fuse is the *nontamperable* fuse. It is fitted with an adapter that disengages from the main fuse as the fuse is installed. The

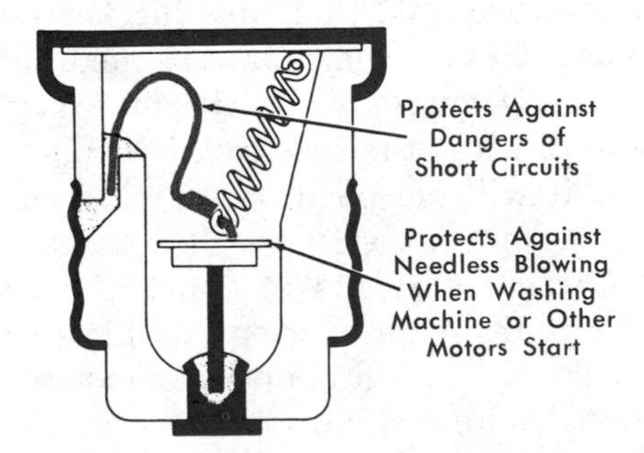

Courtesy Bussman Mfg. Co.

Fig. 7-38. Cross-section of a time-lag fuse.

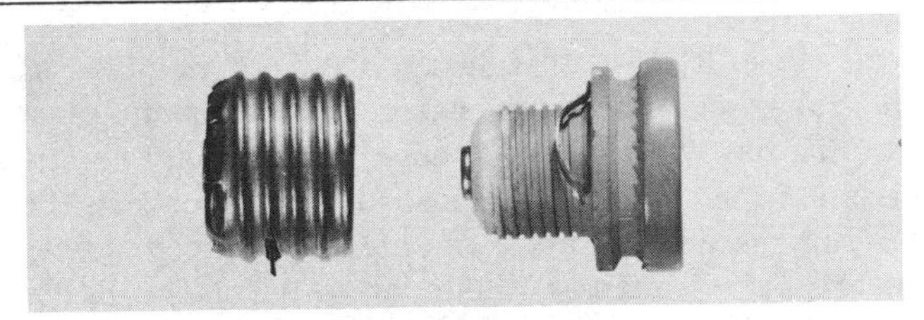

Fig. 7-39. A tamper-proof fuse.

adapter will accommodate a fuse of proper size or smaller only. The adapter is so designed that it is difficult to tamper with or to bridge the fuse. The fuse and its adapter are shown in Fig. 7-39.

Fig. 7-40 shows several incorrect mounting conditions that will often be found with cartridge fuse clips. Fig. 7-41 shows several ways in which fuse links may blow.

Fig. 7-40. Damaged fuse clips.

If the window in a fuse plug is clear and if the strip is melted in two, a light overload probably blew the fuse. A badly blackened window, due to a violent blowing of the fuse, usually indicates a severe overload or short circuit. Whenever a fuse blows, all possible causes in the circuit should be checked before the fuse is replaced. Frequent blowing of fuses usually indicates an overloaded circuit. If so, another circuit and set of fuses should be installed.

When looking for short circuits and grounds, see that each light on the circuit is turned off and each plug removed from any outlets in the circuit. If the trouble clears up, one of these devices

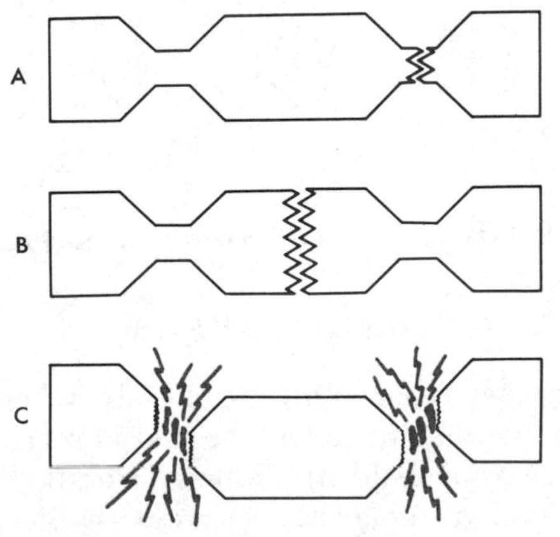

Fig. 7-41. Burned-out cartridge fuse links.

was at fault. If a test lamp, inserted in place of the fuse, will still burn after all equipment on a circuit has been disconnected, there is a short circuit in the wiring. Almost always, shorts in the wiring system will be found at poorly taped splices in the outlet boxes. Whenever some of the lights in a circuit are excessively bright and others are very dim, the cause is likely to be a blown neutral fuse on one of the older nonpolarized wiring installations.

AUTO AND EQUIPMENT FUSES

Other types of fuses have been designed for use in automobiles to safeguard the wiring and the different pieces of equipment against short circuits and overloads. There are fuses also used by the electronics industry to protect such things as radios, tv receivers, meters, and power supplies.

In this group of fuses (see Fig. 7-42) are found two different types, the "Slo-blo" and the standard. Their construction is basically the same in having a hairlike metal ribbon inside a glass tube with metal caps on each end. However, the "Slo-blo" type contains a spring and a small resistor as well, which provides for planned line surges in certain kinds of equipment.

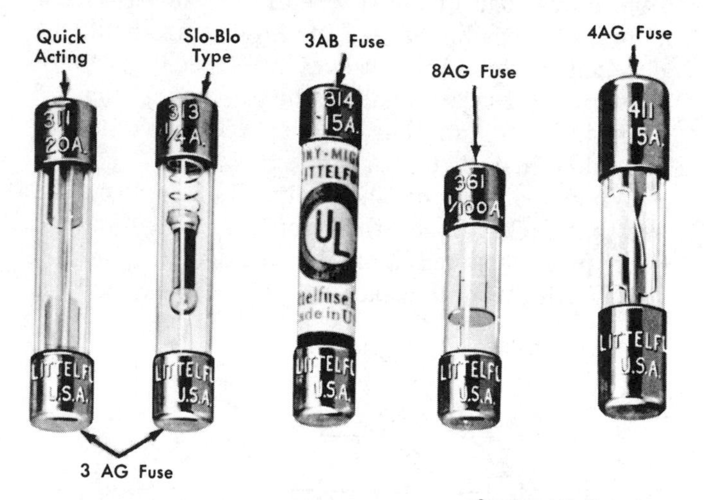

Courtesy Littelfuse, Inc.

Fig. 7-42. Fuses come in many types and sizes.

DELAY FUSES

Sometimes fuses blow needlessly when motors on washing machines, oil burners, refrigerators, and other household appliances are started. Such motors use a moderate current while running normally, but from the instant a motor is first started until it has accelerated to about normal operating speed, it draws a current far in excess of its normal operating current. During this short period of time in starting, it often blows a fuse. This is especially true, unless the fuse is of a much higher rating than necessary for protection of the motor under normal running condition.

Courtesy Bussman Mfg. Co.

Fig. 7-43. External view of a common Fusetron fuse.

Fuses are available which will not blow when a momentary overload such as that imposed by starting a motor occurs in the circuit. Such a fuse is shown in Fig. 7-43. It is a combined fuse element and thermal cutout. It protects, as does an ordinary fuse, against short circuits and overloads, but it will not blow on intermittent excess currents, such as momentary overloads produced by motors. It blows only when the overload persists.

Delay fuses may be used as replacements for ordinary fuses to eliminate needless fuse blowing. They help keep lights burning and appliances operating and often save the expense and trouble of calling a service man.

Fig. 7-38 shows a detailed cross section of a delay fuse. From the outside, this protective device looks like an ordinary fuse, but the inside is different. It has not only a fuse-link element but also a thermal cutout element.

Excessive current resulting from any overload causes the thermal cutout to heat, and if the overload persists, the solder of the thermal cutout softens until it permits the spring to pull out the end of the fuse link, thus opening the circuit.

Because it takes some time to melt solder, even with a heavy current, the thermal cutout is not fast-acting, while the fuse link is made sufficiently heavy so it will not open on starting current of a motor; therefore, the fuse will not open on heavy motor-starting currents of short duration. When a short circuit or an overload as high as 500 percent occurs, the fuse link opens in exactly the same manner as in an ordinary fuse.

Referring to Table 7-1 it can be seen how the delay fuse hangs on. It gives motors plenty of time

Table 7-1. Opening Time of Fuses (in Seconds)

Current Passing Through Fuse (Amps)	Delay Fuses				Ordinary Fuses	
	Fuse Ratings (Amps)				Fuse Ratings (Amps)	
	15A.	20A.	25A.	30A.	15A.	30A.
30A.	38.0 Sec.	120.0 Sec.	— Sec.	— Sec.	3.0 Sec.	— Sec.
45A.	11.0	23.0	42.0	80.0	0.7	12.0
60A.	4.7	10.3	15.2	22.0	0.2	3.8
75A.	1.3	5.1	8.2	12.0	0.1	2.2
90A.	0.8	2.1	5.3	6.0	—	1.3

to start, but observe how quickly an ordinary fuse opens on ordinary motor-starting currents. Actually, the 15-ampere delay fuse carries starting current like that of a 30-ampere ordinary fuse, and blows as does a 15-ampere ordinary fuse on a prolonged overload.

The small screw-type fuse plug is inadequate for handling the heavy currents encountered in the generation and distribution of electricity. This is also true in thousands of industrial and commercial applications of electricity. For this reason, fuses are designed and manufactured in an almost endless number of styles, types, and sizes to meet all the exacting requirements for fuse protection.

The number of electrons required to do even the simplest job staggers the imagination. To keep a 100-watt lamp burning requires a flow of six-billion-billion electrons—not per day, hour, or minute, but every second! Six-billion-billion—six with eighteen zeros after it! Yet, the individual electrons are so small that all this vast horde weighs next to nothing at all.

Now this insignificant weight of the electron is the secret of one of the greatest advantages of electricity—its speed. When you throw an electric switch, something happens right now! You must spin the starter to start your automobile; you have to wait for steam to build up to start a steam engine; but electric power is always right there, poised on its toes, ready to go. It is instantaneous because the electrons have practically no cumbersome weight to get moving, no inertia to overcome. They begin to flow the instant a path is completed for them, as soon as an operator throws the switch.

The exceedingly common and little-thought-of electric switch is after all the master, directing and controlling the countless millions of horsepower of surging electrical energy in industry, transportation, even the turning on and off of our lights or television.

Among the many types of heavy-duty fuses, one that is found quite adequate for general protection and utility is the renewable link-type fuse, which is shown in Fig. 7-36. The insulative protective case has been removed in order to show the interior structure and arrangement of the fuse parts. Renewable link fuses are designed for easy replacement of blown links.

Another type of fuse for heavy-duty use is the nonrenewable fuse shown in Fig. 7-44. The rigid girder assembly keeps the fuse from twisting or coming loose from its contacts in the holder. Fig. 7-45 shows the common cartridge-type fuse which comes in a variety of amperage ratings. There are literally hundreds of styles, types, and sizes of fuses in addition to these described.

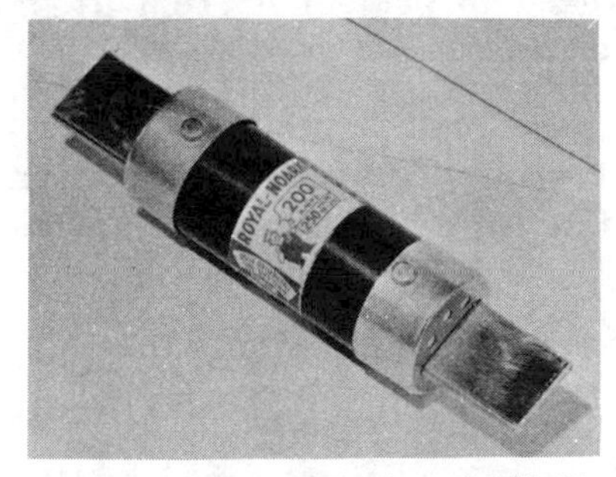

Courtesy ITT, Wire and Cable Div.

Fig. 7-44. Rigid-girder, nonrenewable-type fuse.

CIRCUIT BREAKERS

The *circuit breaker* protects electrical equipment from damage by opening the circuit whenever the current rises above a certain level. Circuit breakers come in two types—electromagnetic and thermal. Some circuit breakers must be reset manually, whereas others reset themselves.

Should the current rise very rapidly, as in a short circuit, the circuit breaker should open the circuit as quickly as possible, yet should allow the necessary overloads when motors and other equipment are starting. The magnetic-type circuit breaker, which is a sealed tripping unit, acts quickly for a short circuit and has a delayed action for harmless overloads. In the magnetic-type circuit breaker, a magnetic field is set up by the current flowing through the circuit. A change in the strength of the field will cause the unit to trip and to open the circuit. There is no unnecessary waiting period before a magnetic relay can be reset. After it has opened on an overload or short, it can be reset immediately, provided the overload or short no longer exists.

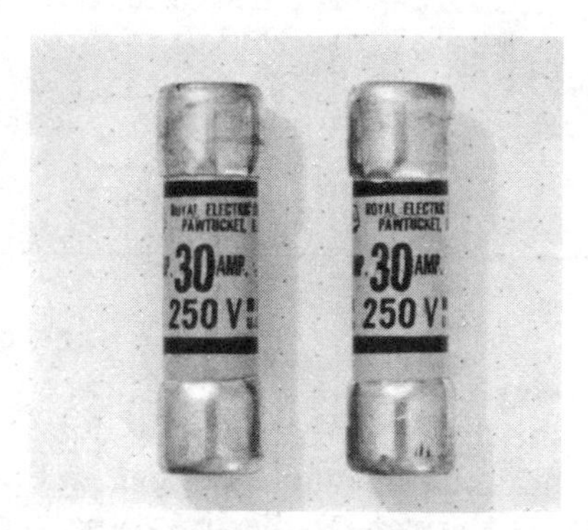

Courtesy ITT, Wire and Cable Div.

Fig. 7-45. Cartridge-type fuses are made in many amperage ratings.

Courtesy Heinemann Electric Co.

Fig. 7-46. A single-pole circuit breaker.

A *thermal* circuit breaker, after opening on a severe overload or short circuit, requires time for the heated parts to cool off before it can be reset. A single-pole circuit breaker is shown in Fig. 7-46.

Summary

A fuse is a safety device. Every wiring system must be fused to protect it from fires and from damage to equipment.

When excessive current flows through a fuse, the soft metal melts and opens the circuit.

There are renewable and nonrenewable fuses.

Screw-in socket fuses are used in branch circuits that do not exceed 30 amperes.

Time-lag fuses look like any other fuse, but will withstand a big overload for a fraction of a minute and will blow instantly on a direct short.

Circuit breakers operate like fuses and open the circuit when it becomes overloaded.

Questions and Problems

29. Would a 15-ampere fuse blow on a 120-volt circuit with a 1900-watt load?

30. What is the purpose of a fuse?

31. If a nontamperable fuse in a circuit blows, can it be replaced with a regular screw-in socket fuse?

32. What happens when an excessive current flows through a fuse?

33. What is the main advantage of a time-lag fuse?

34. What kinds of fuses are used in most houses?

35. What kind of fuse protection would be used in a circuit that has several appliances run by electric motors? Why?

36. Is it necessary to inspect blown fuses? What would you look for?

8

DISTRIBUTION

CONDUCTORS

A conductor is a material, usually a metal, which will allow an electric current to pass through it. The atoms of a good conductor have loosely held valence electrons which move as "free electrons" when a small electrical potential is applied. A wire or cable is the most common conductor.

Materials

Silver is the best conductor. For this reason, it is the standard by which other conducting materials are compared. However, since silver is relatively expensive, it is used only where good conductivity is necessary, such as in contacts and other small parts.

Copper is the next best conductor and is the best *practical* conductor. It is used for most of the electrical wiring in our homes and for much of the wiring of industrial electric equipment.

Because copper is rather heavy, aluminum is sometimes used. Aluminum is not as good a conductor as copper. Hence, a larger diameter wire is needed. In spite of the larger diameter wire, however, an aluminum conductor weighs only a little more than half as much as a comparable copper wire.

Resistance of Conductors

From Ohm's law and from the definition of a conductor, you can see that a good conductor must offer little resistance to current flow. The purpose of a conductor is to allow a large quantity of electrons (which is the same as a large current) to flow with a small electrical potential. For this to happen, the conductor must not offer too much resistance.

All metals used as conductors have different resistance values. Table 8-1 gives the resistances of different sizes of copper wire. In the first column in Table 8-1 you will notice that the wire size is given as a *gauge* number. This is the most common way of stating the diameter of wire. Fig. 8-1 shows a Brown and Sharpe wire gauge (B&S) or American Standard Wire Gauge (AWG), which is a standard gauge for measuring wire sizes. This gauge has the gauge number on one side and the decimal equivalent of the gauge number on the other side. The decimal equivalents are

Fig. 8-1. A wire gauge.

shown in the second column in Table 8-1. The third column lists the resistance of each wire size. Note that the resistance decreases as the wire size increases.

Measurement of Conductors

Since a larger wire has less resistance than a smaller wire, it follows that a larger wire will carry more current than a smaller one. Table 8-2 lists some common wire sizes and their current-carrying capacities. If the wire is too small for the current, it will resist the current and become overheated. How much current a wire will carry depends upon its cross-sectional area. This is stated in *circular mils*. To find the area in circular mils, square the diameter (in mils) of the wire.

The square mil is another term referring to the area of a conductor. It is used for square or rectangular conductors, such as *bus* lines. A bus line is a conductor, usually square or rectangular, that carries heavy electrical currents to distribution points.

Table 8-1. Gauge Equivalents Showing the Resistances of Standard Annealed Copper Wire

B. & S. American Wire Gauge No.	Diameter In Inches	Area Circular Mils	Ohms at 68° Fahrenheit Per 1000 Ft.	Per Mile	Per Pound
0000	0.460	211600.	0.04906	0.25903	0.000077
000	0.40964	167805.	0.06186	0.32664	0.00012
00	0.3648	133079.	0.07801	0.41187	0.00019
0	0.32486	105534.	0.09831	0.51909	0.00031
1	0.2893	83694.	0.12404	0.65490	0.00049
2	0.25763	66373.	0.1563	0.8258	0.00078
3	0.22942	52634.	0.19723	1.0414	0.00125
4	0.20431	41743.	0.24869	1.313	0.00198
5	0.18194	33102.	0.31361	1.655	0.00314
6	0.16202	26251.	0.39546	2.088	0.00499
7	0.14428	20817.	0.49871	2.633	0.00797
8	0.12849	16510.	0.6529	3.3	0.0125
9	0.11443	13094.	0.7892	4.1	0.0197
10	0.10189	10382.	0.8441	4.4	0.0270
11	0.090742	8234.	1.254	6.4	0.0501
12	0.080808	6530.	1.580	8.3	0.079
13	0.071961	5178.	1.995	10.4	0.127
14	0.064084	4107.	2.504	13.2	0.200
15	0.057068	3257.	3.172	16.7	0.320
16	0.05082	2583.	4.001	23.	0.512
17	0.045257	2048.	5.04	26.	0.811
18	0.040303	1624.	6.36	33.	1.29
19	0.03589	1288.	8.25	43.	2.11
20	0.031961	1021.	10.12	53.	3.27
21	0.028462	810.	12.76	68.	5.20
22	0.025347	642.	16.25	85.	8.35
23	0.022571	509.	20.30	108.	13.3
24	0.0201	404.	25.60	135.	20.9
25	0.0179	326.	32.2	170.	33.2
26	0.01594	254.	40.7	214.	52.9
27	0.014195	201.	51.3	270.	84.2
28	0.012641	159.8	64.8	343.	134.
29	0.011257	126.7	81.6	432.	213.
30	0.010025	100.5	103.	538.	338.
31	0.008928	79.7	130.	685.	539.
32	0.00795	63.	164.	865.	856.
33	0.00708	50.1	206.	1033.	1357.
34	0.006304	39.74	260.	1389.	2166.
35	0.005614	31.5	328.	1820.	3521.
36	0.005	25.	414.	2200.	5469.
37	0.004453	19.8	523.	2765.	8742.
38	0.003965	15.72	660.	3486.	13772.
39	0.003531	12.47	832.	4395.	21896.
40	0.003144	9.88	1049.	5542.	34823.

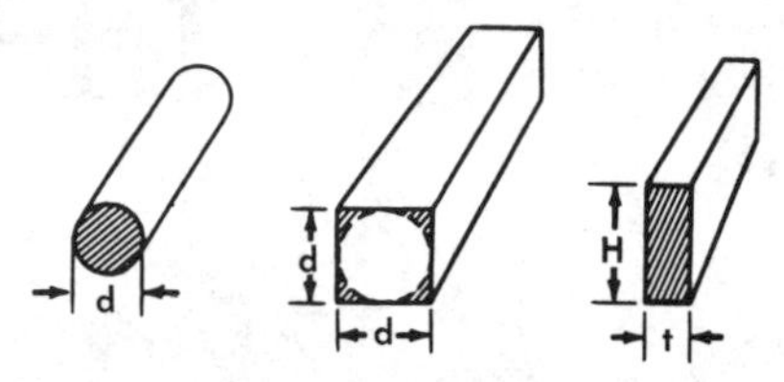

Fig. 8-2. Typical bus lines.

The cross section of a bus line is shown in Fig. 8-2. Obviously, a one-inch diameter round conductor has a smaller *cross-sectional area* than a one-inch square conductor. A round conductor has .7854 times the area of a square conductor of the same size. Therefore, to convert circular mils to square mils, multiply by .7854. Divide by .7854 to convert square mils to circular mils.

The length of a wire also affects its total resistance and, therefore, the amount of current it can safely carry. Table 8-1 gives the resistance

Table 8-2. The Diameter, Area, and Current-Carrying Capacity of Various Sizes of Wire

B. & S. Gauge	Diam. of Solid Wires in Mils	Area In Circular Mils	Table A Rubber Insulation Amperes
18	40.3	1,624	3
16	50.8	2,583	6
14	64.1	4,107	20
12	80.8	6,530	25
10	101.9	10,380	40
8	128.5	16,510	55
6	162.0	26,250	80
4	204.3	41,740	105
3	229.4	52,630	120
2	257.6	66,370	140
1	289.3	83,690	165
0	325.	105,500	195
00	364.8	133,100	225
000	409.6	167,800	260
0000	460.	211,600	300

per 1000 feet and per mile for various wires. For instance, the resistance of No. 14 wire is 2.504 ohms per 1000 feet. Therefore, 100 feet would have a resistance of .2504 ohm. Doubling the length of the wire would double its resistance, making it .5008 ohm.

Types of Conductors

There are numerous kinds of conductors. Some of the more common ones are shown in Fig. 8-3. You will notice that different insulation is used for different applications. Also note that some of the conductors are *solid* and that others are *stranded* for more flexibility.

Choice of the type and size of wire is simplified by the requirements set forth in various electrical codes. These codes are established by local, state, and national groups to insure safe electrical installations. For instance, the National Electrical Code requires that wires for feeder lines be large enough to cause no more than a 1% voltage drop. In addition, they must have a current-carrying capacity as large as the maximum possible load.

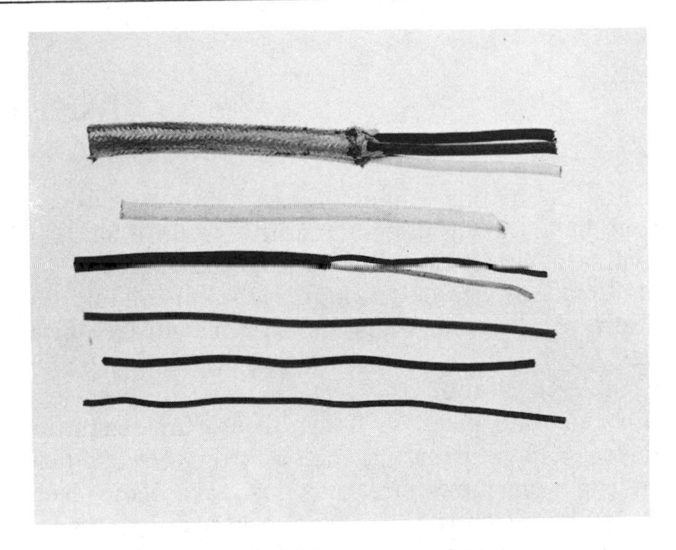

Fig. 8-3. Some common conductors.

No conductors smaller than No. 14 gauge may be used in branch circuits. A 15-ampere circuit requires No. 14 wire; a 20-ampere circuit, No. 12; and a 30-ampere circuit, No. 10. Electrical codes also give methods for finding the load in a circuit.

Summary

Copper is one of the better conductors. It is used for most industrial and residential wiring.

Silver is the best conductor. However, it is too expensive for ordinary usage.

Aluminum also is a good conductor. It is used where a lightweight conductor is required.

The current-carrying capacity of a conductor is based upon its cross-sectional area, which is expressed in *circular mils* or in *square mils.*

The resistance of a conductor depends upon the material, the area, and the length of the conductor.

Questions and Problems

1. What is the ratio between the square mil and the circular mil?
2. List the three factors which determine the resistance of a conductor.
3. Define a conductor.
4. If a certain circuit in a house has a total load of 1870 watts, what size wire should be used?
5. The measurement of the diameter of a wire is .229 inches. What gauge is it?
6. What is the cross-sectional area (in circular mils) of the wire in Question 5?
7. Why are some metals, such as copper, considered to be better conductors than other metals, such as iron?
8. Where would you find the information about the wire size needed to install an electric range?

INSULATORS

Just as metals and some nonmetals let the current flow through them readily, so do other materials obstruct the flow of current. These latter materials are called *insulators*. An insulator, unlike a conductor, is a material with no free electrons in it. In fact, the electrons are tightly bound within each atom. However, a *very* high voltage *can* cause an electron flow through an insulator. It does this by exerting such a strong electric field on the atom that electrons are torn loose from their attraction to the nucleus and are then free to become a part of the current.

Insulators guide the current along the path we wish it to follow and prevent it from wandering off into places where it is not wanted. For example, copper wire usually has a thin covering of rubber over it (Fig. 8-4). When thus covered with an insulation, the conductor may touch grounded metal without causing a *short circuit*. The layer

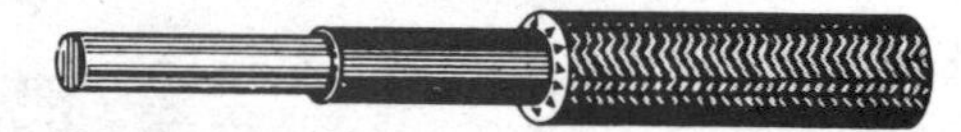

Fig. 8-4. Rubber and fabric insulation on a solid conductor.

of rubber keeps the conductor from touching an unwanted electrical contact, such as another conductor. There are other types of insulation besides rubber. The most common ones are glass, mica, asbestos, dry wood, porcelain, Bakelite, dry air, and paper.

Fig. 8-5. Glass insulator.

Within recent years, plastic insulation has been developed. A thin plastic covering provides better protection than a thick coat of rubber. This has resulted in smaller and lighter wiring, which is necessary wherever many wires must be passed through the same passageway or bundled in a single cable (like telephone wires, for example). Typical insulating plastics are polystyrene, nylon, teflon, and polyvinyl chloride.

A new type of insulation is flexible, woven glass cloth. Because it is a better insulator than rubber or plastic, glass is used in coal mines and powder plants, where a spark might set off an explosion. Many motor windings are insulated with glass. This permits the rotors to operate at much higher temperatures than before and makes smaller and more powerful motors possible.

Fig. 8-6. Porcelain insulator.

Glass is the insulator we see most often on telephone and electric poles along our streets and roads. A relatively inexpensive glass insulator is shown in Fig. 8-5.

Table 8-3. Typical Insulating Materials

Material	Dielectric Strength (Volts per Mil)
Air	31
Bakelite (paper base)	250-585
Glass (window)	760
Mica	2030-5080
Nylon	285-470
Paper (Paraffin coated)	1170
Polystyrene	300-710
Porcelain	200-400
Rubber (hard)	470

The porcelain insulator in Fig. 8-6 is used on power lines carrying very high voltages. These insulators are built to resist the effects of snow, sleet, and moisture.

Dry air is one of our most important insulators. It is fortunate that air—which is free—is an insulator.

Oil is another insulator. Power transformer coils are submerged in oil. Not only does the oil provide a satisfactory insulation between the coils, but it also cools the transformer. Because the current flowing through the transformer coils is quite large, the coils may become overheated. The oil seals any break in the insulation between the coils (Fig. 8-7).

Insulating materials are sometimes called *dielectrics,* and this term may be used to describe some of the characteristics of a material. For example, the ability to withstand applied voltages without breaking down is called the *dielectric strength.* Table 8-3 lists some common insulating materials and their dielectric strengths. These strengths are given in volts per mil of thickness. For example, the dielectric strength of glass is given as 760. This means that a piece of glass 1 mil in thickness will withstand 760 volts without breakdown. A piece 2 mils thick would withstand twice that amount, or 1520 volts, and so on.

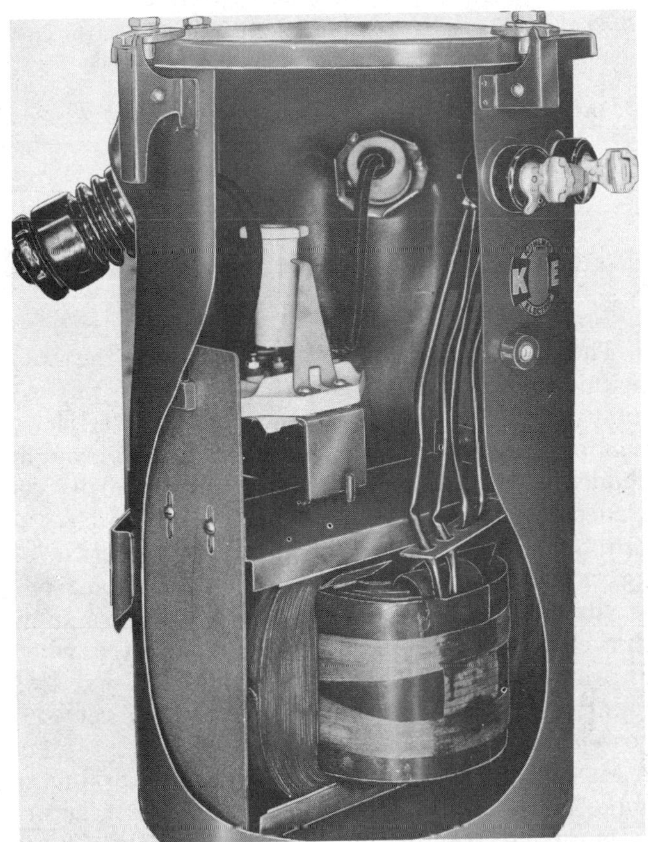

Courtesy Kuhlman Electric Co.

Fig. 8-7. Cutaway view of an oil-insulated, self-cooled, outdoor type of power transformer.

Summary

Materials that are poor conductors are called insulators.

Insulators have no free electrons in their atomic structure.

Porcelain, glass, oil, and rubber are excellent insulators.
Thin plastic insulation will usually provide better protection than thick rubber insulation.

Dry air is one of the most important and most common insulators. It is the standard against which all other insulators are compared.

Dielectric is another term for insulator.

Dielectric strength is the amount of voltage a dielectric can resist before it allows current to flow through it.

Questions and Problems

9. Is Bakelite a good insulator? Why?

10. What type of insulation has been used recently on motor windings?

11. What is the dielectric constant of porcelain?

12. Is porcelain a good conductor?

13. What is the dielectric strength of hard rubber?

14. Why will an arc jump from a high-voltage transmission line to the tower when ice has formed on a porcelain insulator supporting the line?

15. Compare the atomic structure of a dielectric with that of a conductor.

16. Will doorbell wiring require a different insulation from that of an automotive spark plug wire? Explain.

17. Is glass a better insulator than hard rubber? Explain.

18. Since window glass is a good insulator, would the inside of a greenhouse be a relatively safe place to stay during a thunderstorm? Explain.

WIRING

There is a vast difference between the way primitive man viewed electricity and the way we view it today. To the primitive man, electricity was a fearsome, unknown thing that brought sudden death and destruction. Evidence of its destructive power was everywhere after an electrical storm. Ignorance made him fear the crash of lightning. Since then, electricity has become one of our most faithful and useful servants. We know that electricity is dangerous and destructive when improperly handled. But we also know that this giant can be harnessed to give us light, heat, and power.

Many persons have worked and are working to make electricity safe. The electrical manufacturers support large testing laboratories. These laboratories are constantly testing and improving the equipment that has made electricity one of our most valuable allies.

Let's now examine the wiring system, which distributes the electricity in our homes, farms, and factories.

WIRES AND CABLES

There are several methods of wiring that are recognized and approved by the National Board of Fire Underwriters, a laboratory which tests electrical apparatus of all kinds to determine whether they are electrically safe. This organization is responsible for the National Electrical Code. The NEC will be discussed at a later point in this chapter.

Open wiring was used in homes constructed in the early 1900s. In this method the wires were strung on porcelain knobs and cleats, unprotected by an enclosure. Although the open wiring method is still approved by the National Electrical Code, its use has been discouraged. In fact, some localities have banned it.

Armored cable consists of two or more insulated conductors enclosed within a heavy steel armor. A length of armored cable is shown in Fig. 8-8.

Fig. 8-8. Armored cable and a fiber bushing used to protect the conductor insulation.

The insulated conductors are tightly bundled into a heavy covering of tough insulating paper. The thick paper is covered by an interlocking spiral of steel armor which makes the cable flexible and thus easier to handle.

The armored cable can be fastened to metal switch boxes, junction boxes, and convenience outlets by means of special connectors. Some of these connectors are shown in Fig. 8-9.

Nonmetallic cable consists of two or three conductors covered with a thick wrapping of paper, as shown in Fig. 8-10. The protective paper is, in turn, covered by a covering of tough plastic material which is highly resistive to wear and moisture.

Wiring with nonmetallic sheathed cable is done much the same as it is with armored cable. The conductors in a nonmetallic cable do not have as much protection as the ones in an armored cable. Quite often the extra protection is not needed anyway. The nonmetallic cable is connected to

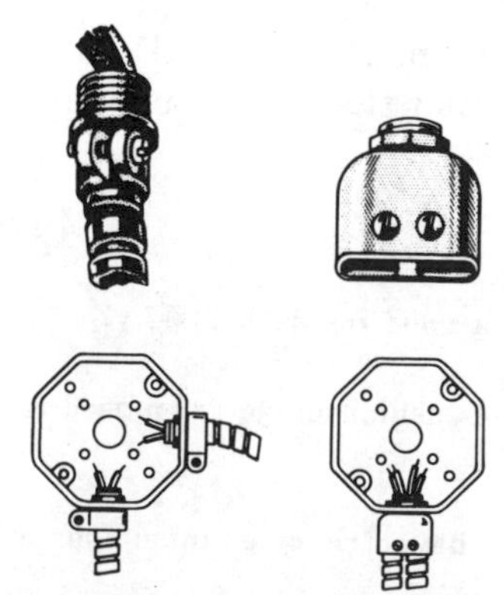

Fig. 8-9. Armored-cable connectors and their uses.

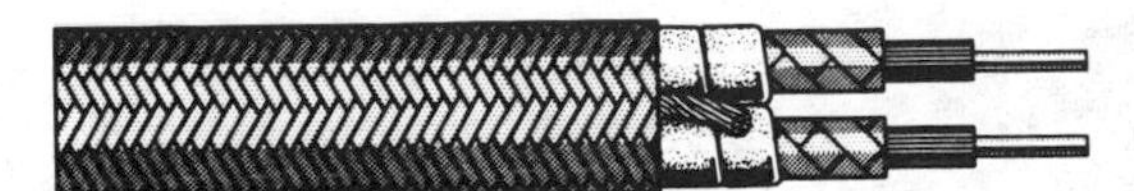

Fig. 8-10. Nonmetallic sheathed cable.

outlets and switch boxes by connectors similar to those used with armored cable. Many of the switch boxes and outlets have a provision for clamping the cable inside them, making special connectors unnecessary.

Metal raceway is an ornamental metal tubing through which conductors can be run when the wiring must be above the surface. The raceway can be installed on walls, baseboards, ceilings, and workbenches. There are several well-known trade names for raceway, one of which is *Wiremold.* Fig. 8-11 shows some typical cross sections and methods of working Wiremold metal raceway.

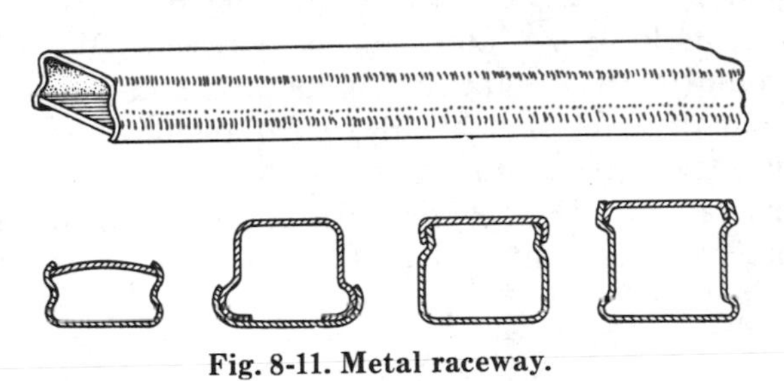

Fig. 8-11. Metal raceway.

Special fittings, such as elbows and tees, can be used between lengths of metal raceway to make junctions and to form bends. Special boxes, switches, and outlets also are available.

Rigid electrical conduit is sold in 10-foot lengths. It is galvanized on the outside to delay corrosion and is varnished or lacquered on the inside for added conductor insulation.

The rigid conduit is connected by locknuts and bushings to switch boxes, fuse cabinets, and other electrical fixtures (Fig. 8-12).

Thinwall conduit (electrical metallic tubing) is available in the same sizes as rigid conduit. Thinwall conduit is much thinner than rigid conduit—

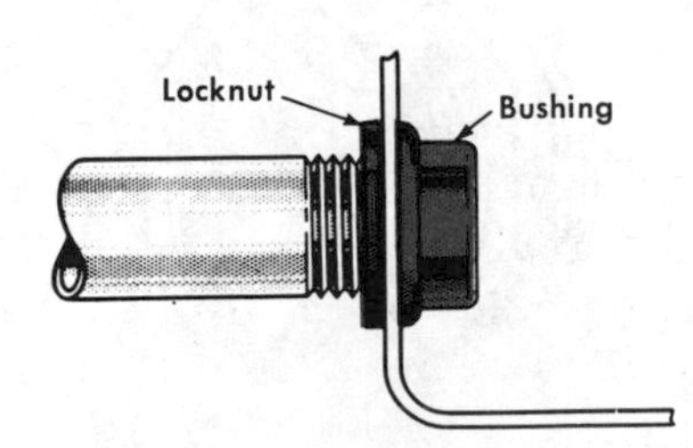

Fig 8-12. A locknut and bushing used to fasten rigid conduit to an outlet box.

so thin, in fact, that the National Electrical Code will not permit it to be threaded. Thinwall conduit is fastened to switch boxes and electrical fixtures by friction connectors like the ones in Fig. 8-13. The angle connnector is used to make right-angle connections to a box. The coupling is used to connect two lengths of conduit for one continuous length. The connector is used to make a straight connection into a metal enclosure.

Thinwall and rigid conduit look like water or gas pipes. With a few exceptions, thinwall conduit can be substituted for rigid conduit. However, some local codes do not permit thinwall conduit to be used outdoors, nor where corrosive fumes are present. Because thinwall conduit is easy to handle, it has practically replaced rigid conduit wherever its use is not prohibited.

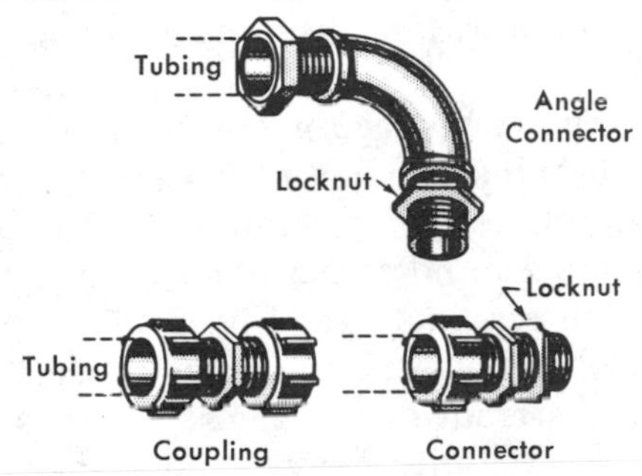

Fig. 8-13. Thinwall conduit fittings.

ENTRY SERVICE

Electricity is brought into a building by either a two- or a three-wire system. A typical three-wire entry service is shown in Fig. 8-14. The meter is usually mounted on the outside. It must be enclosed in glass, as shown in Fig. 8-15, or in

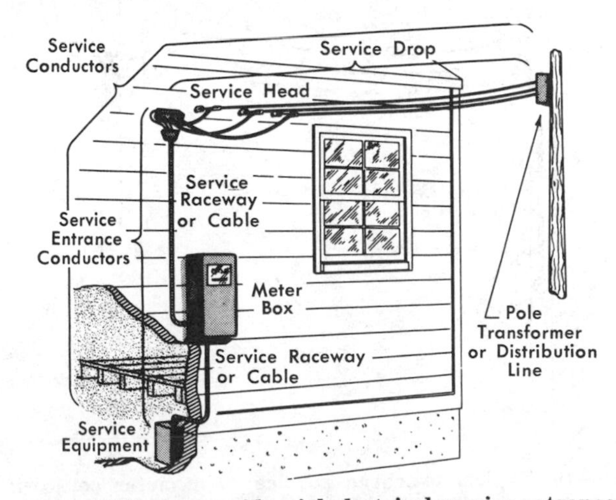

Fig. 8-14. Typical residential electrical service entrance.

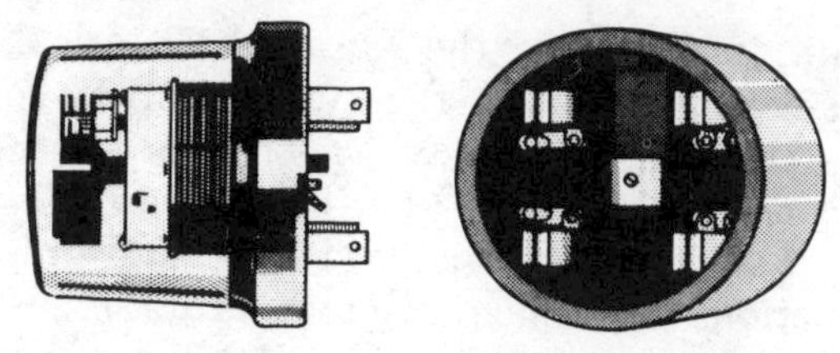

Fig. 8-15. Glass-enclosed outdoor meter without a protective cabinet.

a metal cabinet. The meter may also be mounted on a power pole, as shown in Fig. 8-16. The conduit is fastened to the meter base with threaded connections, making a watertight junction. The service conductors are enclosed by the service head (at the top of the conduit in Fig. 8-16). The service head protects the insulation of the service conductors, at the point where they emerge from the conduit, and keeps rain or snow out of the conduit.

The entry service *cabinet* is usually on the inside. It contains the main fuses, plus a switch to control all electrical power to the building. The main fuses and the branch circuit fuses are usually in one cabinet (see Fig. 8-17).

A special entry service cable is used in many electrical distribution systems. It is a braided insulated cable, quite heavy, and larger than the largest automobile battery cable. The outer braid is usually greenish-gray. Fig. 8-18 shows the construction of the cable.

At the center of the cable is a pair of heavy insulated conductors, which rarely are smaller

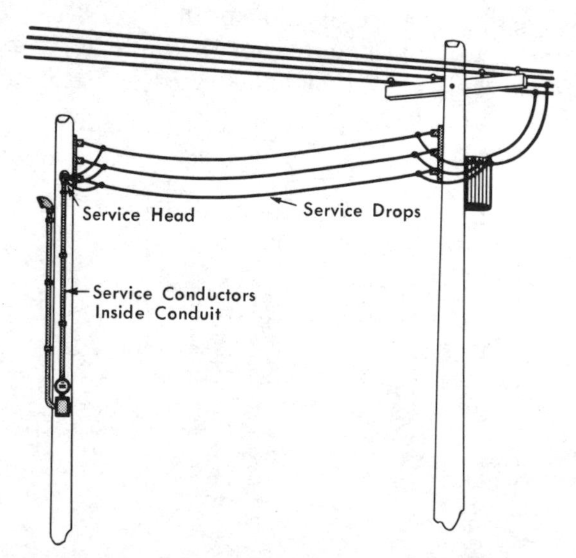

Fig. 8-16. Typical overhead service to a meter pole, with service conductors, service drops, and the service head.

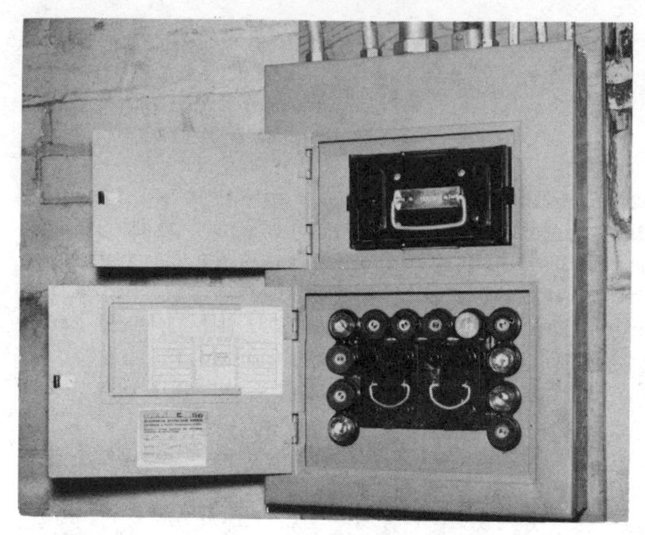

Fig. 8-17. A main service installation for a residence.

than No. 6 because they must be heavy enough to handle large amounts of current. Wrapped around the two inner conductors and protected from them by a heavy layer of insulating paper or fiber is a twisted layer of wires. These wires protect the inner conductors and their insulation and are the neutral conductor of the three-wire system.

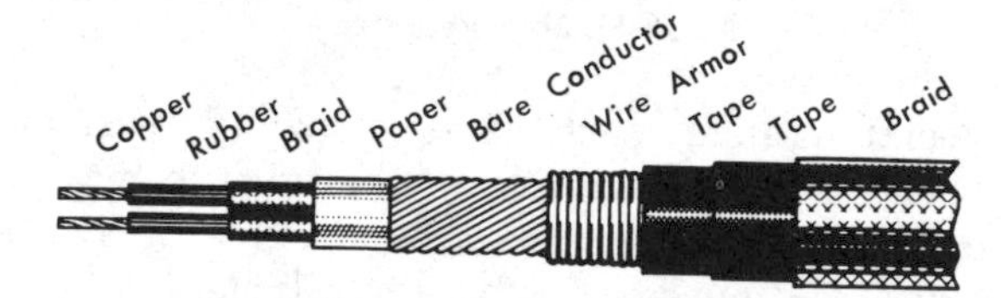

Fig. 8-18. Service-entrance cable construction.

Any entry service cable can be connected to the service head by a watertight connecter like the one in Fig. 8-19. The cable will pass through the hole in the rubber bushing inside the connector. When the packing nut is tightened, the bushing will pack tightly around the cable and

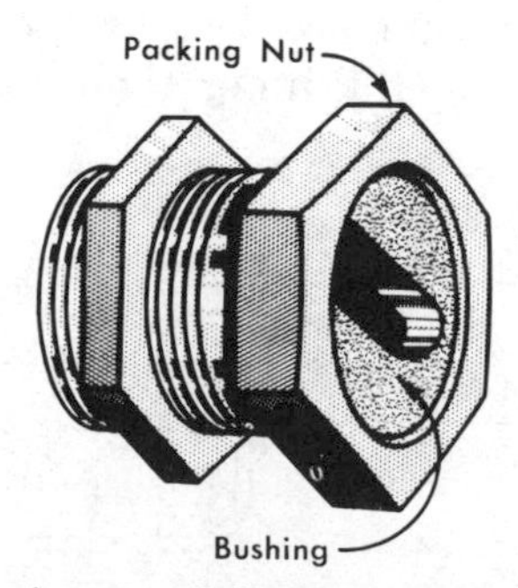

Fig. 8-19. A water-tight fitting for outdoor use with service-entrance cable.

make a watertight fit. A similar connector is used at the point where the service cable enters the meter base or the cabinet of the meter box.

Two-Wire Entry Service

In the two-wire entry service, one wire is grounded (connected to earth by being fastened to a cold-water pipe or to a metal rod driven deep into the earth). The ground wire in an entry service cable is covered with a *white* insulation; the "hot" wire, with a *black* insulation. In other words all the white wires are connected together to provide a ground for every branch circuit, and all the black wires are "hot" throughout the system.

Three-Wire Entry Service

You may have noticed that the electrical power supply wires are usually strung from utility poles to homes in sets of three wires. If these wires are lettered "X," "Y," and "Z," connections to "X" and "Z" have a potential (voltage) of 230 volts; those to "X" and "Y" have 115 volts, and those to "Y" and "Z," 115 volts. The white center wire ("Y" in our example) is the *neutral* or ground wire. The other wires are "hot."

The electrician who wires a residential service attempts to estimate the load which may be required of the various circuits within the house and connects these circuits in the service entrance switch (main distribution panel) so that half the total 115-volt load will be connected to "X" and "Y" and the other half to "Y" and "Z." If the home has an electric water heater where 230 volts are required, "X" and "Z" would be used.

In the city, the ground wire can be connected to a cold-water pipe, or to a water main for the best possible ground. In rural areas where deep wells supply the water, an electrical system grounded to a water pipe can be hazardous. If the water pumping equipment should be removed, the electrical system will no longer be grounded. A long copper rod driven deep into the earth provides the best *ground* for most rural electrical systems.

The grounding wire is fastened to the water pipe or ground rod by a *grounding clamp*. The grounding clamp provides a solid electrical and mechanical connection to the ground pipe or rod. Fig. 8-20 shows two types of pipe clamps, plus a ground strap for tying the grounding wire to the ground pipe.

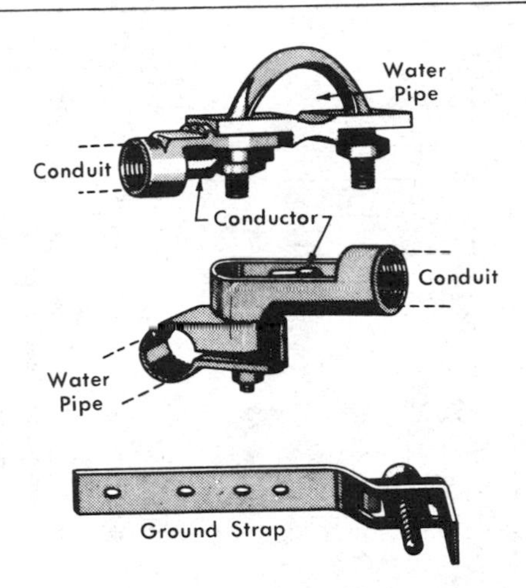

Fig. 8-20. Typical ground clamps and straps.

OUTLET WIRING

A good home wiring system is a necessity. Even though a full supply of electricity is available from the power company, the extent to which it can be used is determined by the *interior wiring* in the home. Starting from the service entrance where the electricity enters, the wiring system must deliver a *full* supply of electricity to each outlet.

In a properly planned system, each circuit and outlet serves a specific purpose, and controls (switches) are located with both convenience and safety in mind.

The first step in wiring a system is to prepare plans detailing how it must function and showing all elements in the system.

A floor plan of the building should be marked with the location of each box, switch, and outlet. The amount of wiring material needed can then be calculated. Fig. 8-21 shows a typical house floor plan with the location of the center lights, switches, and outlets marked. The ceiling outlet boxes can be used as *junction boxes,* in addition to providing a place to install the light fixtures. Fig. 8-22 shows a method of using an outlet box for a light fixture and of running wires to another outlet. It was previously stated that all white wires were tied together and grounded. But the cable entering the light switch has white and black wires, both tied in the "hot" circuit. Therefore, the white wire is "hot." This situation occurs only when switches are wired. Fig. 8-23 shows how the switch wires are connected to an outlet box.

Outlets are a convenient way to supply current to lamps, motors, radios, and other electrical ap-

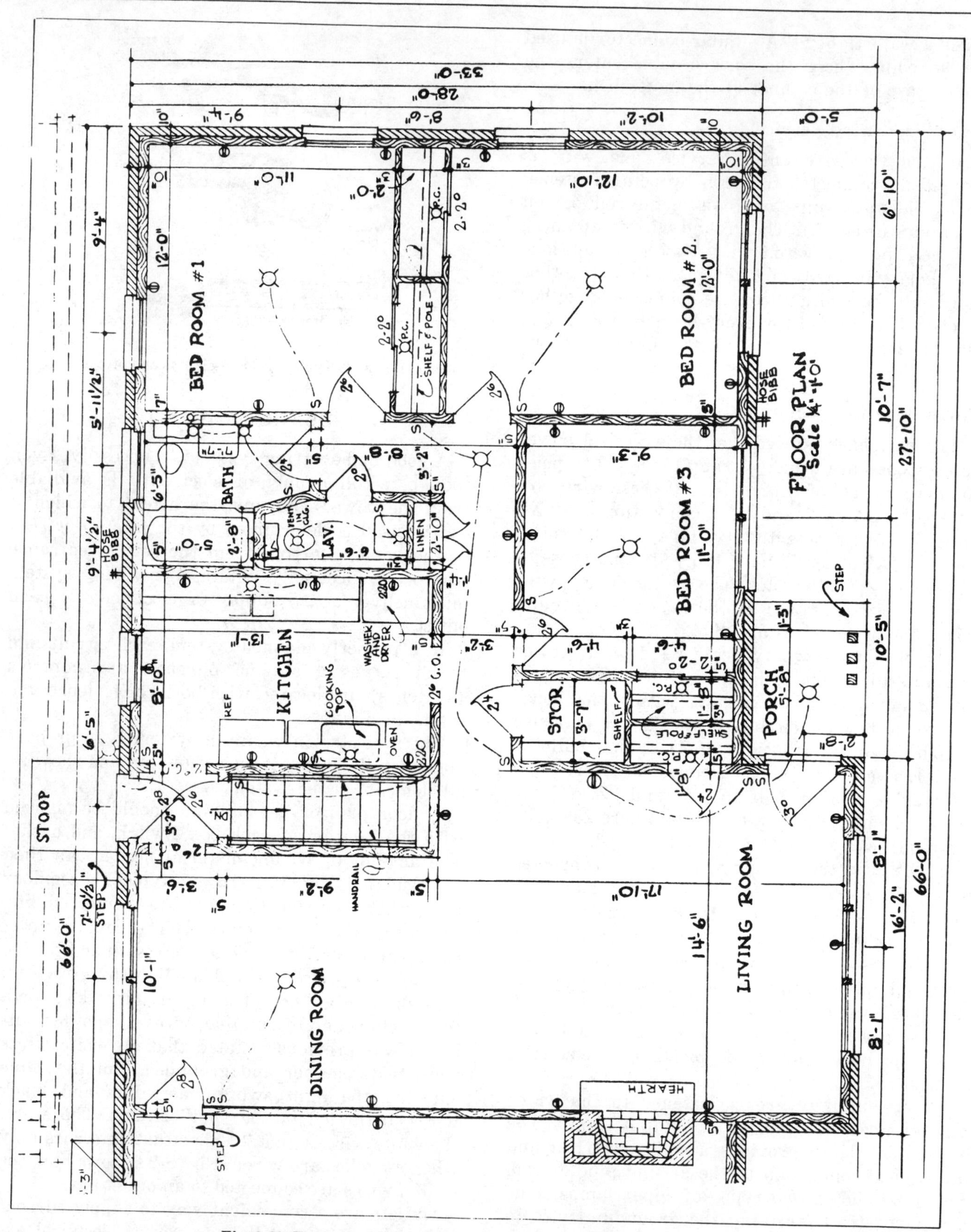

Fig. 8-21. Floor plan of a house, showing the wiring plan.

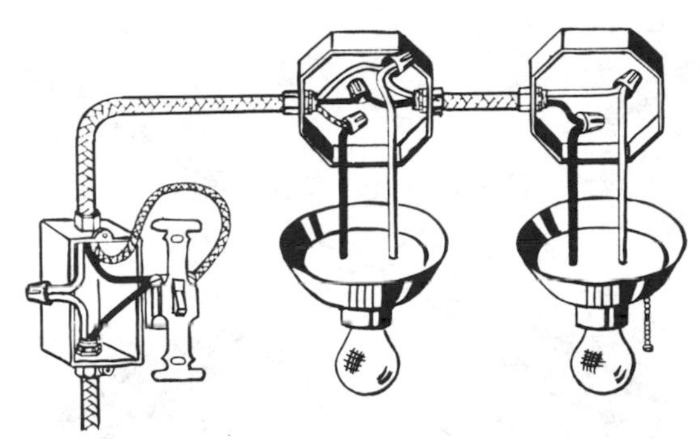

Fig. 8-22. Installation of two ceiling lights on the same supply line; one controlled by a switch.

pliances (which is why they are called convenience outlets). Most outlets are placed in the wall, near the floor. The outlets in bathrooms, kitchens, basements, and garages are usually placed on the wall, approximately four feet above the floor. In a factory, the outlets are often on bench tops.

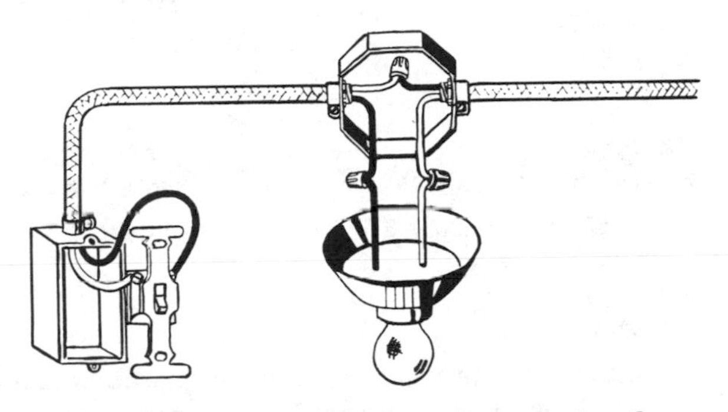

Fig. 8-23. Ceiling light connected to a single-pole wall switch.

Fig. 8-24 shows an outlet with a third terminal which is connected to ground. This type of outlet is used for operating electrical hand tools. The ground wire assures the operator that the tool is properly grounded. Fig. 8-25 shows an adapter plug with a grounded connection.

SPLICES AND JOINTS

When conductors are joined, the connection must be strong and must be electrically "tight" or solid. A "loose" connection reduces efficiency by introducing added resistance in the circuit.

A *splice* joins the ends of two wires together. A wire joined at right angles to another continuous wire is called a *tap.*

The first step in making any splice or tap is to *strip* (remove the insulation from) the ends of the wires. This is done by cutting at a slant—as

Fig. 8-24. Grounded outlet receptacle for power tools.

in sharpening a pencil—because nicks can reduce the conducting area and thus weaken the wire. (See Fig. 8-26.) The tin coating should not be removed.

The pigtail or rat-tail splice, the Western Union splice, the tee or common tap, the knotted tap, and the fixture splice are all different types of splices that can be used.

Fig. 8-25. Special three-prong attachment plug for grounded power tools.

Pigtail Splice

The pigtail splice (also known as a rat-tail splice) is shown in Fig. 8-27. Three inches of insulation are removed from each wire. The wires are crossed one inch from the insulation and are

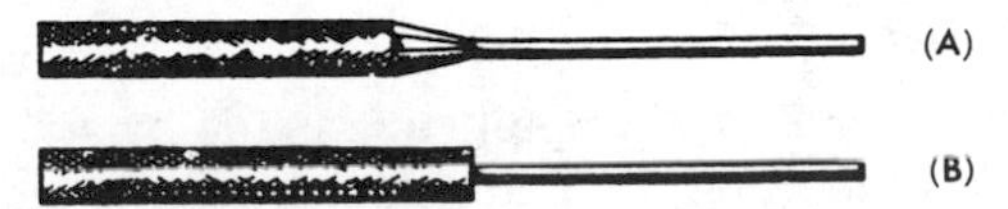

Fig. 8-26. Correct (A) and incorrect (B) removal of insulation from a rubber-covered single conductor.

twisted—with the fingers and pliers—at least five turns. The wires are then bent back, as shown in Fig. 8-27C, to prevent their puncturing the tape that will be added to insulate the splice. Because this splice is not very strong, it is often used in outlet boxes and at other junctions where there is no strain on the wires.

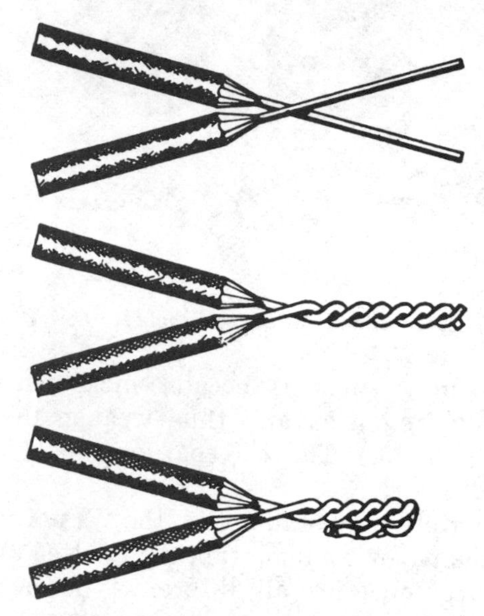

Fig. 8-27. Making a pigtail splice.

The bare wires are then coated with soldering paste (flux) and heated with a soldering iron. Solder is next melted into each crevice in the splice. After the wires are soldered, they are insulated by wrapping the splice with *rubber* tape and covering that with *friction* tape. The latest type of insulation is *plastic* tape. It does the work of both the rubber and the friction tape, and is waterproof and acidproof.

Western Union Splice

The construction of a Western Union splice is shown in Fig. 8-28. The wires are twisted and wrapped, then soldered and taped. A double Western Union splice may be used to join a pair of parallel wires. The joints are staggered, as shown in Fig. 8-29. The Western Union splice is often used for splicing straight runs of wire. It is a good, strong splice.

Tap Splice

A tee or common tap splice is made by stripping three-fourths inch of insulation from the *main* or *running* wire and three inches from the end of the tap wire. (See Fig. 8-30.)

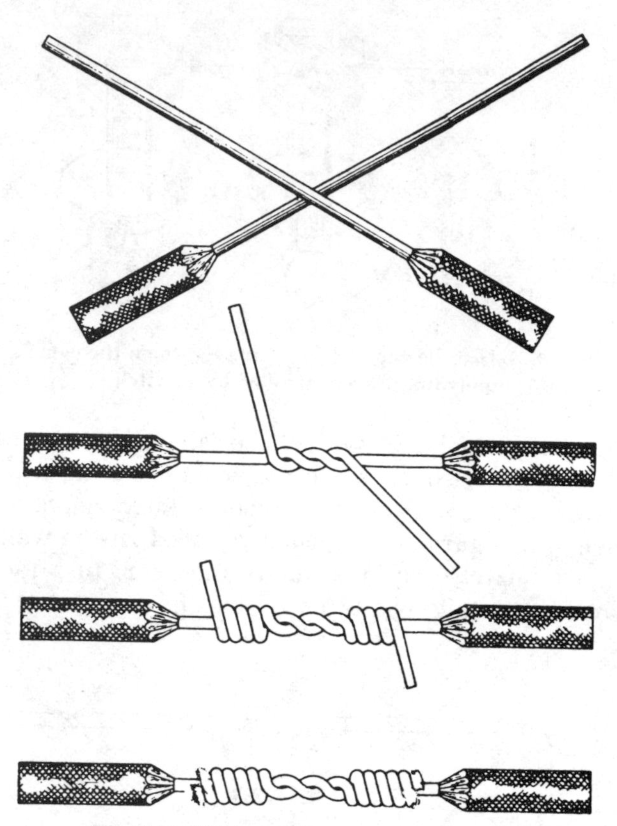

Fig. 8-28. Making a Western Union splice.

Fixture Splice

A fixture wire is smaller than the branch circuit wire to which it is connected. The steps required to make a fixture splice are shown in Fig. 8-31. The fixture splice is often used to fasten together two wires of different sizes.

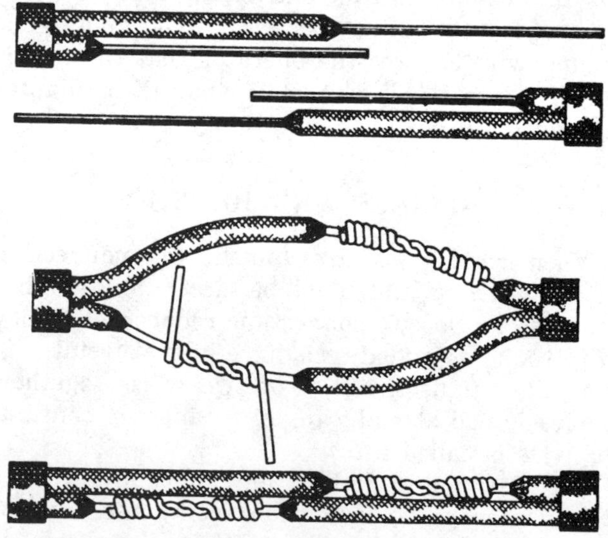

Fig. 8-29. Making a double Western Union splice.

The *wire nut* or *solderless connector* eliminates the need for soldering joints. Since it is made of insulating material, the wires need not be taped.

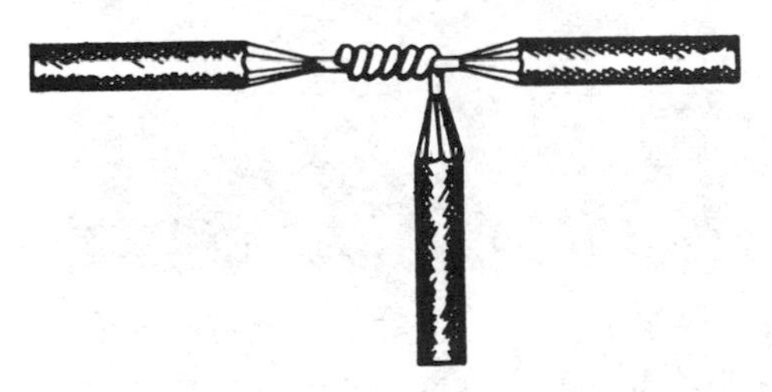

Fig. 8-30. Making a common tap or tee splice.

The solderless connector can be used on a pigtail splice only. It is screwed over the wires, as shown in Fig. 8-32.

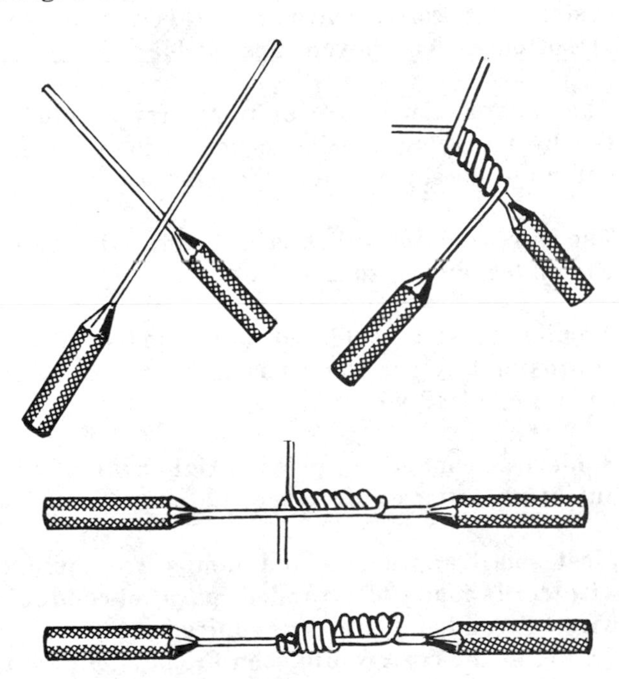

Fig. 8-31. Making a fixture splice.

WIRING SOCKETS AND PLUGS

The conductors we have discussed so far carry large amounts of current. However, there are other conductors, in the home, that carry smaller amounts of current to the lights and appliances. One of these is the *stranded wire conductor*. It is used with small electrical appliances, such as lamps, small motors, clocks, radios, and toasters.

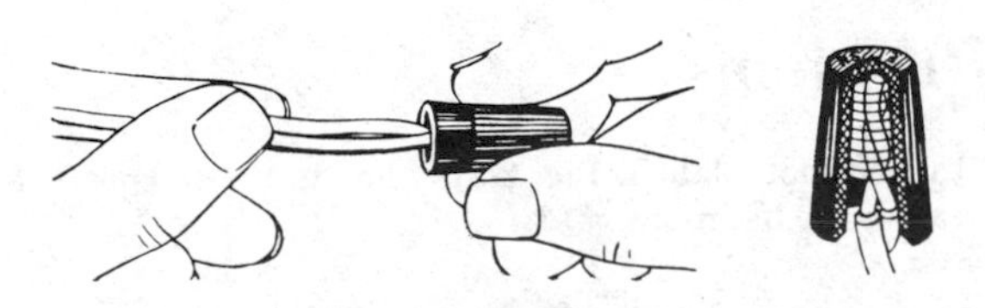

Fig. 8-32. Using solderless connectors.

Stranded wire, usually made of copper, actually is a number of small wires twisted together to provide a conductor with a larger diameter and more flexibility than a single, solid wire. The stranded wire is better able to withstand repeated flexings without breaking. In most lamps and other small appliances, a twin-conductor No. 18 lamp cord is used, insulated with plastic, rubber, or cloth-covered rubber.

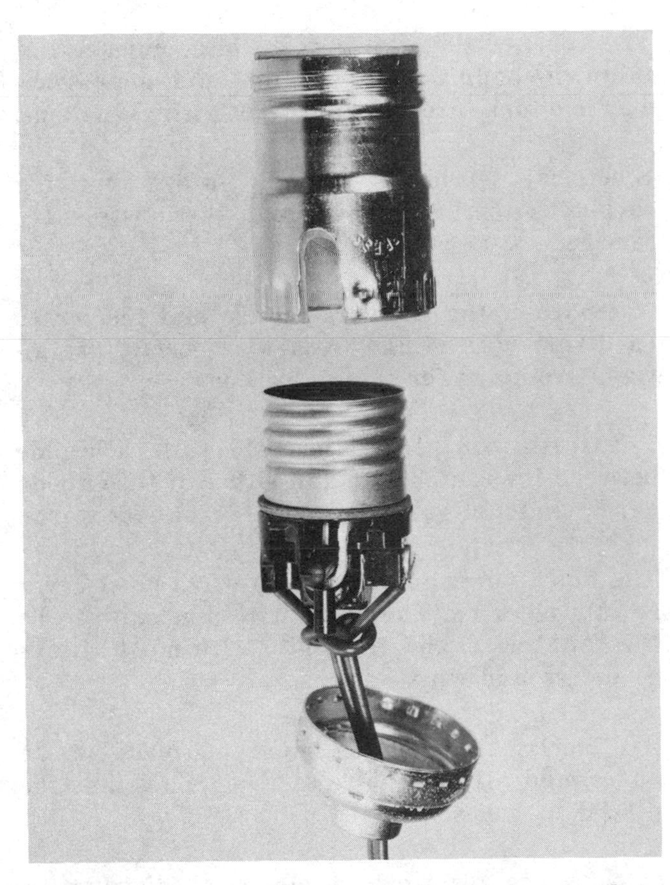

Fig. 8-33. Using an underwriters knot with a keyed socket.

Extension cords are a useful addition to any home lighting system, and are easy to make. There are two types of extension cords—a heavy cord for the garage, home shop, or basement; and a lighter one for inside the home, where the problem of wear is not as great. However, the same wiring procedures apply to both.

The brass socket must be carefully connected to the extension cord. For a safer and stronger connection, an underwriter's knot should be used (Fig. 8-33) because it relieves the cord of strain.

The attachment plug is connected to the end of an extension cord after a *single* knot is tied in the cord to protect the terminal connection (see Fig. 8-34). A single knot is used because a small plug normally does not have enough space for an underwriter's knot. When an attachment plug on a lamp or other small appliance is damaged, a new attachment plug can be installed in the same manner.

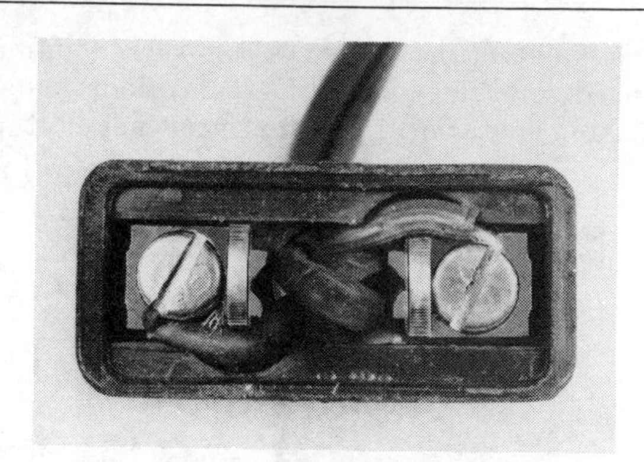

Fig. 8-34. Using a single knot to protect the terminal connection of an attachment plug.

Summary

The open-wiring, armored-cable, nonmetallic, thinwall-conduit, rigid-conduit, and metal-raceway methods are all approved wiring methods.

Electrical service is brought into a building through either a two-wire or a three-wire entry service.

One wire of the two-wire service and the center (neutral) wire of the three-wire service are always grounded for safety reasons.

The three-wire service has 115 volts available between the center wire and either of the outside wires, and 230 volts between the outside wires.

The black wires and the white wires must never be interchanged. The black wire must always be the "hot" wire, and the white wire must always be the ground wire.

All electrical junctions must be made inside boxes, and all switches and receptacles must be placed in boxes.

Outlets are installed in electrical distribution systems for convenience. For this reason, they are often called convenience outlets.

The ends of a wire are prepared for a splice or tap by removing the insulation with a *slanting* cut, as when sharpening a pencil.

The more common splices are the pigtail, Western Union, tee, and fixture splices.

A splice must be soldered and taped to provide approximately the same strength and insulation as an unspliced wire.

Solderless connectors provide tight splices without being soldered or taped.

Most small appliances and lamps are equipped with cords made of stranded, parallel-conductor wire. Stranded wire is required because the flexing of the cord would soon break a solid wire.

The *underwriter's knot* is used for attaching a socket to an extension cord, where strain could pull the cord from the socket. An extension cord has a *single knot* to prevent an attachment plug from pulling free of the wire.

Questions and Problems

19. What is the principal advantage of a three-wire entry service over a two-wire entry service?

20. What is the purpose of a service head in an electrical entry service?

21. In the floor plan in Fig. 8-21, what is the purpose of the switch in bedroom #1?

22. Is a receptacle connected across both service wires or across only one wire?

23. What are the requirements of a good splice?

24. When is the pigtail splice usually used?

25. What type of splice is employed to join wires of different sizes?

26. Why are appliance cords made of stranded wire?

THREE-WIRE CIRCUITS

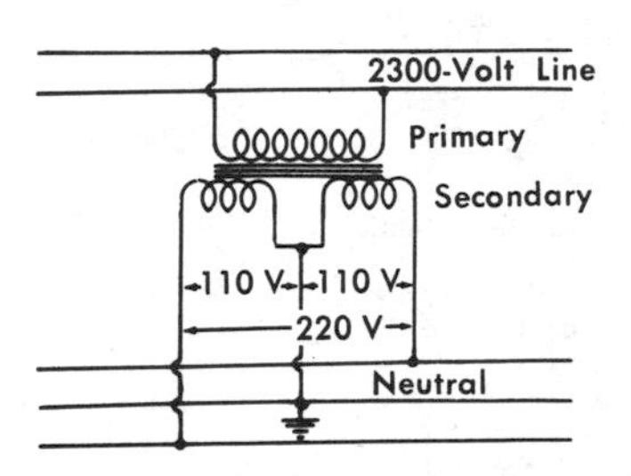

Fig. 8-35. A three-wire ac circuit supply.

The three-wire circuit may be used with either ac or dc, although the original Edison three-wire circuits were dc only. (See Figs. 8-35 and 8-36.)

The three-wire system provides a saving of over 50 percent in copper wire. It also has the advantage of making two different voltages available. Let's examine the importance of this matter of copper wire saving by considering a certain situation. For example:

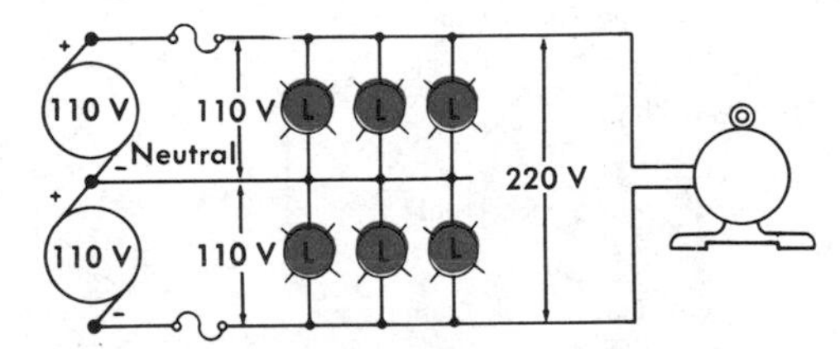

Fig. 8-36. A three-wire dc circuit supply.

A load of 12,000 watts at 120 volts at 1000 feet (Fig. 8-37),

$$I = \frac{12{,}000 \text{ watts}}{120 \text{ volts}} = 100 \text{ amps}$$

#1 Wire Cap = 100 amps

253 lbs per 1000′

253 × 2 = 506 total lbs copper

In Fig. 8-38 the same 12,000-watt load has been connected to a *three-wire* supply.

$$I = \frac{6000 \text{ watts}}{120 \text{ volts}} = 50 \text{ amps}$$

$$I = \frac{12{,}000 \text{ watts}}{240 \text{ volts}} = 50 \text{ amps}$$

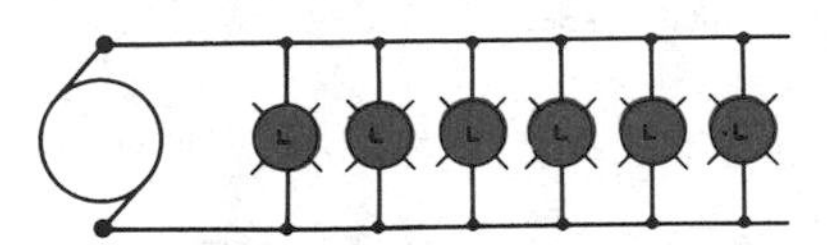

Fig. 8-37. Six lamps connected to a two-wire current supply.

#6 Wire Cap = 50 amps

79.5 lbs per 1000′

79.5 × 3 = 238.5 lbs copper

506 − 238.5 = 267.5 lbs copper saved

Polarization refers to the identification of a wiring system by the color of the wire. Dark-colored insulation indicates a "hot" (ungrounded) wire; a lighter color indicates a neutral (grounded) wire. This method of identification is specified in Article 200 of the National Electrical Code (we will learn more about this code in a later section). Normally, the neutral (ground) wire is white or gray. The code clearly states that it shall be of such a color that it can be easily distinguished from all other conductors in the wiring system.

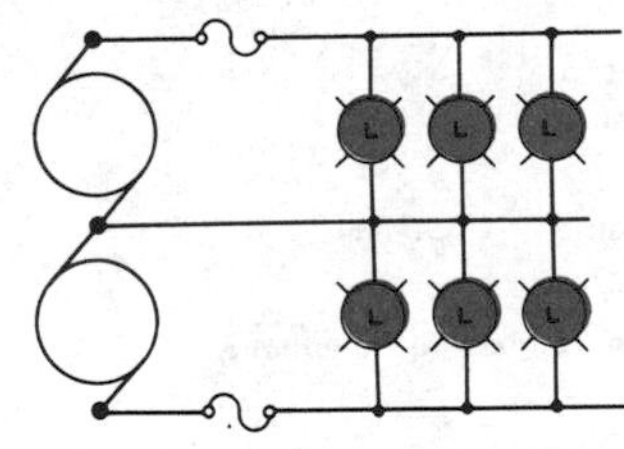

Fig. 8-38. Six lamps connected to a three-wire supply.

When the electrical system is grounded (by connecting the neutral wire directly to ground), no part of the system will ever be more than 120 volts above or below ground. Thus, it is possible to have 240 volts of electrical power available, and yet have no voltage more than 120 volts from ground. Fixture or socket terminals are identified by brass or dark screws for "hot" wires, and white metal screws for neutral wires.

Balanced Three-Wire Circuits

In a perfectly balanced three-wire system, each outside wire carries the total load—the neutral

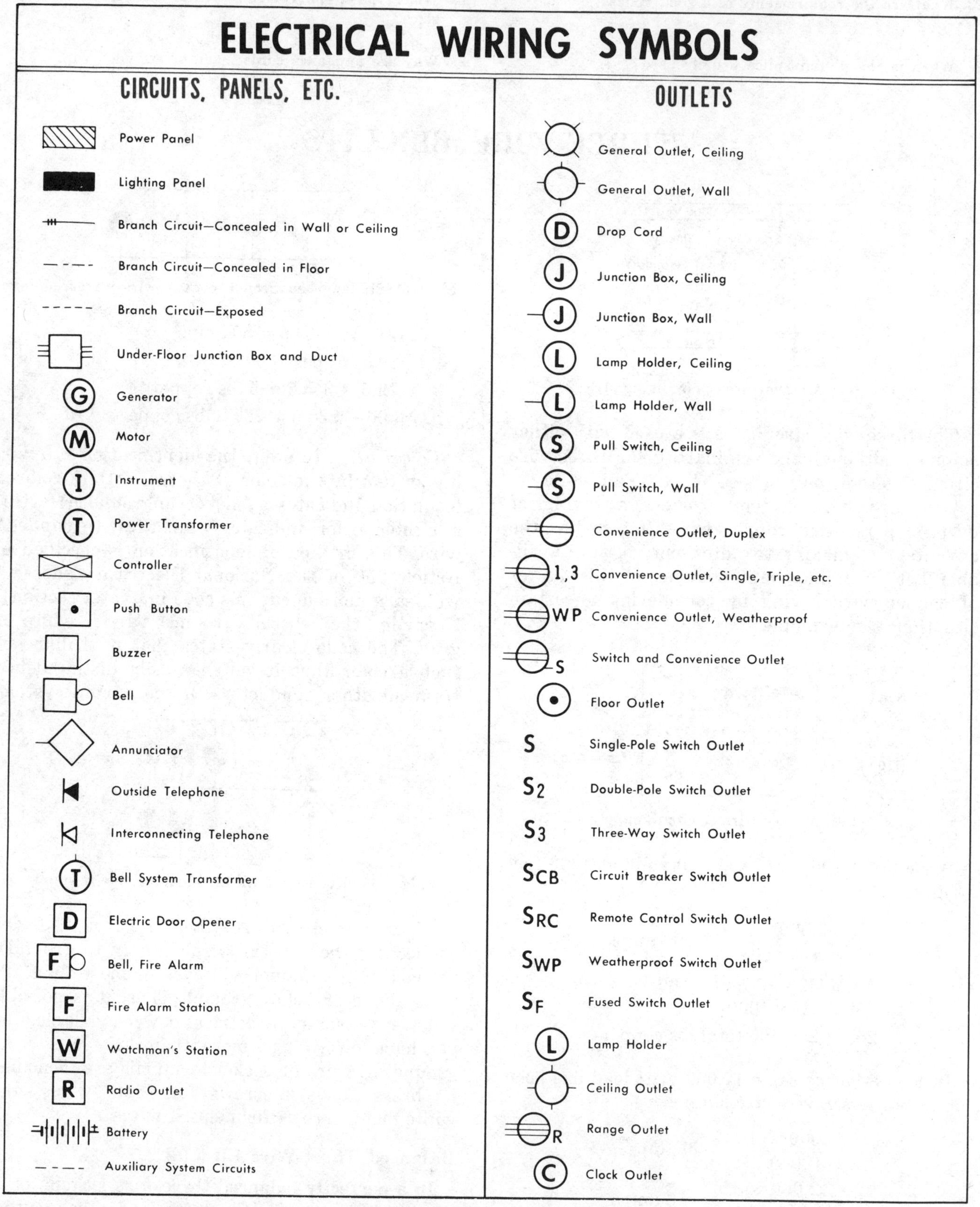

ELECTRICAL WIRING SYMBOLS

CIRCUITS, PANELS, ETC.

Power Panel
Lighting Panel
Branch Circuit—Concealed in Wall or Ceiling
Branch Circuit—Concealed in Floor
Branch Circuit—Exposed
Under-Floor Junction Box and Duct
G Generator
M Motor
I Instrument
T Power Transformer
Controller
Push Button
Buzzer
Bell
Annunciator
Outside Telephone
Interconnecting Telephone
T Bell System Transformer
D Electric Door Opener
F Bell, Fire Alarm
F Fire Alarm Station
W Watchman's Station
R Radio Outlet
Battery
Auxiliary System Circuits

OUTLETS

General Outlet, Ceiling
General Outlet, Wall
D Drop Cord
J Junction Box, Ceiling
J Junction Box, Wall
L Lamp Holder, Ceiling
L Lamp Holder, Wall
S Pull Switch, Ceiling
S Pull Switch, Wall
Convenience Outlet, Duplex
1,3 Convenience Outlet, Single, Triple, etc.
WP Convenience Outlet, Weatherproof
S Switch and Convenience Outlet
Floor Outlet
S Single-Pole Switch Outlet
S_2 Double-Pole Switch Outlet
S_3 Three-Way Switch Outlet
S_{CB} Circuit Breaker Switch Outlet
S_{RC} Remote Control Switch Outlet
S_{WP} Weatherproof Switch Outlet
S_F Fused Switch Outlet
L Lamp Holder
Ceiling Outlet
R Range Outlet
C Clock Outlet

Fig. 8-41. Electrical wiring symbols.

wire carries no load (see Fig. 8-36). In a three-wire system in which the load is unbalanced, one of the outside wires carries the total load. The remaining two wires, one of which is neutral, divide the total load between them, and the division may or may not be equal. (See Fig. 8-39.)

Unbalanced Load Circuits and Open Neutral

Fig. 8-40 shows an unbalanced three-wire system. If the neutral wire breaks, the lightly loaded side must carry all the current that the heavily loaded side requires, and the load will be damaged because of the *surge* current.

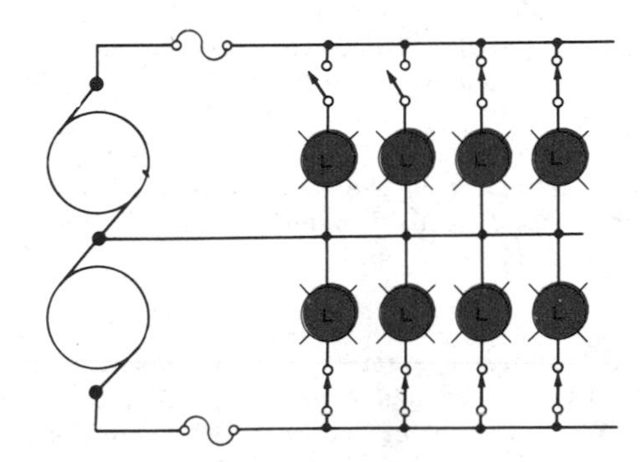

Fig. 8-39. An unbalanced three-wire system with a solid neutral.

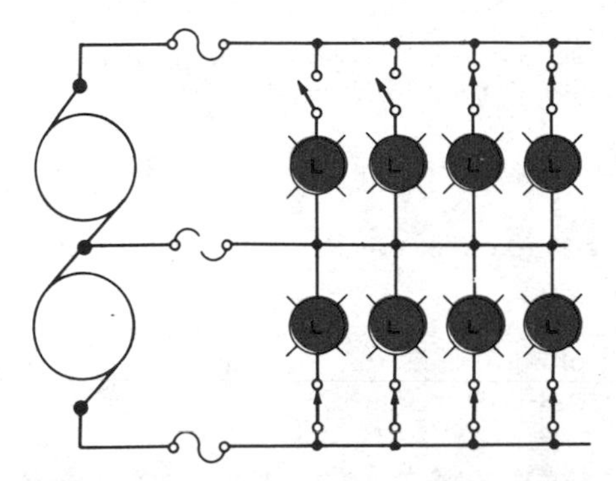

Fig. 8-40. An unbalanced three-wire system with an open neutral.

Resistance of each lamp = 100 ohms

$$\text{Resistance of two upper lamps} = \frac{100}{2} = 50 \text{ ohms}$$

$$\text{Resistance of four lower lamps} = \frac{100}{4} = 25 \text{ ohms}$$

$$R_t = 50 + 25 = 75 \text{ ohms}$$

$$L_t = E/R = \frac{200}{75} = 2.93\text{A}$$

$$E = I \times R = 2.93 \times 50 = 147$$

$$2.93 \times 25 = \underline{\ 73\ }$$

$$220 \text{ volts}$$

Nonpolarized systems often cause injury to personnel working on a "hot" system. If all three-wire systems, new or reconditioned, are properly installed and marked, fires and other mishaps cannot occur.

A number of common symbols used in electrical wiring blueprints are shown in Fig. 8-41.

In most power stations supplying dc, the three-wire system is used to distribute the current. Thus, lamps can be operated at one potential, and motors can be operated at the same or twice this potential.

THREE-WIRE CURRENT SOURCE

The three-wire system dc generator supplies the two potentials from a single dynamo. It is called a *three-wire generator*. Such a 220-volt generator may be equipped with slip rings and a commutator. An iron-core reactor usually is connected across the slip rings to establish a tap for the neutral wire. This type of *bipolar* machine is shown in Fig. 8-42.

The iron-core reactor has a *high* inductance and a *low* resistance. It is bridged across the armature at opposite (180°) points on the winding. Therefore, the center tap of the reactor has a voltage halfway between the voltages of the brushes on the commutator. The voltage across the reactor is *alternating*. Because of the high *reactance* of the winding, however, the ac through the winding is small. When the load is *balanced* on both sides of the system, no dc flows in the neutral wire or in the reactor. But when the loads are *unbalanced*, there is dc in one direction or the other through the neutral wire and through half of the reactor. Because of the low resistance of the reactor, the voltage drop (due to the dc) across the reactor is small.

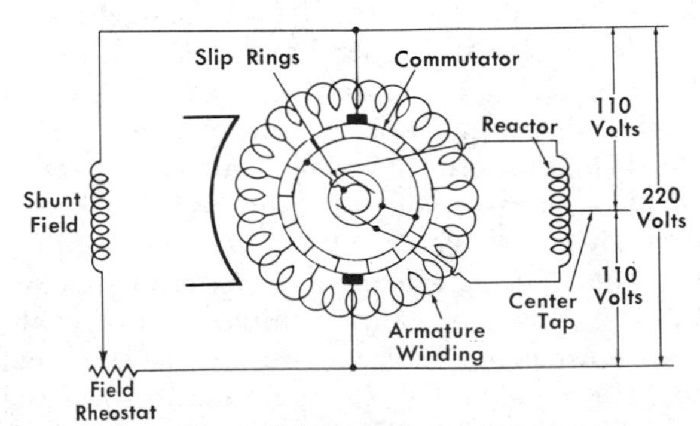

Fig. 8-42. Three-wire generators.

Summary

The three-wire system provides two different supply voltages, and can carry a much larger load than a two-wire system.

It is best to plan the wiring to get a balanced load from the system, so that each of the outside wires carries only one-half the total load.

A three-wire system dc generator supplies two potentials from a single dynamo.

Questions and Problems

27. What are the two advantages of a three-wire system over those of a two-wire system?

28. What is the symbol for a three-way switch?

29. What is the symbol for a waterproof switch?

30. In Fig. 8-40, does the top bank of lamps have higher or lower voltages than the bottom bank of lamps when the neutral fuse is blown?

31. Should the white wire in a service entrance installation be connected to ground?

32. How are the neutral connections (terminals) of an electrical fixture usually identified?

33. What will happen if the neutral wire of a three-wire system is accidentally broken off between the load and the generator?

34. A dc distribution system requires 10 tons of copper wire to carry a certain load operated at 115 volts with two wires. What amount of copper will be required to operate the system at 230 volts with three wires, assuming the same load and the same percentage of voltage drop?

NATIONAL ELECTRICAL CODE

Electricity has become one of man's most faithful and useful servants, even though it can be destructive—and a killer—if not handled properly. No one knows how many billions of dollars worth of property have gone up in smoke from fires caused by improper electrical wiring. Thousands of persons have lost their lives from this invisible force.

From facts like these, one might think that electricity is too dangerous to use and that we should turn to other sources for our light, heat, and power. Such reasoning ignores one most important fact about electricity—its danger stems from careless and improper handling.

To make electricity safe for everyone, a test laboratory is supported by large sums from electrical manufacturers. This laboratory tests electrical apparatus of all kinds to determine whether they are electrically safe, so that they will not be a fire hazard or a potential danger to persons. Fire insurance companies are probably the most concerned. They and the electrical manufacturers have banded together to form the National Board of Fire Underwriters, with offices in principal cities all over the United States. The Underwriter's Laboratories' stamp of approval is the user's best guarantee that an electrical apparatus is as safe as science can make it. No product can carry the Underwriter's Laboratories' stamp of approval unless it has been tested and approved by the laboratory.

The Underwriter's Laboratories have done more than this. In cooperation with state and municipal electrical inspectors, fire prevention bureaus, and electrical manufacturers, they have borne most of the expense of preparing an electrical code telling how electrical apparatus should be made and how wiring should be installed. In addition, they keep the code up-to-date as new knowledge and equipment comes into existence.

The National Electrical Code is the standard for electricians throughout the United States. Its value is emphasized by the fact that many states and cities have incorporated it into their statutes, word for word.

The National Electrical Code is a code of minimum standards. That is, any wiring or electrical apparatus that does not meet its standards is rejected. Nevertheless, there is no reason why a city or state cannot adopt even stricter standards than the ones in the code.

The important points to be considered in any electrical wiring job are:

1. The proper size wires for the amount of current.
2. The proper insulation according to the voltage.
3. Proper mechanical support and protection for the runs of wire.
4. Secure and permanent splices and connections.
5. Protection against any danger of fire and shock.

The National Electrical Code has one purpose—to safeguard persons and property from electrical hazards wherever electricity is used. It standardizes and simplifies the rules of good wiring and provides a reliable guide for electrical construction men. This code originally was prepared in 1897. It is revised frequently to meet the changing conditions of improved equipment and materials. Many electrical engineers, electrical equipment manufacturers, insurance companies, and architects all deserve credit for contributing to its success.

There are at least four codes governing electrical work:

1. National Electrical Code.
2. State codes.
3. Local (city) codes.
4. Central station rules.

Wiring Methods

The National Electrical Code recognizes the following seventeen types of wiring for light and power systems.

1. Rigid metal conduit.
2. Flexible metal conduit (Greenfield).
3. Armored cable (BX and BXL).
4. Electrical metallic tubing (steel tube, and thinwall conduit or Thinwall).
5. Surface metal raceway (Wiremold).
6. Cellular metal floor raceways.
7. Underfloor raceways.
8. Wireways.
9. Busways.
10. Nonmetallic sheathed cable (Romex or Loomflex).
11. Nonmetallic waterproof (rubber-sheathed) cable wiring.
12. Nonmetallic surface extensions.
13. Service entrance cable.
14. Concealed knob-and-tube work.
15. Open wiring on insulator (cleats).
16. Bare-conductor feeders.
17. Underplaster extensions.

Wiring Systems

After eliminating those methods which apply only to special installations, we can divide the remaining wiring systems into the following six groups:

1. Conduit (rigid) or electrical metallic tubing (Thinwall).
2. Armored cable.
3. Metal raceway.
4. Nonmetallic sheathed cable.
5. Knob and tube.
6. Cleats.

The first three are known as "metallic systems" and the latter three as "nonmetallic systems." The usual local city codes permit the use of either metallic or nonmetallic systems.

Branch Circuit Requirements

A branch circuit is defined by the National Electrical Code as the part of a wiring system beyond the final fuse box protecting the circuit. Branch circuits carry electrical current to appliance outlets, light fixtures, and switches.

The code recognizes four branch circuits with ratings of 15, 20, 30, and 50 amperes. The 50-ampere branch circuit is recommended for ranges and water heaters only. The 30-ampere branch circuit is usually used for appliances or lights in nonresidential applications. The 20-ampere branch circuit (as well as the 30- and 50-ampere branches) can have only heavy-duty lampholders of the *mogul* or porcelain keyless type when used for lighting purposes. Also, the fixture wire to the lampholder must be no smaller than No. 14. The only branch circuit suitable for general lighting is the 15-ampere branch circuit. Any type lampholder may be used on the 15-ampere branch circuit, and it may be used for appliances as well as for lighting.

For the 15-ampere branch circuit, the National Electrical Code specifications are:

1. No wire smaller than No. 14.
2. No fuse larger than 15 amperes.

If conductors of a larger diameter, such as No. 12, are used to reduce the voltage drop, the rating of the circuit would not be changed. The 15-ampere fuse would still be the largest permitted.

Branch Circuits Required

The branch circuit requirements for homes vary with local codes, but most are similar to those of the National Electrical Code. For instance, the code recommends one 15-ampere circuit for lighting each 500 square feet. This will supply approximately three watts of artificial light per square foot.

In addition to the lighting circuits, a branch circuit must be provided for the small appliances in the kitchen, laundry, pantry, dining room, and breakfast room. A minimum-sized (No. 12) conductor is specified by these branch circuits. They may be protected by 20-ampere (maximum) fuses. However, 15-ampere fuses are usually employed because most appliances are equipped with only No. 18 conductors.

Table 8-4. Maximum Current-Carrying Capacity of Rubber-Covered Wire

Wire Size	In Cable or Conduit	Knob and Tube Work	Weatherproofed Wire
14	15 amps	20 amps	20 amps
12	20 amps	26 amps	30 amps
10	25 amps	35 amps	35 amps
8	35 amps	48 amps	50 amps
6	45 amps	65 amps	70 amps
4	60 amps	87 amps	90 amps
2	80 amps	118 amps	125 amps
0	105 amps	160 amps	200 amps

Receptacle Outlets Required

The National Electrical Code requires receptacle outlets in the kitchen, dining room, breakfast room, living room, parlor, library, den, sun room, recreation room, and bedroom. One receptacle outlet shall be provided for every 12 feet of the total distance around the room. At least one receptacle outlet shall be installed for the connection of laundry appliances. This receptacle shall be a 3-pole type designed for grounding. All outlets should be divided as equally as possible among the branch circuits to avoid overloads.

The maximum current load for each wire (conductor) size, as recommended by the National Electrical Code, is shown in Table 8-4. Not more than four outlets can be wired onto one circuit.

Types of Insulation

Different types of insulation are shown in Table 8-5. The letter designation of the Underwriter's Laboratories is given in the table for each wire, together with the trade name by which it is commonly known. The letter represents an abbreviation of the material used for insulation. For example, *R* indicates rubber insulation. This is the kind of wire used for interior wiring. It is known also as *Code-Grade Type R* because it meets the requirements of the National Board of Fire Underwriters.

Rubber conductors are flexible and inexpensive. For this reason, they are found around the home in lamp wiring, extension cords, and small appliance wiring.

Type RH is used for appliances which operate above 122° F because it is more heat-resistant than *Type R* wire.

Summary

The National Board of Fire Underwriters is supported by electrical equipment manufacturers and fire insurance companies.

The National Electrical Code is a set of minimum requirements and standard practices for the installation of electrical wiring systems and devices. The code was prepared by the Underwriter's Laboratories in cooperation with equipment manufacturers, fire prevention bureaus, and many state and local electrical inspectors.

A good wiring system must provide for:

1. **The proper size wires for the amount of current.**
2. **The proper type of insulation to protect the system according to the voltage.**

3. Good mechanical support and protection for the runs of wire.
4. Secure and permanent splices and connections.
5. Protection against shock and fire.

Two approved systems of wiring are the metallic and nonmetallic.

The 15-ampere branch circuit is used in house wiring for lighting and small appliances.

Table 8-5. Required Conductor Insulation for Current-Carrying Circuits Under 6000 Volts

Trade Name	Type Letter	Insulation	Outer Covering	Use
Code	R	Code Grade Rubber	Moisture-Resistant Flame-Retardant Fibrous Covering	General Use
Moisture-Resistant	RW	Moisture-Resistant Rubber	Moisture-Resistant Flame-Retardant Fibrous Covering	General Use, Especially in Wet Locations
Heat-Resistant	RH	Heat-Resistant Rubber	Moisture-Resistant Flame-Retardant Fibrous Covering	General Use
Latex Rubber	RU	90-Percent Unmilled Grainless Rubber	Moisture-Resistant Flame-Retardant Fibrous Covering	General Use
Thermoplastic	T and TW	Flame-Retardant Thermoplastic Compound	None	T—General Use TW—in Wet Locations
Thermoplastic and Asbestos	TA	Thermoplastic and Asbestos	Flame-Retardant Cotton Braid	Switchboard Wiring Only
Asbestos and Varnished Cambric	AVA	Impregnated Asbestos and Varnished Cambric	Asbestos Braid	Dry Locations Only
Asbestos and Varnished Cambric	AVB	Same as Type AVA	Flame-Retardant Cotton Braid	Dry Locations Only
Asbestos and Varnished Cambric	AVL	Same as Type AVA	Asbestos Braid and Lead Sheath	Wet Locations
Slow Burning	SB	3 Braids Impregnated Fire-Retardant Cotton Thread	Outer Cover Finished Smooth and Hard	Dry Locations Only
Slow Burning Weatherproof	SBW	2 Layers Impregnated Cotton Thread	Outer Fire-Retardant Coating	Open Wiring Only

Questions and Problems

35. What agency prepares and publishes the National Electrical Code?

36. Are all electrical codes identical to the National Electrical Code? Explain your answer.

37. If you were going to add an outlet receptacle in a bedroom, what size conductors could be used, to comply with the code?

38. What kind of receptacle is specified by the National Electrical Code for laundry appliances? Why is this necessary?

39. What size conductor is used for lighting circuits in house wiring?

40. If No. 10 or No. 12 wire were used for the lighting circuit described in Question 39, could a 20-ampere fuse be used to protect the circuit?

41. List four of the precautions to be considered when planning an electrical wiring system for a new home.

42. Name three examples of materials used for metallic wiring systems.

43. What size wire is recommended for an outlet branch circuit in a kitchen in which small appliances will be used?

9

HEATING AND LIGHTING

HEATING EFFECTS

Heat has always been one of man's primary concerns. He has needed it, in one form or another, to keep him warm. At first, the sun supplied this warmth. Later, the discovery of fire enabled him to keep warm and to cook his food. Today, electricity enables us to produce heat in nearly any desirable form and to control this heat with extreme accuracy.

Many of the electrical appliances in your home depend on the heating effects of an electric current. A few coils of high resistance wire, which glow red-hot when a current passes through them, are found in toasters, coffee makers, steam and dry irons, and other appliances.

Heat from electricity can be produced by the resistance, induction, and dielectric methods.

Resistance Heating

This is probably the most common method of producing heat from electrical energy. As current flows through a resistance, the electrical energy changes to heat. This principle is used in most heating appliances, such as stoves, ovens, heat lamps, and electric blankets. Arc welding is a form of resistance heating. The arc takes place through a column of vaporized metal. A very large current, flowing in a small area, melts the metal at the ends of the arc. Spot welding is another example. The electrodes are pressed against each side of the metals to be joined, and the combined high current and pressure fuses the metals together.

Heat can be transferred from a heated electrical element by *conduction, convection,* and *radiation.* An example of each is shown in Fig. 9-1. The electric stove transfers the heat to a pan by contact (*conduction*) with the stove elements. A soldering iron is another good example of conduction heating. The heat is conducted from the element to the copper tip and then to the metal to be soldered.

Convection is the second method of transferring heat. A medium such as air, gas, or liquid conveys heat to the desired location. The fan-type heater in Fig. 9-1 is one form of this type of heating. The air near the resistance element is heated, and the fan blows the heated air into the room.

Radiation is the third form of heat transfer. The heat lamp in Fig. 9-1 radiates heat energy.

Conduction, convection, and radiation are the three methods of moving heat from one location to another.

Induction Heating

Electrical conductors in a changing magnetic field become heated due to induced eddy currents. Eddy currents are set up when a rapidly changing magnetic field cuts through a metallic object. The high currents flowing inside the metal cause it to heat. Another way to view induction heating is to think of the metal as a shorted secondary winding of a transformer.

The amount of heat produced by induction heating depends upon the strength of the field, the rate

(A) *Conduction.*

(B) *Convection.*

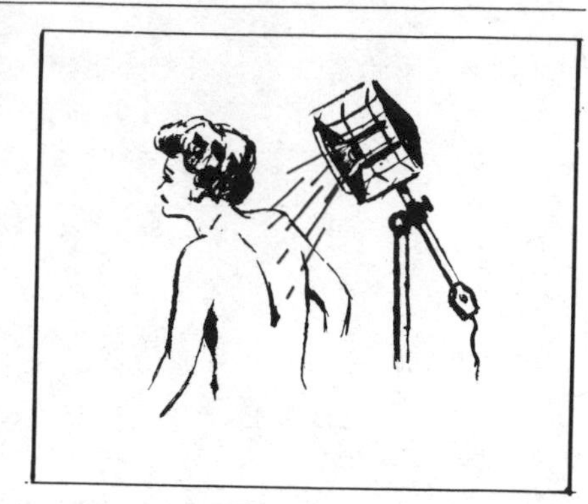

(C) *Radiation.*

Fig. 9-1. Three methods of heat transfer.

at which the field changes, the specific resistance of the metal being heated, the area of the metal, and its position with respect to the magnetic field, among others.

Dielectric Heating

If we rapidly bend a piece of wire back and forth until it breaks and then touch the point where the break occurs, we will feel heat. The heat is caused by the stress we placed on the metal by bending it back and forth. A similar stress is produced when a *dielectric* (insulator) is placed between two oppositely charged plates, and this stress produces heat. (See Fig. 9-2.)

The dielectric process is used in the manufacture of some plywoods. The separate sheets of plywood, with glue between them, are placed between metal plates, and a rapidly changing charge is applied to the plates. The stress produced in the wood produces heat to cure the glue.

The efficiency of dielectric heating depends upon the type, size, and thickness of the dielectric and upon the frequency and voltage of the electric charge.

Mechanical Equivalent of Heat

Heat is measured in calories or in BTUs (British thermal units). A *calorie* is defined as the amount of energy required to raise the temperature of one *gram* of water through one degree *Celsius*. The *BTU* is defined as the amount of energy required to raise the temperature of one *pound* of water through one degree *Fahrenheit.*

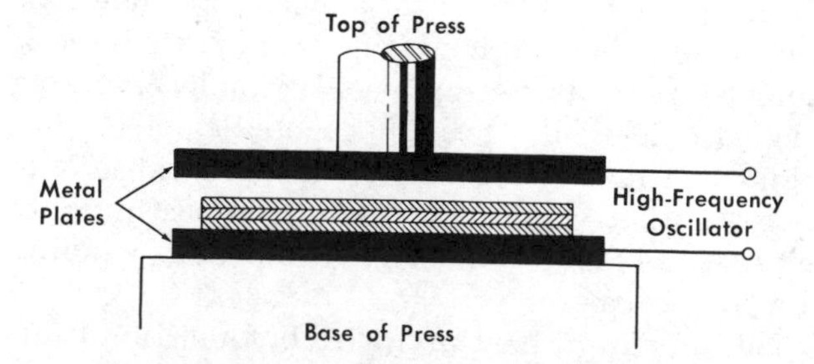

Fig. 9-2. Dielectric heating used in plywood fabrication.

In electrical terms, the *joule* is a unit of work equal to one watt-second, and the calorie is equal to 0.239 joule. One BTU is equal to 252 calories. The relation of these terms to each other is expressed by the following formulas:

$$\text{Joules} = Pt = I^2Rt$$
$$\text{Calorie} = 0.239 \text{ joule}$$
$$\text{Joule} = 4.185 \text{ calories}$$
$$\text{BTU} = 252 \text{ calories}$$
$$\text{Heat} = I^2Rt \text{ joules}$$
$$= 4.185\ I^2Rt \text{ calories}$$

where,

P is the power in watts,
I is the current in amperes,
R is the resistance in ohms,
t is the time in seconds.

Therefore, the heat generated in a conductor is proportional to the *time*, the *resistance*, and the *square of the current.*

Heat can be produced in many ways. However, the production of heat by electricity is the most versatile and the most easily controlled method. From the formula, $J = I^2Rt$, we can see that the amount of heat can be controlled by changing the voltage, the current, the resistance, the time, or any of these.

Energy Relations of an Electric Current

We have learned that the *power* (rate of doing work) is the product of the voltage drop between the ends of a conductor and the *current strength.* If the voltage drop (*potential difference* or P.D.)

is expressed in volts and the current in amperes, the power is given in watts; thus:

$$\text{Watts} = \text{Volts} \times \text{Amperes}$$

The *energy* of our electric current usually is measured by watt-hour meters, the common unit being kilowatt hours. One kilowatt hour is the amount of energy produced in one hour by a current with a power (rate of spending energy) of one kilowatt (one kilowatt equals 1000 watts).

Summary

Heat can be produced electrically by the resistance, induction, and dielectric methods.

Heat can be transferred by conduction, convection, and radiation.

Heat is measured in calories or in British thermal units (BTU).

Electrically produced heat can be controlled exactly.

The calories of heat developed by an electric current in any number of seconds equals $4.185I^2Rt$.

The heating effect of current is used in many household appliances.

Questions and Problems

1. By what means does an infrared lamp transfer heat?

2. What is the name given to the electric current in a block of metal being heated by induction?

3. Name three methods of transferring heat.

4. What type of material is heated by induction heating?

5. What type of material is heated by dielectric heating?

6. The heat produced by electrical heating is measured in what units?

7. What are some advantages of electrical heating?

8. List some common applications of the three methods of electrical heating:
   (a) Resistance.
   (b) Induction.
   (c) Dielectric.

9. Define the term *joule*.

10. If an automobile storage battery has an emf of 12 volts and furnishes a momentary current of 200 amperes while starting the engine, what is its power (rate of spending energy) in watts? In kilowatts?

11. An aquarium contains 16 kg of water heated by an electric heater that draws 0.7 ampere through a 160-ohm element. How many *minutes* will it take for this heater to raise the water temperature from 10°C to 25°C?

LIGHTING EFFECTS

Artificial lighting probably was discovered accidentally. Perhaps one of our cave-dwelling ancestors, grasping the end of a burning fagot, realized he held a crude torch with which he could light dark corners of the cave. Since then, artificial lighting has taken many forms. Oil lamps, gas lights, candles, carbon arc lamps, and Edison's incandescent lamp are still in use today, together with such more modern devices as fluorescent lamps and cold-cathode lights.

INCANDESCENT LIGHTING

The incandescent lamp marked the beginning of efficient electric lighting. Since that time, the *lumens-per-watt* efficiency has been steadily increasing, accompanied by a decrease in the cost of electric lighting. The following list shows this progress.

1897—Edison's first commercial lamp. Carbon filament—1.4 lumens per watt.

1893—Carbonized cellulose filament—3.3 lumens per watt.
1905—Metallized carbon—4 lumens per watt.
1906—Osmium and tantalum—4.8 lumens per watt.
1907—Pressed or squirted tungsten—7.9 lumens per watt.
1911—Mazda drawn-tungsten wire, Type B—10 lumens per watt.
1915—500-, 300-, 200-, 100-watt sizes—12.6 lumens per watt.

Today, the larger incandescent lamps have an efficiency of more than 26 lumens per watt.

The early development of more efficient lamps created a new problem—increased glare. The first attempt to overcome glare was by diffusing (reflecting) the light. Unfortunately, much of the light also was lost. Later the undesirable glare was reduced, without any loss of light, by frosting the *inside* of the lamp bulb.

Incandescent lamps range from the 0.2-watt "grain of wheat" lamp used by surgeons to the 50,000-watt Mazda made by General Electric for movie production in Hollywood. The bulbs (glass envelopes) come in a variety of types, such as PS, F, G, S, P, A, C, and T. These code letters refer to the *shape* of the bulb. Numbers are added to the letters to indicate the *diameter* of the bulb in eighths of an inch (e. g., T-12 is a tubular bulb one and one-half inches in diameter). Fig. 9-3 depicts a few of the bulb shapes and their corresponding letter designations.

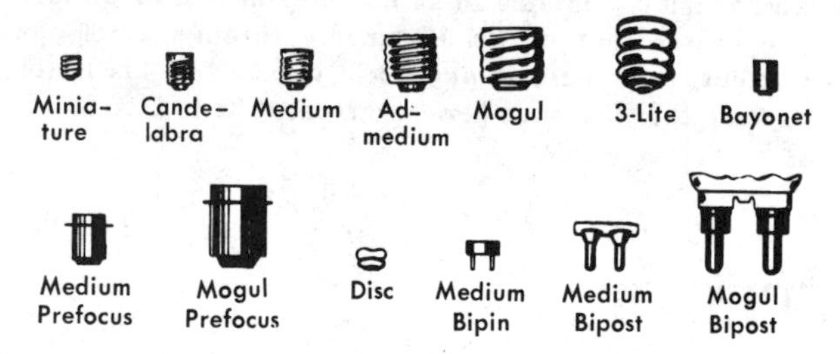

Fig. 9-4. Lamp bases.

Lamp bases for Edison (screw-shell) lamps are made in several sizes, such as miniature, candelabra, medium, admedium, mogul, and "3-lite." These are shown in Fig. 9-4, along with some special types of bases, including the bayonet, disc, and prefocus.

The incandescent lamp is used in table lamps, floor lamps, chandeliers, and wall-mounted fixtures. Fixtures provide *direct* and *indirect* lighting. Direct light rays pass directly from the light source. Indirect light rays are diffused, that is, are reflected or are sent through a translucent material.

The *reflectance,* or reflection factor, is the percentage of light reflected from a surface. Reflectance may vary from 7 percent for certain dark green materials to 87 percent for certain white materials. The light not reflected is absorbed by the surface of the reflector.

Illumination Units and Terms

Photometry refers to the measurement of the intensity of light. Light, which is a form of radiant energy, is evaluated according to its capacity to produce visual sensation. The amount of illumination is a measure of the density of light falling upon a surface. When light strikes a surface, the surface is illuminated.

The intensity of a light source is measured in *candle power* (cp). The amount of light we receive from a light source depends upon the intensity of the source and upon our distance from that source. This amount is measured in *foot-candles.* One foot-candle is the amount of illumination received from a source of one candle power (one standard candle) on a surface that is one foot from and at right angles to the source, as shown in Fig. 9-5. A person seated one foot from a 30-candle power lamp receives 30 foot-candles of light. If he is sitting two feet from such a lamp, however, he receives only 7½ foot-candles. The reason is that the amount of light an object receives is *in-*

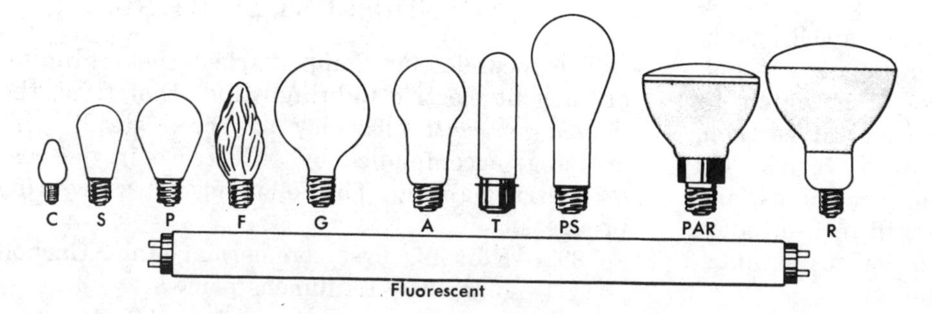

Fig. 9-3. Lamp sizes and shapes.

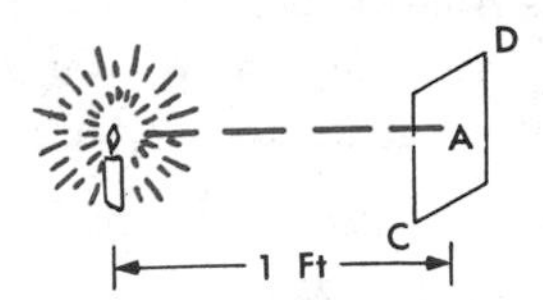

Fig. 9-5. One foot-candle of light (illumination) falls on a surface one foot away from a standard candle.

versely proportional to the square of its distance from the source of light. (See Fig. 9-6.)

$$\text{Illumination} = \frac{cp}{d^2} = \frac{30}{2^2} = 7.5 \text{ foot-candles}$$

where,

cp is the intensity of the light source, in candles,
d is the distance in feet from the source to the object.

The amount of light falling upon a given point is measured with a *light meter* (Fig. 9-7). The light meter shown is calibrated to provide camera exposure settings. A light meter can also be calibrated to measure the amount of light falling on a work area, in which case the meter is calibrated in candle power or *lumens.*

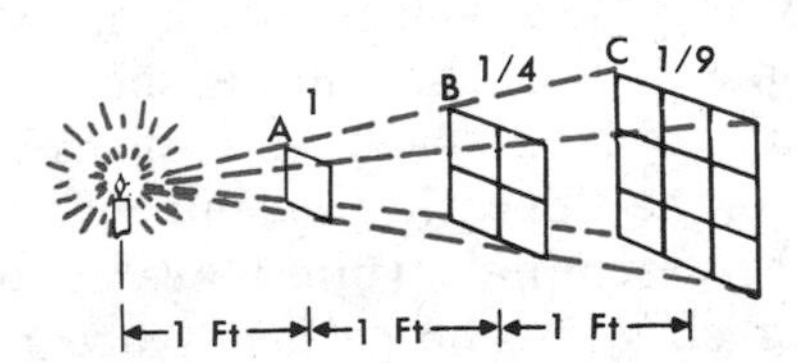

Fig. 9-6. The inverse square law for light. One-fourth as much illumination is provided at a two-foot distance and one-ninth as much at a three-foot distance.

The lumen is the unit for measuring light quantity, that is, the total amount of light actually given off by a source. If we place a standard candle at the center of a sphere of one foot radius, the amount of light falling upon one square foot of the inner surface of the sphere will be one lumen. The inner surface area of such a sphere is 12.57 square feet. Since one candle power would be distributed over that area, you can see that one candle power is equal to 12.57 lumens. The output of Mazda filament is rated in lumens. For example, a 100-watt lamp is rated at 1620 lumens and a 1000-watt lamp, at 21,000 lumens. A 40-watt, white fluorescent lamp is rated at 2320 lumens.

FLUORESCENT LAMPS

The development of the fluorescent lamp in 1938 opened up a new era in lighting which was as revolutionary and far-reaching as Edison's incandescent lamp in 1879. Each major development in lighting has been characterized by one all-important feature—more light. Fluorescent lighting has

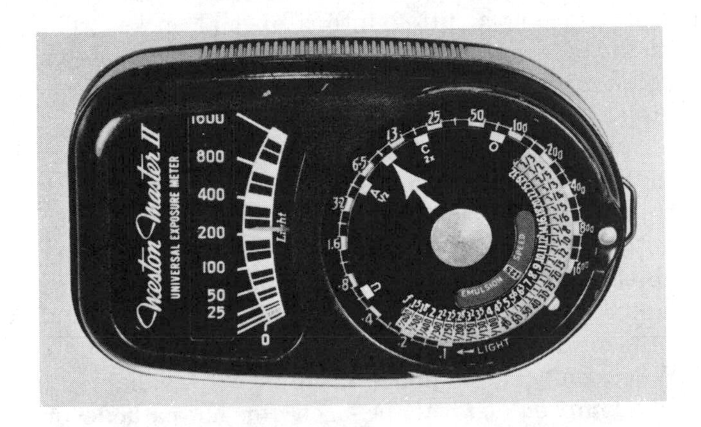

Fig. 9-7. A photographer's light meter.

contributed to better artificial light in our homes, offices, factories, and hospitals, where our eyes rely on artificial lighting for safe and effortless seeing.

Efficiency and Advantages

A fluorescent lamp operates at less than 40 percent of the cost of operating a tungsten lamp of equivalent candle power. Fig. 9-8 compares the output in lumens of the standard candle, a 100-

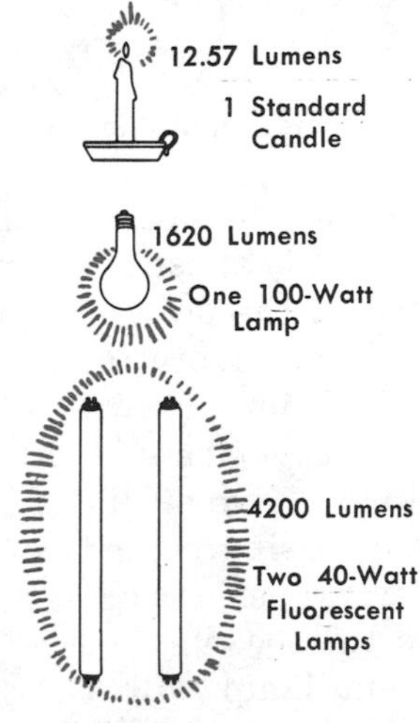

Fig. 9-8. Comparative output, in lumens, of a standard candle, a 100-watt lamp, and two fluorescent lamps.

watt incandescent lamp, and two 40-watt fluorescent lamps.

Fluorescent lamps are available in four distinct types, depending upon the operating circuit—hot-cathode; preheat-starting; hot-cathode, instant-starting; cold-cathode; and the "rf" lamp, which operates from radio-frequency energy. The fluorescent lamp consists of a glass tube one to one and one-half inches in diameter and 18 inches or more in length, fitted into a special socket. The inside of the tube is coated with chemicals known as *phosphors,* which glow when exposed to ultraviolet light.

When operated on ac, the fluorescent light actually goes out 120 times each second. This flicker is usually not noticeable. However, it may be noticeable as a stroboscopic effect when a moving object comes into view. To minimize flicker, the lamps may be operated in pairs on two-phase circuits or in sets of three on three-phase circuits. When only single-phase circuits are available, a two-lamp *ballast* can be used for hot-cathode or cold-cathode lamps. This ballast supplies *leading* current to one lamp and *lagging* current to the other. The phase difference is such that the light pulsations of the two lamps compensate each other and eliminate the flicker (stroboscopic effect).

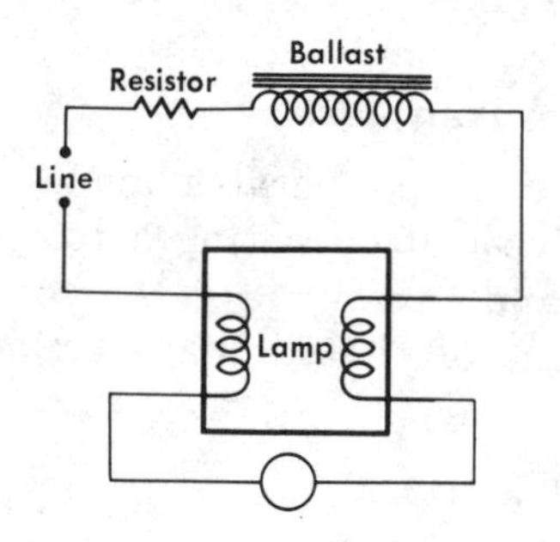

Fig. 9-9. Circuit for dc operation of fluorescent lamps.

Fluorescent lamps may be operated on dc if a resistance is added to the ballast in order to limit the operating current to the designed value. (See Fig. 9-9.) However, lamps more than a few feet long, when operated from dc, will accumulate mercury vapor at the *negative* end. As a result, only part of the bulb will give off light. This fault can be overcome by reversing polarity of the lines feeding the lamps or by periodically turning the lamp end for end in the sockets.

Before a fluorescent lamp will "start," it requires a momentary voltage higher than the line voltage in order to establish the arc discharge through the gas inside the tube. This is done by adding an inductance coil or choke coil (a fluorescent lamp *ballast*) in series with one side of the line.

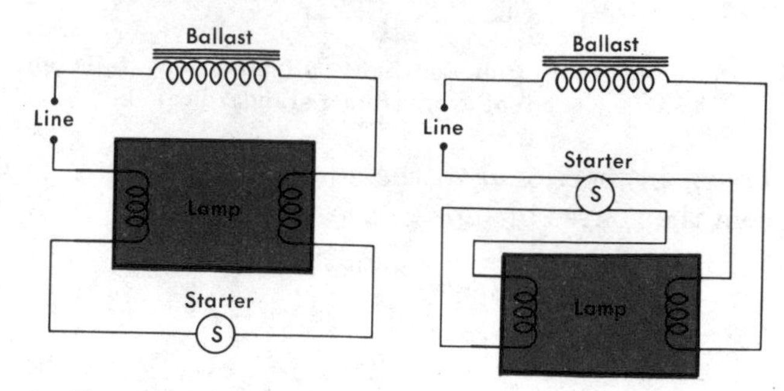

Fig. 9-10. Fluorescent lamp operation, using ballasts.

Fig. 9-10 shows two circuits in which *ballast coils* are used. When the starter is in its closed position, current flows from the line, through the ballast, the cathode, the starter, the second cathode, and back to the line. The current heats the cathodes. When the starter opens, the inductive *surge* from the ballast strikes an arc between the heated cathodes.

Hot-Cathode, Preheat-Starting Lamps

A hot-cathode, preheat-starting, fluorescent lamp as it would look broken open is shown in Fig. 9-11. The operation of the lamp is shown also.

One terminal from each end of the lamp is connected to the operating circuit (Fig. 9-12A) and the other terminals (pins) are connected to the starting circuit. The cathodes are preheated for a short time to start the lamp. An arc is struck because of the inductive surge of current as the starting circuit is opened. (See Fig. 9-12B.)

Hot-Cathode, Instant-Starting Lamp

One type of hot-cathode, instant-starting, fluorescent lamp looks like the preheat-starting lamps. Another type has only a single-pin connection at each end. When there is a two-pin base at each end, the two pins are short-circuited internally and act like a single pin. The other difference between instant-starting and preheat-starting lamps is in the cathodes. The instant-starting cathode must withstand the shock of a high-voltage arc striking between cold cathodes. After the arc is struck, a cathode flow is produced, and the lamp operates like a preheat-starting lamp.

Instant-starting lamps are started by applying, between the cathodes, a voltage high enough to

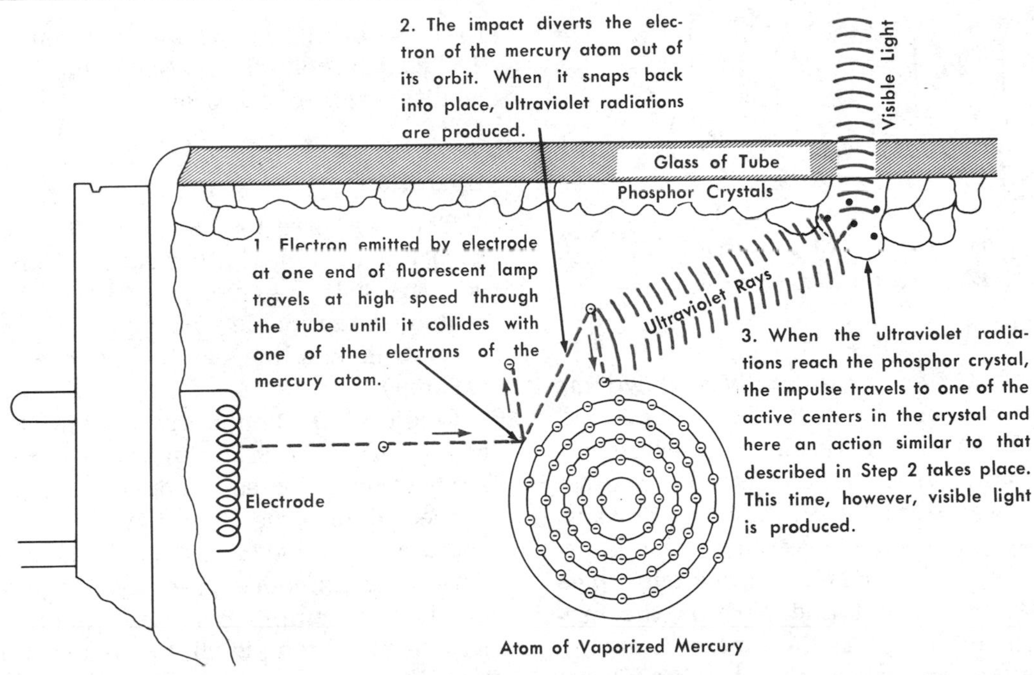

Fig. 9-11. The three-step operation of a hot-cathode, preheat-starting, fluorescent lamp.

"break down" the gas and start the arc. Lamps of this type require starting voltages of 450 to 750 volts. Because of the high voltage, electrical codes require the lamp holders to be so wired that the primary circuit will automatically turn off when a lamp is removed.

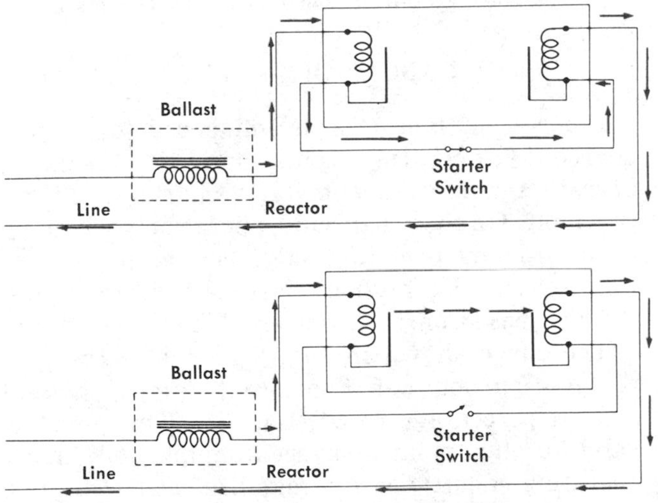

Fig. 9-12. Starting and operating circuits of a hot-cathode, preheat-starting, fluorescent lamp.

Lamp Holders

One style of lamp holder for two-pin lamps is illustrated in Fig. 9-13. The preheat-starting lamps require a starter for each lamp. Only one of the holders will have a starter socket. A typical holder for the single-pin lamp is shown in Fig. 9-14. One holder is used for the high-voltage end of the lamp and the other holder is used for the low-voltage end.

Fig. 9-13. Holders for bi-pin lamps.

STARTERS

Starting switches are available in three principal types: manual, glo-switch, and thermal.

Manual Starter Switch

This type of starter is hand operated. The preheating period is controlled by the operator. A push-button switch is usually depressed until a glow appears at the ends of the tubes. An inductive surge strikes an arc through the tube (bulb) when the push-button is released.

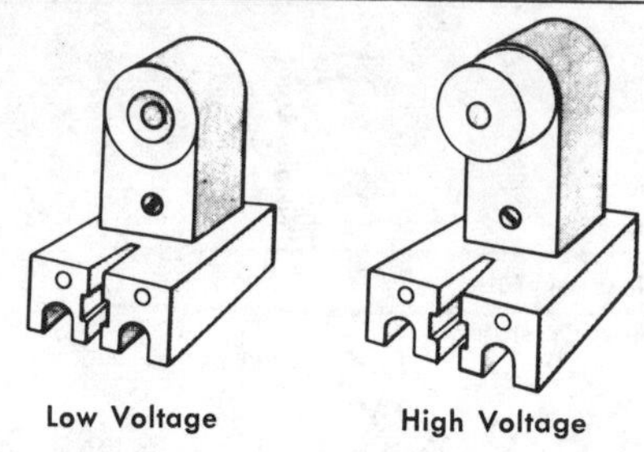

Fig. 9-14. Holders for single-pin lamps.

Glo-Switch Starter

This type of starter is essentially a glow-lamp in which one of the electrodes is a bimetallic, thermostatic blade. When the lamp is energized, the glow discharge around the electrodes releases enough heat to cause the bimetallic blade to warp, thereby touching the other electrode of the glow-lamp. This contact is, in effect, a switch which closes the starting circuit and permits the preheating current to flow until the contact is cooled. When the contact is broken, the starting circuit is opened.

Thermal Starter

This type of starter includes a resistive element which carries a filament current and is, in turn, heated by this filament current. The heater element warms a bimetallic, thermostatic blade, causing the blade to warp, thus opening the filament connection after the filaments have been warmed enough to glow. After the arc is struck in the tube (lamp), the operating current which flows through the heater holds the starter switch contacts in the open position.

COLD-CATHODE LIGHTING

Cold-cathode lighting is often associated with neon lighting. However, cold-cathode lamps are more difficult, hence more time-consuming, to manufacture than neon fixtures. Both the cold-cathode tubing and the neon tubing can be bent into letters and outlines.

Cold-cathode fixtures are available in many sizes and shapes. The four-lamp, eight-foot room fixture may be connected to a standard 110-volt ac outlet, and it has a self-contained ballast that automatically steps up the 110 volts in order to start the lamp. Custom installations can be recessed, so that the lamps will be placed in coves or above a luminous ceiling. A remote high-voltage transformer is used, with its leads running into the cove or ceiling where the lamps are located. The cold-cathode fluorescent lamp furnishes high-quality lighting at low operating and maintenance cost.

Operation

The gas pressure of a cold-cathode lamp is higher than that of a hot-cathode lamp. The cathodes are *hollow* cylinders of pure iron, which are coated on the inside with one of the active oxides of certain alkaline earths (such as barium and calcium).

Cold-cathode lamps are instant-starting. They can be used in either multiple-circuit or series applications. For series operation, the lamps are connected in series, as shown in Fig. 9-15, and operated from a ballast or from a transformer with a high secondary voltage. Depending on the number of lamps in series, the ballast voltage may range from 2000 to 15,000 volts and the lamp operating currents from 12 to 100 milliamperes.

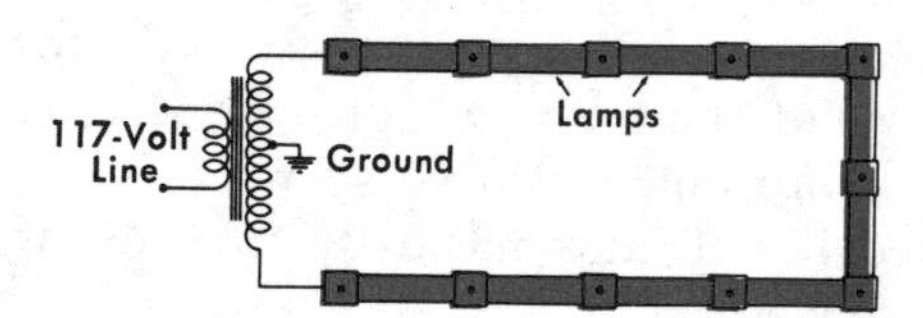

Fig. 9-15. Series operation of cold-cathode lamps.

LAMP MORTALITY

Fluorescent lamp life is influenced to a great degree by operating conditions. For example, operation with frequent starts shortens life appreciably. On the other hand, operation with many burning hours per start lengthens lamp life. For this reason, lamps have several different life ratings, based on the number of hours per start.

The curve shown in Fig. 9-16 shows the percentage burnout of fluorescent lamps, plotted against percentage rated lamp life. This curve is valid for all popular fluorescent lamps, under most operating conditions. Average life ratings for the various lamps and operating conditions can be obtained from fluorescent manufacturers' data sheets.

LUMEN MAINTENANCE

Like other types of lamps, fluorescent lamps depreciate in light output as they age. The initial

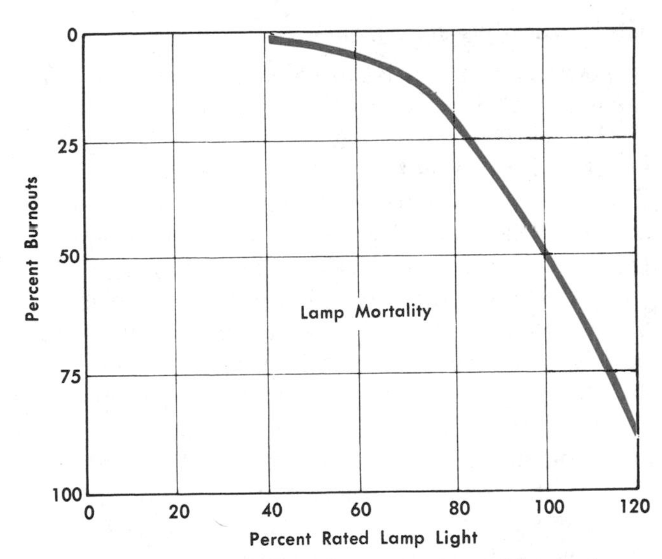

Fig. 9-16. Chart illustrating percentage of burnout of fluorescent lamps.

light output is variable from lamp to lamp, and this value may decrease rapidly during the first 100 hours of operation. The lumen depreciation may amount to as much as 10%. For rating purposes, the 100-hour value is used as the initial value.

The curve shown in Fig. 9-17 shows the typical range of fluorescent lamp depreciation in light output versus time. The vast majority of lamp types depreciate along a line near the upper limit of the curve. Light depreciation is not noticeably affected by the number of burning hours per start.

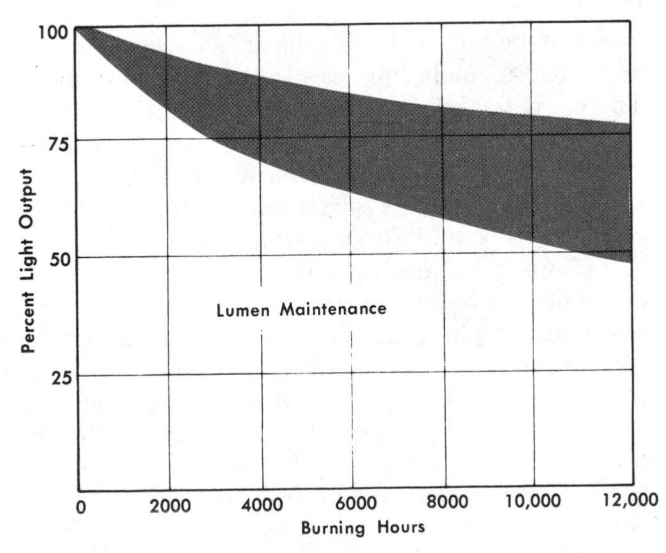

Fig. 9-17. Chart illustrating light output versus burning time.

MERCURY LAMPS

Mercury lamps combine the relatively small size of incandescent lamps with the long life and high efficiency characteristics of fluorescent lamps. These high-wattage light sources produce a considerable amount of light for their size. As a result, mercury lamps offer many advantages to the user, particularly in industrial, street lighting, floodlighting, and other outdoor commercial applications. Mercury lamps are electric discharge lamps. As such, they produce light by passing a current through a gas vapor under pressure, rather than by heating a filament as in incandescent lamps.

Although the first high-pressure mercury lamp was not introduced until 1934, the first mercury vapor lamp was developed in 1901 by Peter Cooper Hewitt. Known as the Cooper Hewitt lamp, it was approximately 4 feet long and produced bluish-green light. The mercury lamp, as it is known today, was first introduced in the 400-watt size with a glass arc tube. Later, other lamps were added, and today the line of general lighting mercury lamps ranges from as low as 100 watts to as high as 3000 watts.

In recent years, several significant improvements in mercury lamp performance have been introduced. The development of color-improved and white phosphor-coated lamps added versatility in light output and color rendition with mercury lamps. Today's 400-watt clear mercury lamp is rated at 20,500 lumens, with an average life in excess of 16,000 hours, and 92% lumens over 16,000 hours. A few years ago, a 400-watt clear lamp was rated at 20,000 lumens, with 3000 hours of life at 83% lumens over 3000 hours.

Fig. 9-18 illustrates the basic parts of the mercury lamp. Nearly all mercury lamps consist of an arc tube enclosed within an outer tube, plus supplementary parts. The arc tube contains the essential operating components—electrodes, mercury vapor, and argon gas. The outer bulb maintains nearly constant arc tube temperature, and protects the arc tube and internal parts from the atmosphere.

The initial light output rating of mercury lamps is based on the average light output after 100 hours of operation. Lamps are rated for operation in a vertical position. Operating a mercury lamp in a horizontal position reduces the light output. In the event of a power interruption or voltage dip lasting for more than one cycle, mercury lamps extinguish and do not restart for several minutes. The exact magnitude of the voltage drop required to cause this condition depends on the ballast design.

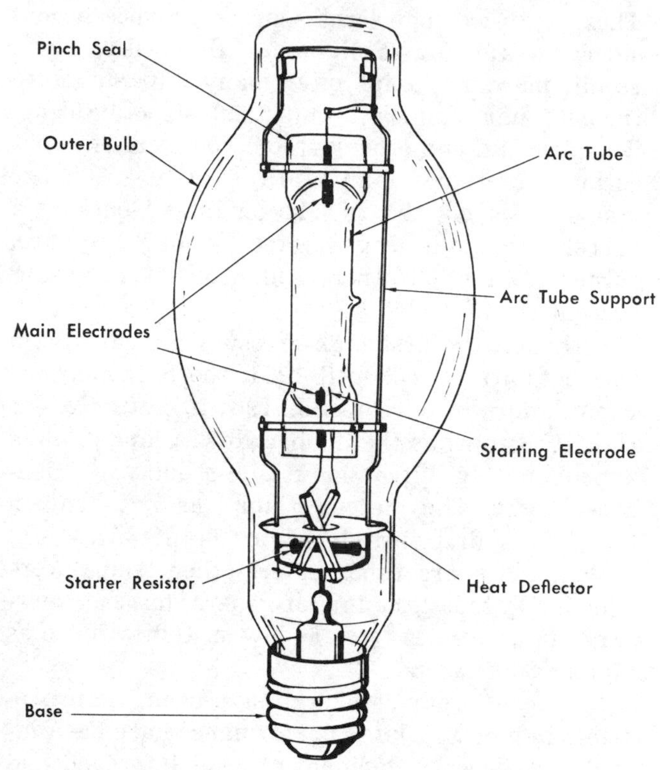

Fig. 9-18. Basic components of the mercury lamp.

Mercury lamps have been operated on frequencies as low as 25 hertz with special ballasts. Lamp flicker is noticeable at 25 hertz. Mercury lamps can operate on direct current, but special ballast circuits are required. Normally light output of mercury lamps is not affected by low temperatures or cold drafts. However, if an exposed lamp is subjected to low temperatures, it may not warm up. This depends on the fixture and ballast design. Once a lamp warms up, the outer bulb isolates the internal arc tube from the effect of ambient conditions.

OTHER LIGHT SOURCES

A number of other sources of light have been developed, and, although most do not put out enough light to make them suited for general illumination purposes, as with the tungsten lamp and fluorescent lamp, they are still of great practical value.

Electroluminescent Panels

These panels consist of some semiconductor material pressed between two electrodes. One of the electrodes is transparent so that the light can be emitted. Materials that have been used to produce light in this manner are cadmium sulfide and zinc sulfide. When an alternating current is applied, the panel emits a soft light of low intensity.

Since the intensity of this light is low, it is used mainly as a night light or as a light on instrument panels that are operated in total darkness. The usable life is very long and power consumption is so low that a night light of this type can be operated continuously at very little cost. Light level can be increased by increasing the voltage and frequency of the applied signal, but at a sacrifice of some of the total life of the light.

Cold Light by Chemical Mixture

A light source that has recently appeared on the market is one that produces light by chemical mixture. The chemicals are stored in a transparent plastic tube but are separated by a barrier. When the tube is twisted or flexed, the barrier is broken and the chemicals are allowed to mix, producing light. The light is not reusable and can not be turned off. The marketing agent states that its chief advantage is as a convenient, easily stored, and reliable emergency light. This method of producing light is said to closely parallel the natural one used by the firefly.

Lasers

A number of materials have been used to produce lasers, including solids, liquids, and gases. The word "laser" is an acronym for Light Amplification by Stimulated Emission of Radiation. Light is produced by a process almost identical to feedback in a resonant electrical circuit. Light is bounced back and forth within the laser, and as the photons of light travel back and forth they cause other photons to be emitted in step with the stimulating photons. This aspect of being "in step" produces light of a very special nature; the light is said to be *coherent*. Ordinary light, such as that produced by an incandescent bulb or a fluorescent lamp, is incoherent—that is, it consists of a great number of wavelengths, more or less out of step with each other.

One property of coherent light is that it can be focused into an extremely narrow beam that is not easily scattered or diffused. Therefore, it can be projected for long distances, and laser beams have been sent to the moon and returned by the retroreflectors placed there by the Apollo astronauts. Lasers have many scientific uses in the field of optics, communications, medicine, and industry.

Light-Emitting Diodes (LEDs)

The light-emitting diode is closely related to one form of laser (the injection laser is a light-emitting diode with a very flat junction and two end mirrors). However, the LED does not emit coherent light. LEDs emit in either the visible-light or the infrared regions of the electromagnetic opectrum. Infrared LEDs are used in optical communicators, intrusion alarms, and many other applications. Visible-light LEDs, because of small size and comparatively low power requirements, are useful as indicator lamps and numeric readouts in electronic calculators.

Fiber Optics

This is not a real light source in itself but a means for transmitting light generated by some of the sources already mentioned. By its use, effects are obtained that are not possible or easily obtained by normal light transmission methods such as direct radiation or optical lens systems. Fig. 9-19 shows how light is transmitted through an optical fiber or "light pipe." After a light ray enters the fiber it eventually strikes the fiber/air interface. If the angle between the light ray and the side of the fiber is smaller than a certain critical angle of refraction, the ray is reflected back into the fiber rather than passed through the fiber wall into the air. This process is repeated at other points along the fiber until finally the ray emerges at the opposite end of the fiber. There are some light losses because of absorption by the fiber and some leakage through the fiber walls.

Fiber-optic cables can be assembled with hundreds, or even thousands, of individual fibers and if each fiber maintains its relative position with the other fibers throughout the cable, images can be transmitted from one end of the cable to the other. A single fiber, or several, can be epoxied to a LED for efficient transmission of its light output. Light can be piped into otherwise inaccessible or hazardous locations, or a fiber-optic cable can be used for viewing under the same type of location conditions.

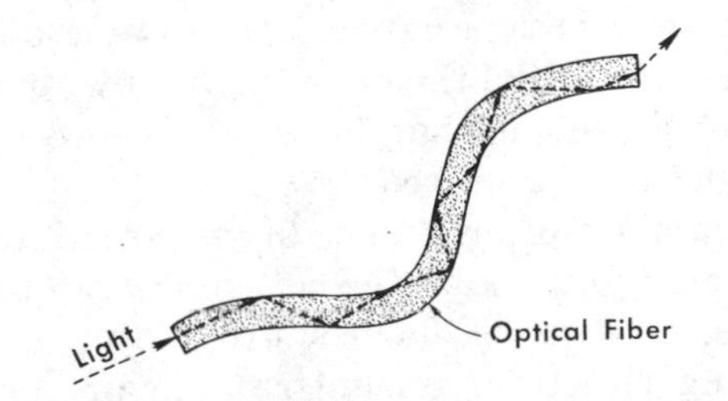

Fig. 9-19. Light transmission through an optical fiber.

ELECTRIC SIGNS

The design of electric signs is an art in itself. Their installation and maintenance can be both interesting and profitable. Although advertising signs were used thousands of years ago, the first *electric* sign of any consequence did not appear until 1893. It was located at Broadway and 23rd Streets in New York City, and advertised "Heinz' 57 Varieties." Today, lower power costs, improved sign flashers, and greater lamp efficiencies have made the electric sign industry so important that there are few communities of any size without at least one "spectacular," as the elaborate signs are called.

Electric signs are classified as exterior illuminated (floodlighting), exposed lamp, concealed lighting, silhouette lighting, and gaseous conductor (neon).

Exterior Illuminated (Floodlighting)

Although sign floodlighting can be done in several ways, ordinary 75- to 100-watt incandescent lamps are normally used. Two familiar examples are illuminated billboards and filling-station signs.

Exposed Lamps

Signs in which exposed bulbs are used can be seen both day and night. They have color, brilliance, and motion. A sign of this type is shown in Fig. 9-20.

Concealed Lighting

This type of sign frequently is used for location markers. It is not effective with color or motion. Such signs usually are stationary (for example, the familiar, back-lighted "EXIT" signs). Although there are many varieties, they consist usually of lamps placed behind translucent glass. (Translucent glass allows some light to pass through, but not enough that objects can be dis-

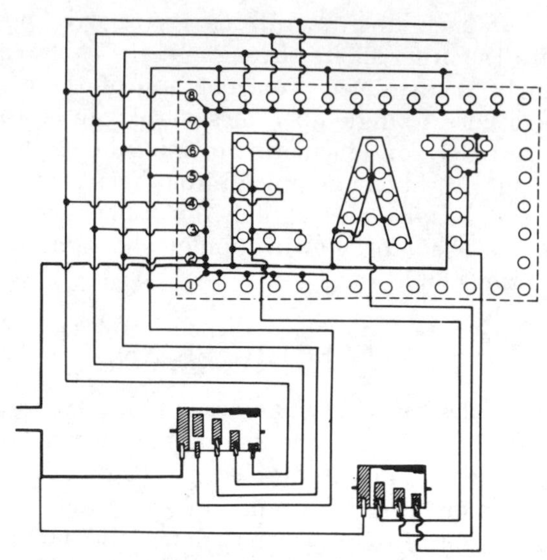

Fig. 9-20. Exposed-lamp sign.

tinguished through it.) A blinker can be used to turn such a sign on and off in order to attract attention to the sign.

Silhouette Lighting

This type of sign consists of an object, such as a letter or a cutout figure in front of a solid background and lighted from behind. Thus, a silhouette is formed of the letter or cutout (Fig. 9-21).

Gaseous Conductor Lighting (Neon) Signs

In this type of electric sign, neon or other rare gases are used to obtain different colors for background and outline effects. These signs are commonly referred to as neon signs because neon was the first type of gas used. The principal parts of a neon sign are shown in Fig. 9-22. Its visible parts

Fig. 9-21. Silhouette-type sign.

consist of glass tubing containing neon (or another gas) which glows when the high voltage from the transformer secondary coil forces current through it. Those parts of the tubing that are not part of the display are painted black. The high-voltage secondary coil of the transformer may supply 600 to 15,000 volts to the electrodes attached at each end of the tube (Fig. 9-23).

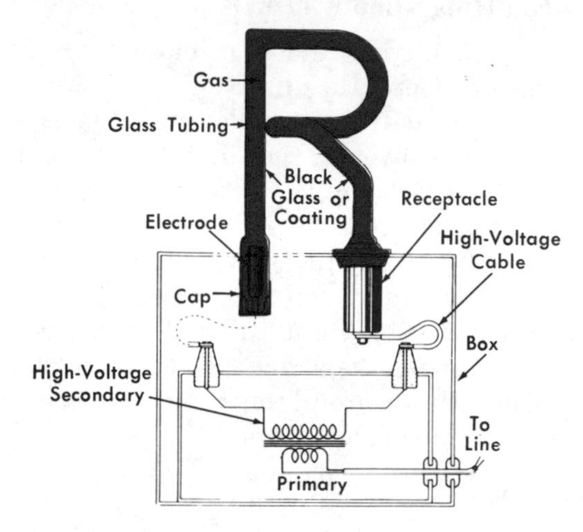

Fig. 9-22. The principal parts of a neon sign.

The tubing of a gas-conductor sign can be filled with neon, helium, mercury, argon, or a mixture of any of these. Pure neon gives off a red light; mercury, a blue light. Mixtures of conducting gases are used to produce a variety of colors. Colored tubing may also be used for the different colors.

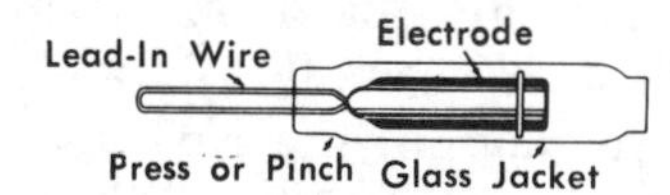

Fig. 9-23. Electrodes in the end of a neon lighting tube.

Animated Signs

Animation effectively attracts attention to an electric sign. This illusion is furnished by flashers, which may be either *thermostatic* or *motor-driven*. The thermostatic type is used for simple off-and-on blinker signs, like the ones in window displays. The motor-driven flasher is used whenever more precise timing is required.

In a number of systems, signs are animated by flashing lights in different combinations and sequences. *On-and-off* flashes provide an on-and-off flashing effect for an entire sign or for only a section of the sign.

The *chaser* effect is used chiefly in sign borders. Only one *set* of lamps is out at any one instant, and the "shadow" can be made to go in either direction.

In the *speller* effect, a lamp or group of lamps is turned on in sequence. This effect forms a letter, a word, or a part of a sign, which gives an add-on effect.

The *script* effect is done by turning on one or more lamps at a time, which gives the impression that the letters are slowly being written.

Summary

Modern incandescent lamps have developed from the inefficient carbon-filament lamp. The fluorescent lamp (tube) is even more efficient than the tungsten-filament incandescent lamps that are in use today.

Diffusion of light means its irregular reflection.

Photometry refers to the measurement of the intensity of light. Light intensity is measured in *candle power*. The amount of illumination is measured in *foot-candles*. The quantity of light emitted is measured in *lumens*.

A foot-candle is the intensity of illumination of a one-cp light at a distance of one foot.

The intensity of light varies inversely with the square of the distance from its source.

The fluorescent tube actually produces an ultraviolet light. The phosphor coating on the inside of the tube converts this ultraviolet radiation into visible light.

In the electric signs used for advertising and information, incandescent or fluorescent lamps are employed in many ways to attract attention.

Many interesting effects are produced by flashing or dimming lights in a predetermined sequence by means of automatic switches.

Long life is one of the performance features of mercury lamps. All general lighting mercury lamps today have an average life in excess of 16,000 hours.

A 400-watt mercury lamp has an increased output of about one third over the same wattage of an incandescent one.

Questions and Problems

12. What is the advantage of having a light fixture at the center of a room and near the ceiling? What are the disadvantages?

13. Five feet from a light source, the light intensity is 100 lumens. What is the light intensity (lumens) falling on a surface 25 feet from the same source?

14. What is the purpose of the phosphor coating on the inside of a fluorescent tube?

15. What is the purpose of a ballast or a ballast transformer in a fluorescent lamp?

16. Sketch a circuit for the connection of a fluorescent tube, a series ballast, and a starter.

17. The amount of light received from a light source depends on what two factors? How can each factor be varied?

18. A student is seated two feet from a study lamp of 20 cp. How many foot-candles of light does he receive?

19. Explain how the flicker, or stroboscopic effect, of fluorescent lights can be corrected.

20. What candle power lamp will give 10 foot-candles of light at a distance of 2.5 feet?

21. Eight foot-candles of light are needed at the work table of a drill press. A lighting fixture with three 25-watt bulbs (20 cp each) is available. How far away must the lighting fixture be placed in order to give the desired light?

10

RADIATIONS AND OTHER MANIFESTATIONS

ELECTROMAGNETIC RADIATION

Man owes much to electromagnetic radiation. No life could exist without the energy transmitted from the sun, in the form of electromagnetic radiation. Radiated heat, light, ultraviolet light, X rays, gamma rays and cosmic rays also are forms of electromagnetic radiation. Another familiar form is the carrier medium for wireless telegraphy, radio, television, and radar and other navigational aids.

Nature of Radiation

Although the exact nature of electromagnetic radiation is not known, it does display wavelike characteristics and often behaves like extremely small particles of projected matter.

Wave Motion

One of the simplest and easiest ways to demonstrate wave motion is to drop a rock into a pool of water. (See Fig. 10-1.) The rock, striking the water, creates a disturbance. This disturbance causes concentric circles of ever-increasing diameter to spread out on the surface. As the circles move from the center, they become less and less noticeable until they disappear. Should these waves meet a solid object, they will be reflected.

The water waves are *radiated* in just one *plane* —along the surface of the water. The water at the center of the disturbance does not move outward. Instead, it is the disturbance itself which moves.

The sound waves we studied in the previous section travel through the air in the same way. However, they travel up and down as well as sideways, in the form of rapidly expanding spherical waves. Again, it is not the air molecules themselves that travel; it is the disturbance. Sound waves also travel in other materials, such as water, wood, and steel. However, they cannot travel through a vacuum.

Electromagnetic radiation, like sound waves, also travels in all directions. However, unlike a sound wave, electromagnetic radiation apparently needs no medium through which to travel. The sun's rays reach the earth through 92 million miles of empty space—most of it empty enough to be considered an almost perfect vacuum by earth standards. Disturbed by this apparent lack of a medium, scientists at one time theorized that there *was* such a medium, although it could not be detected by any scientific means. They called this medium *ether* and assumed that it had weight and that it occupied all of space. There has never been any proof that ether exists. Today, scientists have discarded the ether theory and define electromagnetic fields of force.

Velocity

Electromagnetic waves travel at approximately 186,000 miles, or 300,000,000 meters, per second through space. Through air their speed is only slightly lower.

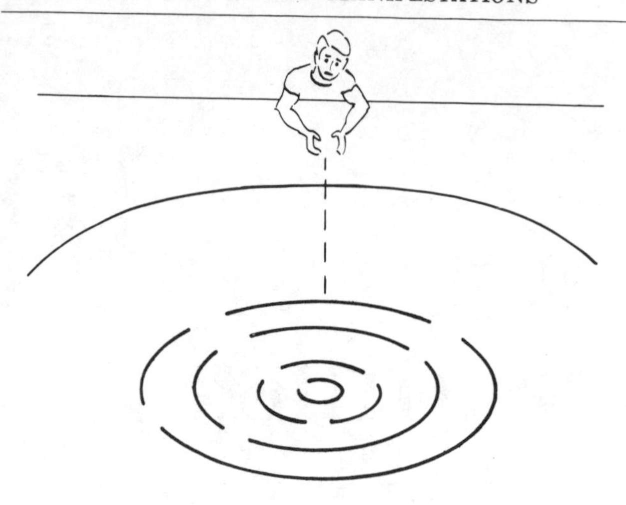

Fig. 10-1. Wave disturbance caused by dropping an object into water.

Wavelength and Frequency

Electromagnetic waves are vibrational phenomena caused by vibrating electrical charges. The vibration is due to man's efforts—when radio waves are broadcast—or to natural causes—when energy is radiated from the sun. The *wavelength, frequency,* and *velocity* of a vibrating waveform are all related to each other.

A *wavelength* is the distance a wave travels during a complete cycle. In Fig. 10-2 it is the distance between any two corresponding points of successive cycles—for example, between A and B or between C and D. Wavelength is represented by the Greek letter *lambda* (λ).

The *frequency* of a repeating waveform is the number of complete cycles occurring in a certain time, usually one second.

If we let v equal the velocity of the electromagnetic waves, f the frequency of vibration, and λ the wavelength, then the following formula applies:

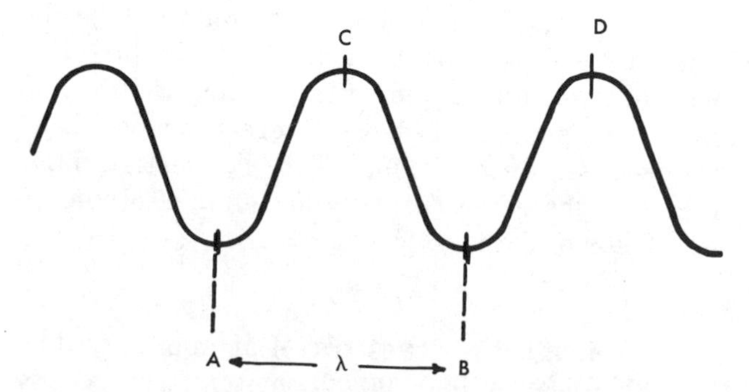

Fig. 10-2. The meaning of the term *wavelength* as applied to a continuous waveform.

$$v = f\lambda$$

The unit for λ must be the same as the one for v. When stated as the speed of light, v is usually given in miles per second. However, it is sometimes more convenient to state λ in meters or centimeters. Therefore, v also must be in meters or centimeters. V is approximately 300 million meters per second in a vacuum.

To illustrate the use of the formula, we will solve a simple problem. Let us find the wavelength, in meters, of a 1-megahertz radio signal. This signal is near the middle of the a-m radio band. The value of v is 300 million meters per second, and the value of f is one million hertz. By using the formula:

$$v = f\lambda$$
$$300{,}000{,}000 = 1{,}000{,}000\lambda$$
$$\lambda = 300 \text{ meters}$$

where,

v is the velocity of the electromagnetic waves in meters per second,
f is frequency of vibration in hertz,
λ is the wavelength in meters.

The Electromagnetic Spectrum

We have said that electromagnetic waves are caused by vibrating electrical charges. The frequency of these vibrations extends over a tremendously wide range—from zero to millions upon millions of hertz. These different vibrations cover the area from the radio frequency waves through infrared rays, visible light rays, ultraviolet light rays, X rays, gamma rays, and cosmic rays. The upper frequency *limit* of such radiations has not yet been determined.

A diagram of the *known* electromagnetic frequency spectrum is shown in Fig. 10-3. The scale is made logarithmic (in multiples of ten) in order to cover the wide range of frequencies in a limited space. Notice that the frequency band for *visible* light is comparatively narrow. The lower limit for visible light is approximately 4.2×10^{14} (420,000,000,000,000) hertz, and the upper limit is approximately 8.6×10^{14} (860,000,000,000,000) hertz. This represents a ratio of two to one. Compare this to the 10 million to 1 for radio frequencies! The *wavelength* of visible light rays ranges from 0.000035 to 0.000070 cm.

The range in wavelength for the radiation spectrum extends from a wavelength of 2×10^6 cm at a frequency of 15,000 hertz, to 1×10^{-12}

(0.000,000,000,001) cm at the frequency of some cosmic rays. The wavelengths for frequencies under 15,000 hertz are longer, of course, and approach infinite length as the frequency approaches zero. The wavelength for the common power-line frequency of 60 hertz is 3100 miles!

The section of the electromagnetic spectrum from about 10 hertz to 1 megahertz contains *subsonic* (below audible sound), *sonic* (audible sound), and *ultrasonic* (above audible sound), frequencies. However, the spectrum still represents electromagnetic radiation at some of these frequencies. The terms are borrowed from sound-wave theory because there are no special electrical terms for these frequencies.

Notice that several of the radiation bands overlap in the spectrum chart in Fig. 10-3. This often occurs throughout the entire spectrum. For example, there is no *distinct* dividing line between ultraviolet light rays and X rays or between X rays and gamma rays.

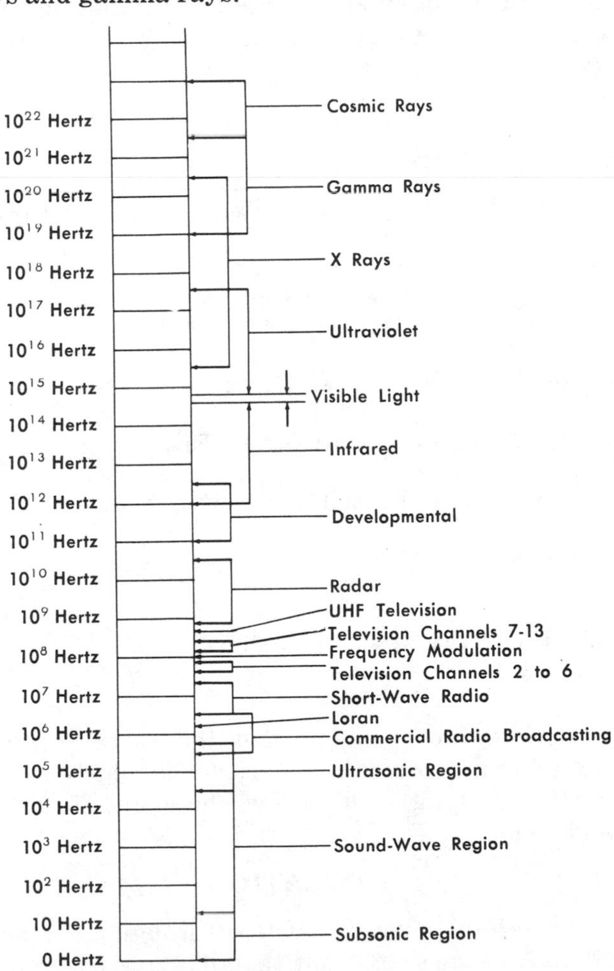

Fig. 10-3. Electromagnetic radiation spectrum chart.

Properties and Uses of Various Radiation Bands

Let's begin at the lower end of the spectrum. Portions of the sound wave and ultrasonic sections are used for *induction heating*. The next group of frequencies is used for commercial communications. This includes both the standard radio and television broadcasting frequencies. *Loran,* an aerial and surface navigational aid, has its waves in this *band* (group). *Radar,* another navigational aid, operates at higher frequencies. Radar is probably most familiar to you as the speed-check device used by law-enforcement officers.

Radio frequencies close to the frequency of light have some of the properties of light. They travel in straight lines and often obey the laws of refraction and reflection.

The sun's radiation is centered about the visible light band and extends into the ultraviolet and infrared regions. Prolonged exposure to ultraviolet radiation can cause sunburn. The earth's atmosphere filters the *short* wavelengths of the ultraviolet rays and thus offers some protection. This explains why people sunburn more easily at high altitudes, where the thinner atmosphere offers less filtering action.

Anything that has a temperature above absolute zero emits infrared radiation. Usable infrared radiation can be produced by special lamps and other devices. It is used for a number of purposes, including cooking, the baking of paints and enamels, the bonding of materials, dehydration, and heat therapy. Infrared rays pass through the atmosphere more readily than visible light rays. For this reason they are useful for long-distance photography.

One of the chief uses of controlled ultraviolet radiation is for sterilizing. Special lamps produce the ultraviolet rays. Since these rays do not pass through ordinary window glass, specially prepared glass is necessary.

X rays are generated when a high-voltage electron stream is projected through a vacuum and strikes a metal target. The resulting splash of energy is converted into electromagnetic radiation of extremely short wavelength—so short, in fact, that it approaches the size of atoms. Apparently, this extremely short wavelength enables X rays to penetrate solid objects. Almost everyone is familiar with the use of X rays to view or photograph the insides of objects. This property is used in both medicine and industry.

Because bones have a different density from the surrounding tissue, they can be seen with X rays, and broken bones will show up clearly. Exposure of living cells to X rays can markedly change the cell growth or structure, and prolonged exposure may kill the cell. This latter effect is used in the treatment of cancer.

X rays are used in industry to detect flaws in metals or other materials and to examine crystalline structures.

Summary

Electromagnetic radiation, in the form of light and heat from the sun, supports all life on the earth.

Man uses controlled radiation in communication, navigation, heating, medicine, and industry.

Although electromagnetic radiation is not clearly understood, it does display some wavelike characteristics and also behaves like projected particles of matter.

Unlike sound, electromagnetic radiation passes easily through a vacuum. Its speed through space is approximately 186,000 miles, or 300 million meters, per second.

The frequency of electronic vibrations in the radiation spectrum varies from zero to more than 10 million quadrillion (10^{22}) hertz.

The wavelengths represented by these frequencies range from thousands of miles at a few hertz to almost atomic smallness at cosmic-ray frequencies.

The velocity of an electromagnetic radiation wave depends upon its frequency and wavelength ($v = f\lambda$).

Questions and Problems

1. How do scientists describe electromagnetic radiation?
2. Does an electromagnetic wave require a medium through which to travel?
3. How fast does electromagnetic radiation travel through space? (Give answers in both miles per second and meters per second.)
4. If a radio station broadcasts at a frequency of 1500 kHz, how long are the waves it radiates?
5. Are ultrasonic waves a form of electromagnetic radiation? Explain your answer.
6. Name the three main portions of the radiation spectrum represented by the radiation from the sun.
7. How are X rays generated?
8. Are X rays easily reflected? Explain.
9. Can X rays be harmful to life? If so, how?

CURRENT FLOW IN GASES

Most gases are poor electrical conductors. Air, for example, is a gas and makes a most effective insulator. However, if two conductors insulated by air are close together and the voltage is high enough, current may still jump the air gap between the conductors. This jumping is called *arcing.*

Lightning is good example of an extremely high potential. Millions of volts are built up between the storm clouds and the earth. Finally the potential becomes so great that current—in the form of lightning—flows between the earth and a cloud.

IONIZATION

Ionization makes conduction through a gas possible. An *ion* is an atom that has gained or lost electrons. An atom that has gained electrons has a negative charge; one that has lost electrons has a positive charge. Ionization is the process by which ions are produced in a gas or electrolyte when a voltage is applied. The atoms of gas can be ionized in numerous ways. Air is ionized somewhat when ultraviolet, gamma, and cosmic rays are passed through it. X rays also have an ionizing effect upon gases. These ionizing rays are all forms of high-energy radiation.

Ionizing by Collision

If a high voltage is applied to a gas, even when there are no ionizing rays, ionization can occur. Fig. 10-4 shows a glass tube containing gas. An electrode is sealed in each end. One electrode (the anode) is *positive*, and the other electrode (the cathode) is *negative*.

A few free ions are always moving about in any gas. The free positive ions are attracted by and move toward the negative electrode, while negative ions will be attracted toward the positive electrode. On their way, they bump into atoms of the gas and *if* they bump the atoms hard enough, electrons will be dislodged from the neutral atoms. These freed electrons may also knock other electrons loose. There will be a continuous flow of electrons and ions within the tube while the potential difference is maintained. When the potential difference is removed, the gas returns to its original state.

Ionization by Concentrated Charges

The ionization of a gas by a *concentrated* electrical charge encourages the ionization of a gas by collision. The completeness of the ionization depends mainly upon the speed of the collision and upon the number of collisions. Increasing the speed of the ions helps to ionize the gas more quickly. In Fig. 10-4 the electrical field about the electrodes attracts and accelerates the nearby ions.

Pointed electrodes will increase the intensity of the electric field. In Fig. 10-5 two pairs of electrodes are shown. One pair (A) has a relatively large area without sharp projections. The charge is spread evenly across the surfaces. The other pair (B) has one electrode with a small, pointed surface. The negative charge is concentrated on this point. Ions or electrons in the gas near such a point are exposed to an intense field. Thereby, they move at greater speeds, and it is easier to ionize the gas.

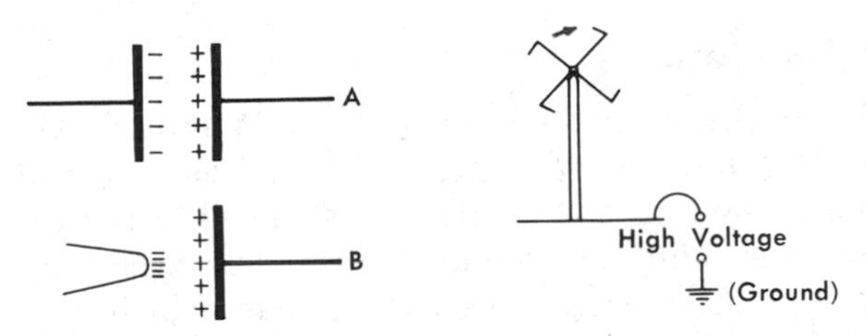

Fig. 10-5. Electrical charges are concentrated on the pointed ends of conductors.

Fig. 10-6. Electrically operated pinwheel showing an ionization principle.

An interesting demonstration can be made from the apparatus in Fig. 10-6. A lightweight vane, similar to a pinwheel, is balanced upon a pivot. The pivot is mounted upon a base that is well insulated from ground. A high electrostatic voltage, inserted between the base and ground, is concentrated at the pointed ends of the pinwheel. The air near the ends of the pinwheel is ionized, and the charge leaks off into the air. The electrical charges exert a force on the surrounding air as they escape, and the pinwheel rotates in the direction shown in the diagram.

Ionization by Heat

If heat is applied to a gas, the speed of the gas molecules will be increased, and ionization will be more likely to occur. This action can be demonstrated by the experiment in Fig. 10-7. When a flame is brought near the terminal of a charged electroscope, the air around the terminal is ionized. The leaves of the electroscope fall together, telling us that the charge has leaked off into the ionized air.

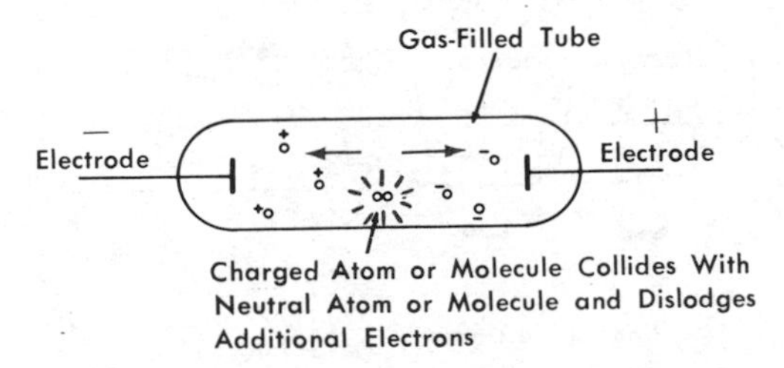

Fig. 10-4. Ionization of a gas within an enclosure by application of an electrical potential.

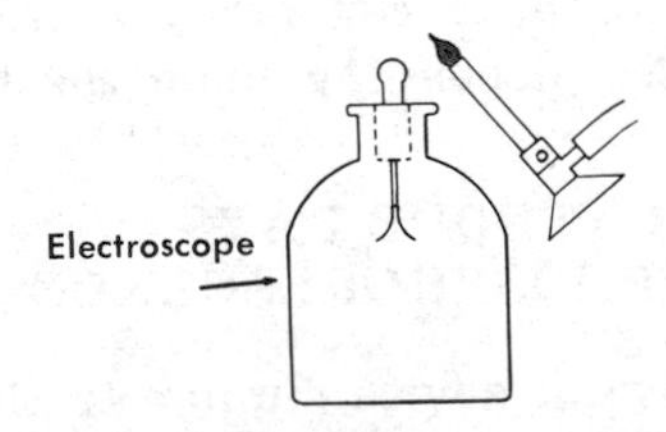

Fig. 10-7. The ionization of air by heating. The electroscope is discharged as air is ionized.

The Hot Cathode

When the cathode in Fig. 10-4 is heated, the electrons move more rapidly. If the heat is great enough, some of the electrons escape into the air or gas surrounding the cathode. The escaped electrons are then attracted by the anode, which is of opposite polarity. In moving toward the anode, the electrons ionize the gas by collision.

Heating the cathode of the device shown in Fig. 10-4 will speed up ionization. The cathode is heated by passing an electric current through it. Another connection point (terminal) must be added. Current can flow either directly through the cathode or through a nearby heater element. The cathode is coated with a material which, when heated, *emits* electrons in large quantities. A tube of this type is called a *hot-cathode tube.* Without the heating feature, such a tube is called a *cold-cathode tube.*

Uses of the Ionization Principle

The ionization of gases is used in the elimination of smoke and dust. This is called *precipitation.* In this process a highly charged electrode ionizes the air around it. The smoke and dust particles become charged and are attracted, or precipitated, to collecting devices such as electronic filters.

The lightning rod is another practical example of gas ionization. In addition to carrying away a lightning bolt, a lightning rod helps *prevent* lightning strokes by *neutralizing* the electrical charge of the clouds passing over. The electrical charge on a passing cloud induces a charge of *opposite* polarity on the buildings and ground beneath it. The induced charge travels up the lightning rod to the tip, where it can leak off because of the concentration on the tip. The ionized air tends to neutralize the charge on the clouds.

By far the widest application of the gas ionization principle is in the electrical and electronic industries. A gas is sealed in a container or *tube,* together with the necessary electrodes. When the gas becomes ionized, it can carry a wide range of current values, depending upon the tube.

CATHODE RAYS AND PHOSPHORESCENCE

Fig. 10-8 shows a device similar to the one in Fig. 10-4, except that an opening with a small tubular neck has been added to the tube. A vacuum pump, connected at this opening, removes most of the gas from the tube. If a voltage too small to cause arcing is applied to the electrodes as the gas is gradually exhausted, the glow within the tube will change colors. At normal atmospheric pressures little current flows. But as the pressure is reduced, the current flow increases. A glow appears around the electrodes, increasing with reduced pressure until a streamer of light crosses the tube from electrode to electrode. The color of the light depends upon the gas.

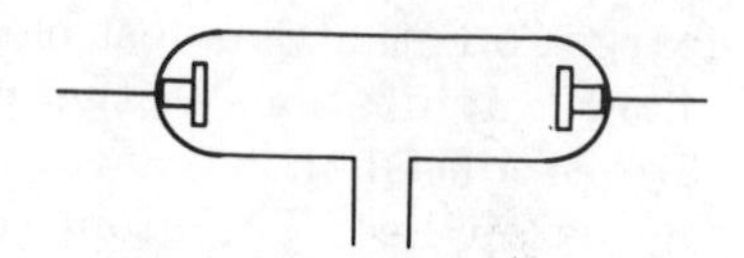

Fig. 10-8. Basic cathode-ray tube.

If the pressure is decreased still further, the glow disappears and the discharge is in the nature of a radiation directed *away* from the cathode—hence the name cathode rays. At extremely low pressures the discharge stops.

When scientists first discovered cathode rays, they concluded that the rays were moving particles, which they called *corpuscles.* Now, the corpuscles are known to be electrons, and the current in a cathode-ray tube is considered to be a high-velocity stream of electrons. A commercial cathode-ray tube is shown in Fig. 10-9 with an undeflected *stream* of *electrons* striking the screen. A cathode-ray tube is evacuated to a very high degree of vacuum. Because the cathode is heated by a filament, electrons are emitted continually even though the gas pressure is low.

When certain materials are placed in the path of the rays, they will glow (*fluoresce*). If the rays strike the glass walls of the tube, they will produce a slight fluorescence there also. Certain phosphors glow brilliantly when struck by cathode rays. The cathode-ray tube with a phosphorescent

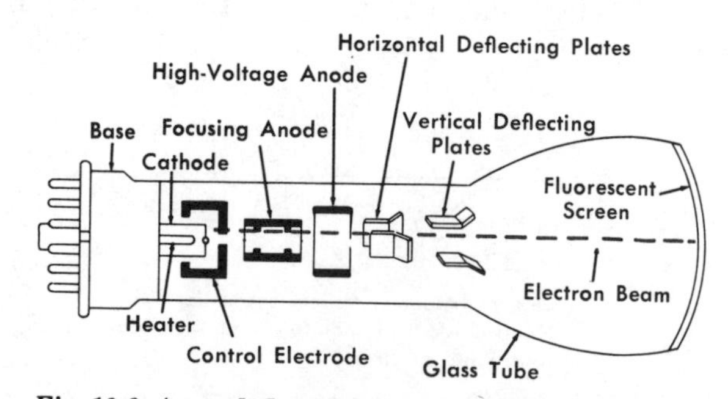

Fig. 10-9. An undeflected beam of electrons flowing in a cathode-ray tube.

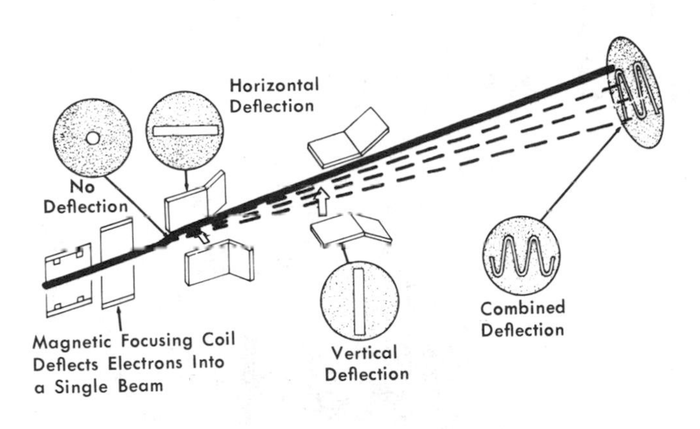

Fig. 10-10. A combination of horizontal deflection and vertical deflection traces a visual pattern of the signal voltage on the screen of the cathode-ray tube.

screen is used in oscilloscopes and television receivers. Many colors of phosphors have been developed, including the red, green, and blue found in color television.

Properties of Cathode Rays

The cathode rays travel in straight lines and cast shadows of objects placed in their path. As the rays strike an object, they produce heat. Their property of exciting phosphorescence or fluorescence has already been mentioned.

Cathode rays can be deflected by electric and magnetic fields. (See Fig. 10-10.) They are attracted by positive fields and repelled by negative fields. Extremely high intensity cathode rays generate *X rays* when they strike an object.

Summary

Most gases are poor conductors. However, they can become good conductors if ionized.

An ion is an atom or molecule that has gained or lost electrons. A positive ion has less than its normal number of electrons; a negative ion has more than its normal number.

Gas atoms and molecules can be ionized by exposure to the high-frequency radiations of ultraviolet, gamma, cosmic, and X rays, and by collision at high speeds with electrons or other ions.

The speeds of moving ions and electrons can be increased by concentrated electric fields, by heat, and by high potentials.

Electrons can be freed from a solid material (such as a cathode element) by heating the material. This makes more electrons available for ionization by collision.

Practical applications of ionization principles include smoke precipitation devices, lightning rods, and gas-filled control tubes.

Cathode rays are high-velocity streams of electrons and ions traveling between the cathode and anode in a partial vacuum. The electrons travel in straight lines, cast shadows, heat objects which they strike, cause some of these objects to fluoresce, and generate X rays if their velocity is high enough. Cathode rays can be deflected by electric and magnetic fields.

Questions and Problems

10. How can gases be changed from poor to good conductors?
11. What does ionization do to an atom or molecule of gas?
12. How does ionized gas conduct current?
13. Name two methods of ionizing gases.
14. Are some ions always in a volume of gas? Explain.
15. How does heat applied to a gas help the ionization process?
16. What are cathode rays?
17. What properties of cathode rays are used in the oscilloscope tube?

ELECTROCHEMISTRY

When we discussed the primary cell, we learned that two dissimilar metals in an electrolyte will produce a voltage. For this voltage to be produced, electrons must travel through the electrolyte to provide an excess of electrons (negative charge) on one terminal of the cell. Fig. 10-11 shows how a primary cell forms positive and negative *ions.* The current flow within the cell is due to these ions.

The cell illustrated is an example of the chemical method of producing current. This type of cell is called a *voltaic cell.* There is yet another cell, an *electrolytic cell,* in which an electrical current produces a chemical action.

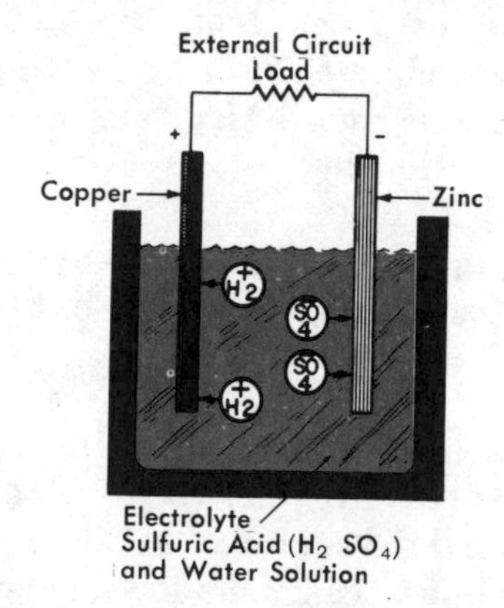

Fig. 10-11. The action of a primary cell.

Electrolysis

The electrolyte of an electrolytic cell consists usually of a water-soluble chemical compound (either salts, bases, or acids) which can form ions. When a current is introduced, as shown in Fig. 10-12, the chemical decomposes, or *disassociates,* in a process called *electrolysis.* Many industrial processes are based upon electrolysis. These processes include electroplating, the commercial production of certain gases, and the refining of metals from their ores.

The electrolytic cell (Fig. 10-12) consists of two electrodes suspended in a solution of hydrochloric acid and water. With the electrodes connected to a source of power, the positive hydrogen ions are attracted to the cathode and are neutralized by the negative charge. These ions are now

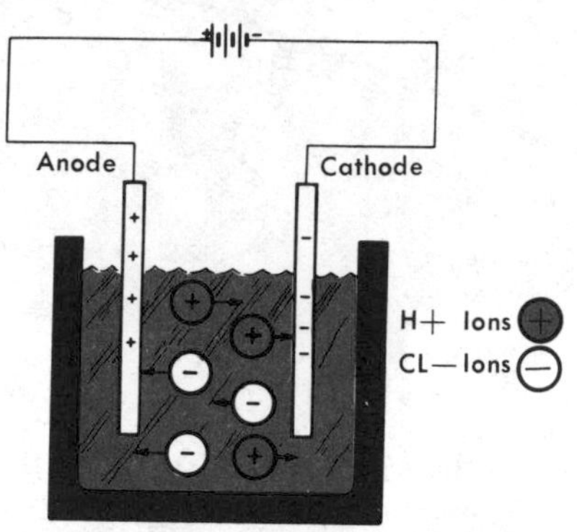

Fig. 10-12. Conduction through the liquid of an electrolytic cell.

neutral atoms of hydrogen, not ions of hydrogen. They are released as hydrogen *gas* at the cathode.

The negative chlorine ions are attracted to the anode where their extra electrons are given up. They then become neutral atoms of chlorine gas. Notice that the flow of current through the cell is due to the ions. Also note that the hydrochloric acid (HCl) is broken up into its hydrogen and chlorine elements.

Electroplating

Another electrolytic cell is shown in Fig. 10-13. This cell has a *copper sulfate* electrolyte. As illustrated, the copper ions (Cu^{++}) and the sulfate ions (SO_4^{--}) disassociate in the solution. The sulfate ions are attracted to the anode. The copper ions are attracted to the cathode, where they gain two electrons and become copper atoms. The copper is thus deposited at the cathode as a *plating* upon the object to be plated. The object to be

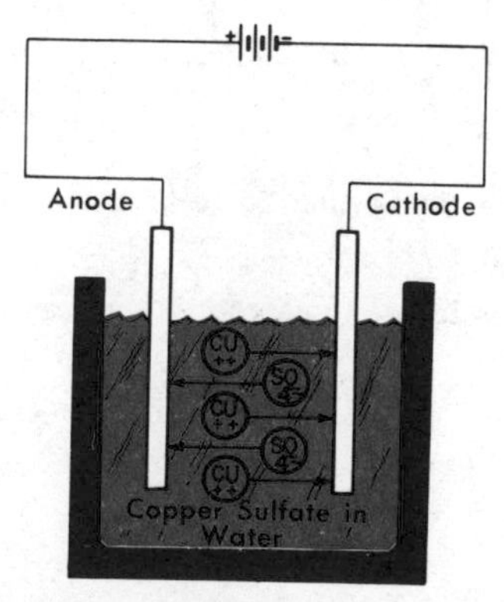

Fig. 10-13. An electrolytic cell for electroplating copper.

plated is the cathode of an electroplating cell. Many metals can be plated with other metals when an electroplating cell is used. The electrolytic solution is a salt of the plating metal, and the anode is a piece of the pure metal to be used for the plating coat.

The thickness of the plating is determined by the *amount of current* and the amount of *time* the current is applied. The higher the current or the longer the current is applied, the thicker the plating becomes.

Some small and intricate parts are made entirely by a plating process known as *electroforming*. A pattern for the part to be produced is used as the cathode. The proper metal is deposited on the pattern until the plate is the required thickness.

Another important commercial application of electrolysis is the Hall process or removing aluminum from ores, chiefly from *bauxite*. Bauxite consists largely of aluminum oxide. An electrolyte of aluminum ore and another mineral known as cryolite are used in the refining process. The cathode is a heavy iron box. The anode of the cell consists of many *carbon* rods which are dipped into the electrolyte of aluminum oxide in melted cryolite. A strong electrical current is sent through the cell and breaks down the electrolyte by electrolysis. The negative oxygen ions move to the carbon rods and the positive aluminum ions move to the cathode (the box). Here they are neutralized to form aluminum atoms. The molten aluminum collects at the bottom of the box.

Several other metals, such as magnesium, sodium, and potassium, are also extracted from their ores by electrolysis. In addition, a number of other metals are refined to a purer state by electrolysis. For instance, copper conductors must be *pure* copper because the impurities would cause a high resistance to current flow. Pure copper, tin, silver, and certain other metals are produced by using the impure metal as the anode of a plating cell. The pure metal is deposited on the cathode of the cell.

The process of extracting ore by electrolysis is known as *electrometallurgy*.

Summary

A *voltaic* cell consists of two dissimilar metals dipped into an electrolyte. This cell produces an electrical current by chemical action.

In an *electrolytic* cell, an electrical current produces a chemical action.

An electrolyte is a solution which can be ionized so that a current can flow due to the charged ions.

Electrolysis is the decomposition of a compound by an electric current. The chemical compound (the electrolyte) is broken up into the elements of the compound.

Many industrial processes, including electroplating and the production of certain metals and gases, are based upon electrolysis.

Questions and Problems

18. Contrast the flow of current through a metal conductor with the flow of current through an electrolyte.

19. What is the difference between a voltaic cell and an electrolytic cell?

20. What chemical compound must be present in the electrolyte of a copper plating cell?

21. Describe the external circuit of a voltaic cell.

22. Describe the external circuit of an electrolytic cell.

23. What type of material is required as an electrolyte (solution) for electroplating?

24. What type of material is used for the anode of an electroplating cell?

25. List three commercial uses of electrolysis.

26. When an electrolytic cell is operating, does a positive ion move toward the anode or toward the cathode of the cell? Why?

27. When a voltaic cell is operating, which ions move toward the cathode of the cell—the positive ions or the negative ions? Why?

11

ELECTRONICS

ELECTRON EMISSION AND CONTROL

All matter contains electrons. In some materials, the electrons are bound into the structure of the atoms and so are not free to move. In other materials, such as metals, some of the electrons are normally free to move from atom to atom. Such electrons are called *free electrons*. Under certain conditions these free electrons can be made to leave the material and enter the surrounding space. This escape of electrons is called *electron emission*. There are four ways electrons can be emitted by a material.

Field Emission

If a high electrostatic potential is concentrated in a small area on the surface of a metal, electrons can be pulled from the surface. The voltage must be very high and the surface very small, so that the voltage per unit square area is very high. This is called *field emission* because the electric *field* supplies the energy needed to enable the electrons to escape.

Photoelectric Emission

Electrons will be released from the surface of certain metals when light of the proper wavelength (or color) falls upon their surfaces. Different metals respond in different degrees to different wavelengths of light.

Thermionic Emission

The molecules in water escape when water is boiled. Similarly, the electrons in a metallic filament will escape if the filament is hot enough (*2700° C* for pure tungsten). The heat causes the electrons to move about rapidly. Eventually, some of them move fast enough to break through the barrier at the surface and escape.

Secondary Emission

The electrons which have been emitted by one of the previous methods can cause further emission. If electrons have been accelerated to a very high rate of speed they may have enough kinetic energy to push one or more electrons out of a material they strike. The electrons emitted by these collisions are called *secondary emission* electrons.

After the electrons have been emitted, their flow must be controlled. Since electrons have a negative charge, they will be repelled by another negative charge and attracted by a positive charge. Electrons which are emitted from a cathode can be influenced by a magnetic field just like the electrons in a conductor.

VACUUM TUBES

Thermionic emission is the source of electrons in most vacuum tubes. There are two types of electron emitters. One is a metallic filament which is heated directly from an ac or dc power source. The other has a sleeve surrounding the filament. The filament heats the sleeve, which emits the electrons. The sleeve is called an indirectly heated *cathode*. It is coated on the outside with an oxide of certain alkaline earths, such as *barium* and *calcium*, which emit large quantities of electrons at comparatively low temperatures.

The Diode

The basic vacuum tube is the *diode* (Fig. 11-1). It contains an electron *emitter* or cathode—and an electron *collector*—or plate. Because electrons are attracted by a positive charge, the anode (plate) will not collect the electrons emitted by the cathode unless the plate is positive. A negative plate repels the electrons and no current can flow.

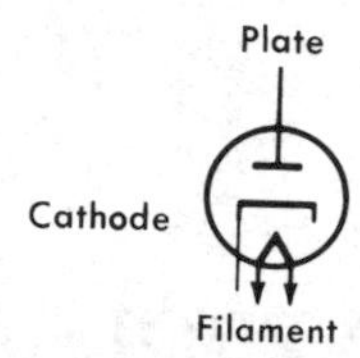

Fig. 11-1. Diode symbol.

The amount of current flowing through the diode can be controlled by raising or lowering the temperature of the cathode. The hotter the cathode, the larger the current. Another way to control the electron flow is to change the plate voltage. As the voltage is increased, the plate becomes more positive and attracts more electrons from the cathode. There is a limit to the amount of electron flow in a diode. When an increase in plate voltage will not increase the amount of current flow the *saturation* point of the tube has been reached.

The Triode

Both methods of control are inefficient and impractical because diodes operate best with certain heater (cathode) and plate voltages. For this reason a third element, called a *control grid,* is placed between the cathode and plate. Since the grid is closer to the cathode than the plate is, the grid exerts a greater control over the electrons than the plate does.

A vacuum tube containing a cathode, a control grid, and a plate is known as a *triode* (Fig. 11-2). The grid is a spirally wound coil of wire placed around the cathode. The turns of the grid wire are

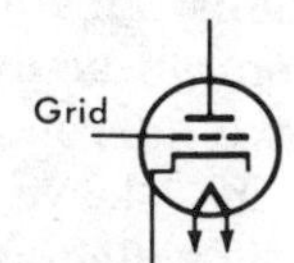

Fig. 11-2. Triode.

widely separated, so that electrons can pass between them to reach the plate. When the grid is made highly negative, however, the electrons will be repelled, and none can reach the plate.

As the grid is made less negative, a few electrons will reach the plate. If the grid is made positive, it attracts the electrons with great force, and huge quantities of electrons pass through the grid to the plate.

The charge placed on the grid is called a *bias.* For most tubes the bias is a small *negative* voltage.

The Tetrode

The cathode of a triode will usually emit more electrons than are required for the plate current. The surplus electrons gather in a "cloud" around the cathode. This cloud is known as a *space charge.* Because the space charge is negative, it hinders the flow of electrons.

To eliminate this disadvantage, a fourth element, the *screen grid,* is placed between the control grid and the plate (Fig. 11-3). The screen grid has a positive charge, which tends to attract the electrons in the space charge. Now the electrons can continue to the plate. Some electrons actually do strike the wires of the screen. However, even though the screen is positive, it does not collect electrons as the plate does.

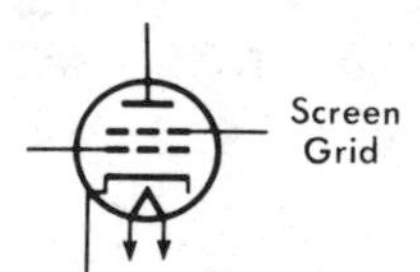

Fig. 11-3. Tetrode.

The screen grid has an extra advantage—it reduces the capacitance between the grid and the plate. This makes the tetrode more suitable for use at high frequencies.

The Pentode

Adding the screen grid greatly increases the speed of the electrons from the emitter to the plate. However, the electrons strike the plate so hard that other free electrons are generated by *secondary* emission. This secondary emission of electrons tends to flow backward in the tube and enter the screen grid, reducing the efficiency of the tube. Therefore, another grid is placed between the plate and the screen grid. This grid has approximately the same negative potential as that of the cathode (Fig. 11-4). The added grid being negative with respect to the plate, repels the secondary electrons back to the plate. Because this grid *suppresses* secondary electrons, it is known as the *suppressor grid.*

There is no limit to the number of elements a vacuum tube can have. Many different types of tubes have been made—far too many to concern ourselves with at this time. However, two other types of tubes should be mentioned—the *phototube* (better known as the "electric eye"), and the *cathode-ray* tube in our television set.

The phototube is a diode. Its cathode is shaped like a half-cylinder. The inside of the cathode is coated with a light-sensitive material, such as cesium. The cesium emits electrons whenever it is exposed to light, just as a cathode does when it is exposed to heat.

The *plate,* or anode, of a phototube is a small metal rod in the center of the half-cylinder formed by the cathode.

Phototubes are used to open doors, to ring a fire or smoke alarm, to control safety interlock switches on machinery, and to count the number of vehicles traveling along a highway.

These tubes are also used to detect changes in light intensity. When a dark, cloudy day results in an early dusk, such a phototube-controlled switch can automatically turn on the street lights before darkness makes city streets dangerous.

The cathode-ray tube contains an electron *gun* consisting of a cathode and several grids. This electron gun "shoots" a beam of electrons toward a "screen" at the front of the picture tube (Fig. 11-5). This screen is coated with a *fluorescent* material that glows when bombarded by an electron beam. The electron beam is focused and swung from side to side and from top to bottom of the screen, by controlled electrostatic and magnetic fields. This produces the pattern for the picture we see on the face of the television picture tube.

The *video* (picture) signal is picked up by the television receiver and sent into the picture tube. This signal varies the strength of the beam of electrons striking the fluorescent screen and "paints" the same picture seen by the television camera. The stronger the flow of electrons, the more the face of the tube will light up. This glow gives us the blacks and whites in the picture.

Special synchronizing circuits keep the beam of the cathode-ray tube in step with the beam of the television camera. The beam must make 525 lines across the face of the tube, thirty times each second. Therefore, the beam is *scanning* across the face of the screen 15,750 times per second.

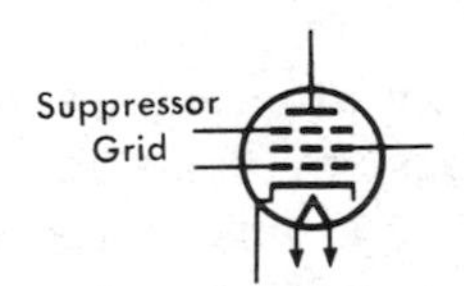

Fig. 11-4. Pentode.

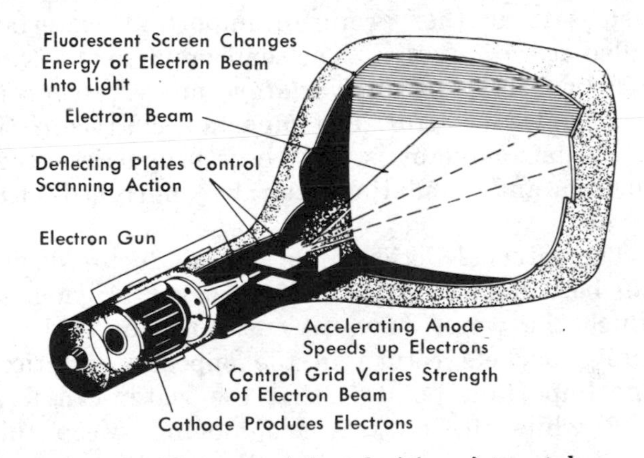

Fig. 11-5. Elements of the television picture tube.

Summary

Field emission is generated by a high electrostatic charge in a very small area.

Secondary emission is generated by the impact of high-velocity electrons.

Photoelectric emission is generated by light falling upon a photo-sensitive material.

Thermionic emission is generated by the application of heat.

Free electrons can be controlled by electrostatic charges and by magnetic fields.

The cathode, plate, control grid, screen grid, and suppressor grid are elements of vacuum tubes.

The picture tube of a television receiver is a cathode-ray tube.

Questions and Problems

1. Define *space charge.* What element of a tube helps overcome space charge?
2. What causes secondary emission?
3. Name the elements in a pentode tube.
4. Name the three ways of controlling electron flow in a triode tube.
5. Name four types of electron emission.
6. Why does the control-grid voltage have more effect on electron flow than the plate voltage does?
7. What must the control grid voltage be to cut off the electron flow, negative or positive?
8. How many elements are there in a diode? What is the purpose of each?
9. Explain the purpose and action of a suppressor grid.
10. Explain the purpose and action of a screen grid.
11. What is the scanning rate of the electron beam in a television picture tube?

TRANSISTORS AND OTHER SEMICONDUCTORS

We are reasonably well-acquainted with conductors and insulators. Yet, between these two extremes is another group of important materials called *semiconductors.* The semiconductor has certain properties of an *insulator* and certain properties of a *conductor.* The important characteristic of a semiconductor is that these properties—conduction and insulation—can be controlled electrically.

The internal action of a semiconductor device can be compared to that of a bucket brigade, in which the water flows in one direction and the empty buckets return in the opposite direction. The important point is that the water can flow only while the buckets are moving. Keep this analogy in mind as you study this section.

Electrons and Holes

Electron is a familiar term associated with electronics and current flow. Current flow through wires, tubes, and other components is generally accepted to be a flow of *negative* particles, or electrons.

The term *hole* is rather new to electronics. A hole denotes a *positive* charge, or the *lack* of an electron—just as *vacuum* denotes the lack of air.

To describe the foregoing more fully, we must briefly review the atomic structure of matter. Atoms consist of a nucleus surrounded by rings of electrons. Each ring of an atom contains a specific number of electrons, depending upon the type of atom. The electrons in the *outer* ring are called *valence* electrons (Fig. 11-6).

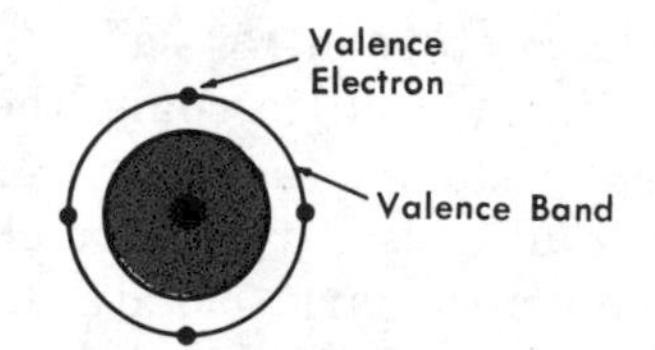

Fig. 11-6. Sketch of an atom and its parts.

Covalence

The electrons of an atom of matter are held in their *orbits,* or rings, by their attraction to the positive nucleus. Two atoms of the same element sometimes *share* electrons by what is known as *covalence.* One or more valence electrons of an atom are held by both its own nucleus and the nucleus of a neighboring atom. (See Fig. 11-7.)

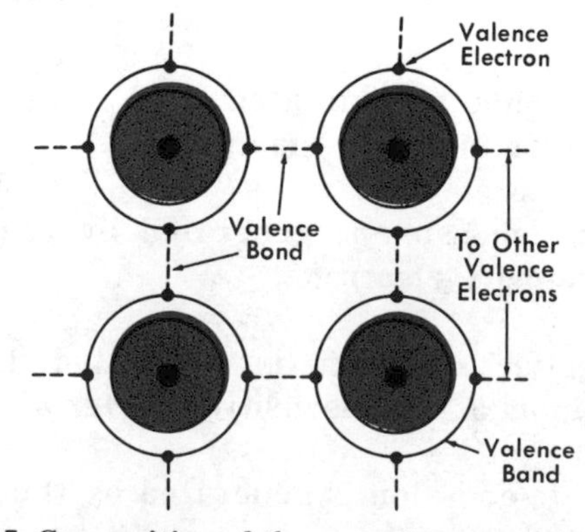

Fig. 11-7. Composition of the crystal structure of atoms.

Covalence can occur in material which has three, four, or five valence electrons. Two different elements also can form covalent bonds.

N-type Semiconductors

If we could add atoms with five valence electrons to the structure shown in Fig. 11-7, four electrons of each atom would be held in a *covalent bond.* In other words, two atoms would be sharing some of the same valence electrons. This would, of course, leave one "extra" electron per atom of impurity, which would not be influenced by either atom. These extra electrons would be *free* electrons, able to move whenever an electrical potential was established.

Germanium is an element with four valence electrons. It is commonly used in semiconductors. Arsenic, with five valence electrons, is often used as an added *impurity.* Extremely small quantities of the impurity are added, in an operation called *doping.*

Germanium and arsenic are elements that form covalent bonds. In such a combination of four valence electrons (germanium) plus five valence electrons (arsenic), there is one "leftover," or *free,* electron. The impurity (arsenic) is called a *donor* impurity because it donates electrons to the semiconductor material. Current can flow through such semiconductor material because of these free electrons, which are the current carriers. Semiconductors of this type are called *n-type* (negative) semiconductors because they have extra electrons. In n-type materials the *electrons* are the *majority carriers.*

Silicon is another element much used for semiconductor material. It is made into n-type material by doping with arsenic or antimony.

P-type Semiconductors

Just as a donor impurity can be added to a semiconductor, so can an *acceptor impurity* be added. Germanium doped with *antimony* or *indium* may be used in this type of semiconductor. Both antimony and indium have three valence electrons.

Germanium (four valence electrons) doped with antimony or indium (three valence electrons) will make a p-type semiconductor. Silicon also makes p-type material when doped with aluminum or boron. When these materials combine, instead of having one extra electron (as in donor or n-types), they lack one electron. Since these materials lack an electron, they are called p-type (positive) semiconductors. The lack of an electron is the hole mentioned previously. The hole, or lack of an electron, causes these materials to *seek,* or to readily *accept,* electrons.

We have learned how free electrons move from one atom to another. In p-type materials the holes move from one atom to another in a similar way. When one atom accepts an electron, a hole is left in a neighboring atom. When such a hole accepts an electron, another hole is created, and so on throughout the material. In the p type, the *hole* is the *majority carrier.* This "transfer" of holes can occur only in solids such as semiconductors; it does not apply to vacuum-tube theory. The study of semiconductor theory is sometimes referred to as *solid-state* electronics.

The main points to remember are that the electrons are negatively charged particles and that the holes are positive charges. Both can move and, as such, can be current carriers. In n-type semiconductors, the electrons are the majority carriers; in p-type semiconductors, the holes are the majority carriers.

Junctions P and N

Transistor operation is based upon the action of the carriers—the electrons and the holes—at the *junction* of p and n materials. The activity of the carriers at a junction can be pictured. Let the blocks labeled *N* and *P* in Fig. 11-8 represent the doped semiconductor materials. The n material is shown as having electrons as majority carriers; the p material is shown as having holes as majority carriers.

The material itself has no charge. Therefore, no current flows between two types of material if they are merely touching each other. The term junction means that the molecules of the materials are fused (melted) together.

When p and n semiconductors are fused together to produce a junction, the majority carriers near the junction move toward each other and cancel each other, as shown in Fig. 11-9. Because of this canceling action, at the junction, a charge

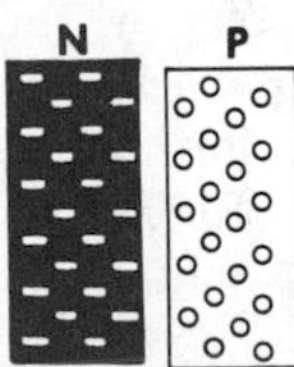

Fig. 11-8. Two types of semiconductor materials and their carriers.

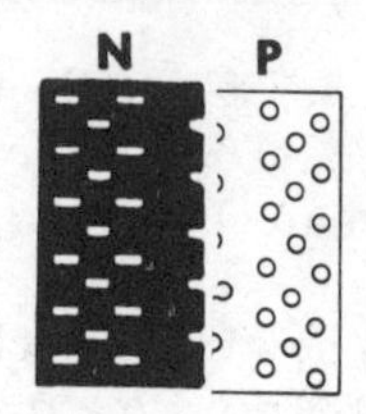

Fig. 11-9. Action that takes place when a junction is produced.

is created between the semiconductor materials. Since some of the majority carriers (electrons in the n type and holes in the p type) have been canceled, the material at the junction assumes a positive charge in the n semiconductor material and a negative charge in the p semiconductor material.

The electrons in the n material are now repelled by the negative charge in the p material, and the holes (positive charges) are repelled by the positive charges in the n material. These majority carriers, therefore, maintain positions away from the junction. The charge and its polarity at the junction are represented by the battery in Fig. 11-10. This charge, or potential, is extremely small—only tenths of a volt—but it does effectively block the current carriers. To pass from one side of the junction to the other, the electron or hole must gain energy equal to this barrier. This energy could be supplied by a power supply or battery.

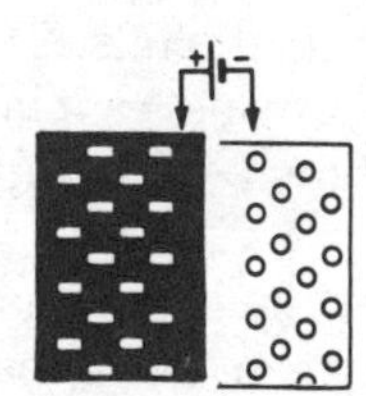

Fig. 11-10. Polarity of the charge at the junction of n and p materials.

Junction Diode

An application of the joined n and p materials is illustrated in Figs. 11-11A and 11-11B. In Fig. 11-11A the battery only adds to the barrier. The electrons are attracted by the positive terminal, and the holes are attracted by the negative terminal. At the junction the holes and electrons are separated even further, making more of a barrier. No current will flow. Connected in this manner, the junction is said to be reverse biased.

In Fig. 11-11B the battery terminals are reversed. The electrons are now repelled by the negative terminal, and the holes are repelled by the positive terminal. In the area of the junction, the electrons (of the n donor material) combine with the holes (of the p acceptor material). This attracts a continuous stream of other electrons from the battery. This flow of electrons, of course, causes the current to flow in the external or battery circuit. Connected in this manner, the junction is said to be forward biased.

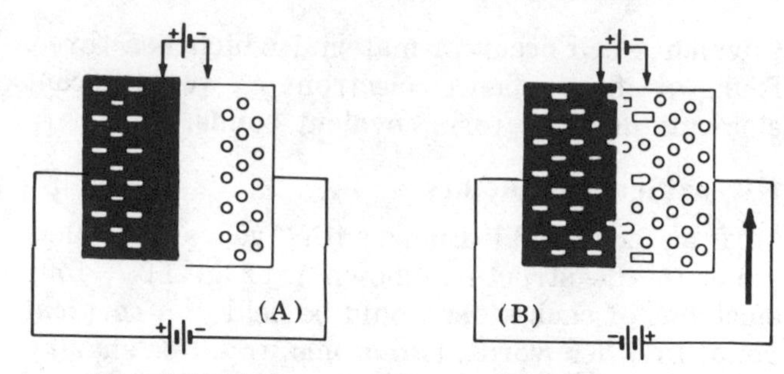

Fig. 11-11. Current flow through a semiconductor.

This type of diode is used in rectification circuit applications. If you can imagine an ac source instead of the battery in Figs. 11-11A and 11-11B, you will realize that only one-half of the ac cycle could pass. This is further illustrated in Fig. 11-12 where the action of a diode tube is compared with a junction diode. The schematic drawing symbol used for junction diodes is shown in Fig. 11-12.

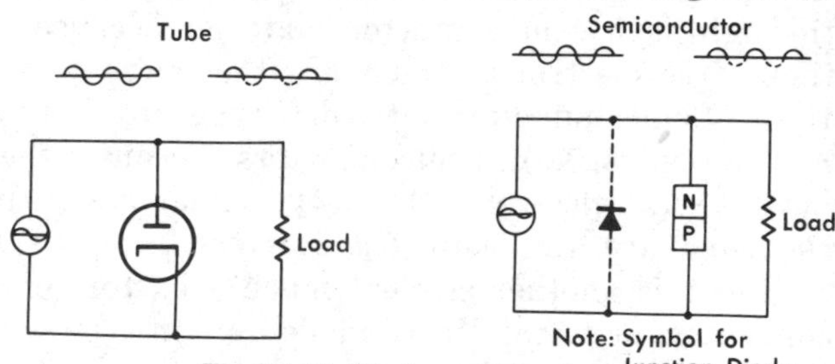

Fig. 11-12. Diode rectification.

THE JUNCTION TRANSISTOR

The addition of a *grid* to a diode tube, making it a triode, makes it possible to control the current passing through such a tube. A small input voltage on the tube grid can control a large output voltage from the tube, resulting in *amplification.*

The junction transistor can function in an electrical circuit to perform many of the same tasks of vacuum tubes. However, you should remember that the *operating principles* of the junction transistor and the vacuum tube are not identical.

A comparison of an npn and a pnp transistor and a triode vacuum tube is shown in Fig. 11-13.

Notice in the tube that the cathode, which emits electrons, is comparable to the transistor *emitter*. The plate of a tube collects the electrons just as the *collector* in the transistor does. The *base* is the control element, similar to the tube grid.

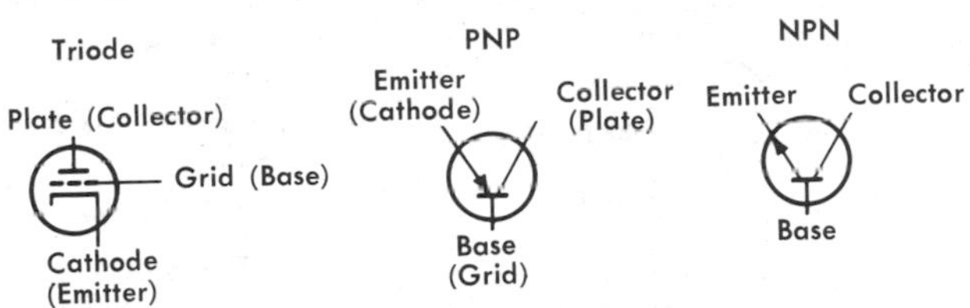

Fig. 11-13. Triode and transistor symbols.

The pnp and npn junction transistors are made as shown in Fig. 11-14, with the n and p materials fused together. The current flow is shown by the arrows. It is controlled by the charge, or *bias*, placed on the emitter-to-base junction. Notice how the two transistors in Fig. 11-14 are biased. In both examples, the emitter-to-base junction is forward biased, and the base-to-collector junction is reverse biased.

Although the diagram of Fig. 11-14 is more or less idealized, it is also a fairly accurate picture of the physical construction of the grown-junction transistor. This was the first type of junction transistor, now largely supplanted by others. New construction methods have been developed with a number of advantages: lower cost, easier fabrication, better performance, etc. Fig. 11-15 shows three of the present methods of construction. Fig. 11-15A is the alloy junction. The starting material is n-type semiconductor, to which p-type semiconductor is alloyed for emitter and collector. Figs. 11-15B and C are diffusion types, in which suitable metals are diffused into semiconductor material to produce the required n and p regions. Masking and etching techniques are used to limit and define the regions.

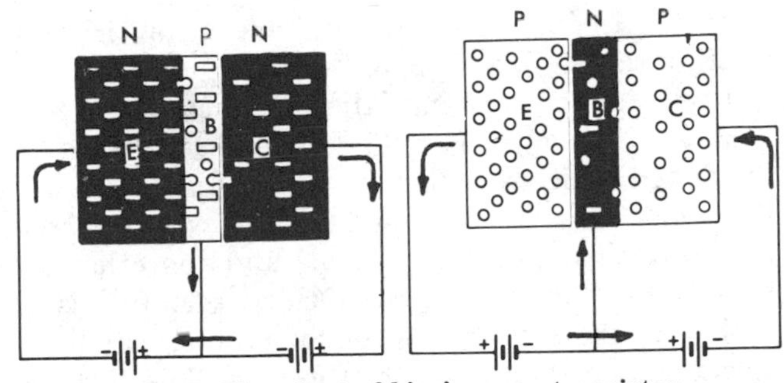

Fig. 11-14. The effect of biasing on a transistor.

Transistors are now used in many circuits because they have important advantages over vacuum tubes. They are much smaller than tubes and require much less power to operate. They are rugged and dependable, and under normal operating conditions they should last indefinitely. Several types of transistors are shown in Fig. 11-16.

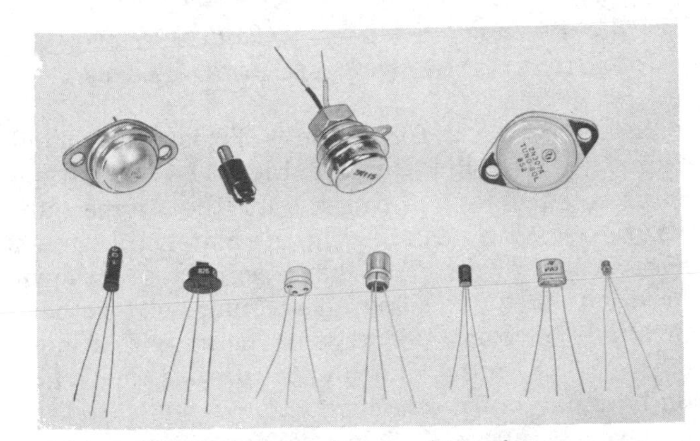

Fig. 11-16. General types of transistors.

FIELD-EFFECT TRANSISTORS (FETs)

The field-effect transistor, like the junction transistor, uses semiconductor material, but its action is quite different. Basically, current is controlled by an electric field, rather than by injected

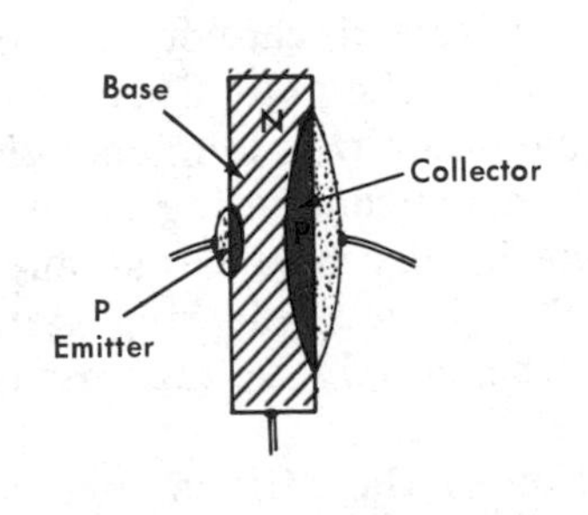

(A) Alloy junction.

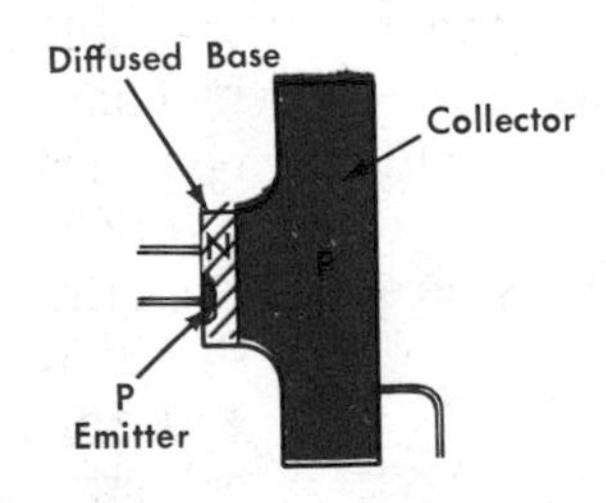

(B) Diffused junction. Mesa construction.

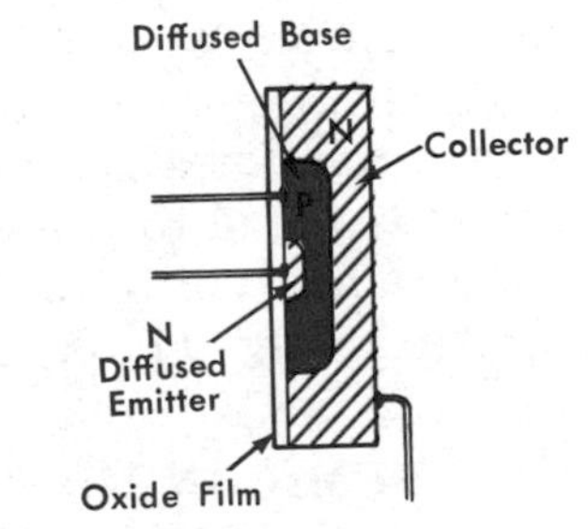

(C) Diffused junction. Planar construction.

Fig. 11-15. Structures of junction transistors.

current. In that respect, its operation is quite similar to the operation of the vacuum tube.

In its simplest form, the FET can be represented as shown in Fig. 11-17A. It is merely a bar of doped silicon, either p or n material, with terminals at each end. The terminal where current is injected is called the *source* and the other terminal is called the *drain*. A third terminal is attached to the side of the semiconductor bar but is electrically insulated from the bar. This terminal is called the *gate*. Current through the bar depends on the voltage applied to source and drain and on the gate voltage.

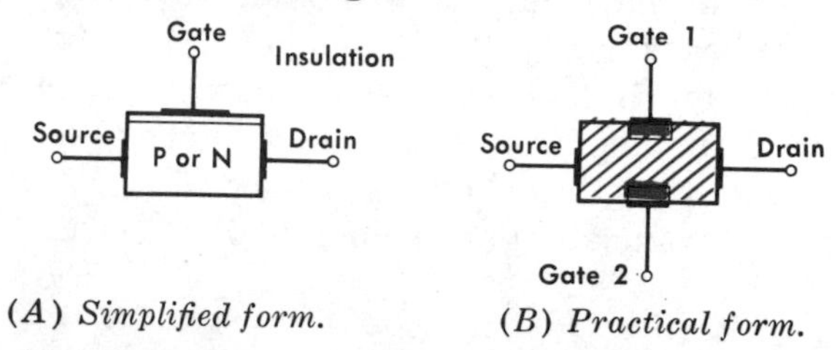

(*A*) *Simplified form.* (*B*) *Practical form.*

Fig. 11-17. Construction of a field-effect transistor.

In its practical form, the FET is constructed more like the diagram of Fig. 11-17B. P-type regions have been diffused into the n-type substrate, creating a *channel* in the material between source and drain. Two gate regions are shown here, but they are sometimes connected together; normally only one side is constructed with a gate, since it is more difficult to diffuse materials on both sides of a semiconductor wafer.

Several types of field-effect transistors have been developed. One is the junction field-effect transistor, or JFET. Another type is the insulated-gate field-effect transistor or IGFET, sometimes called the metal-oxide semiconductor field-effect transistor (MOSFET). In addition, each type can be either n-channel or p-channel.

The input impedance of the FET is high, and its circuit behavior resembles that of the vacuum tube, being voltage controlled rather than current controlled. As a matter of experiment, a FET was substituted for a pentode tube in a certain circuit and was found to work.

A number of circuit symbols for FETs are shown in Fig. 11-18.

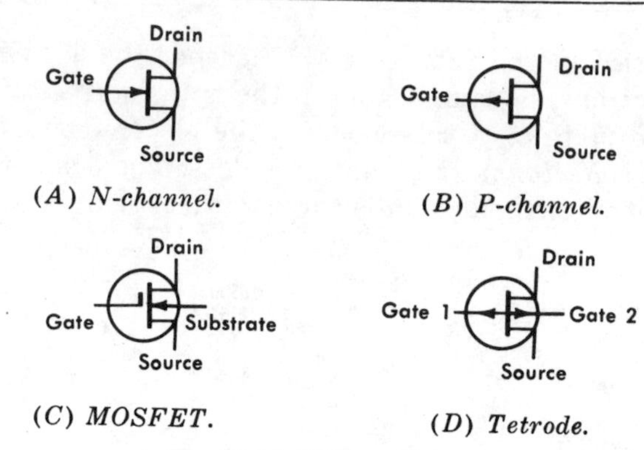

(*A*) *N-channel.* (*B*) *P-channel.*

(*C*) *MOSFET.* (*D*) *Tetrode.*

Fig. 11-18. FET symbols.

SEMICONDUCTOR CONTROLLED RECTIFIERS (SCRs)

The semiconductor controlled rectifier (often called silicon controlled rectifier) is a four-layer semiconductor device made up of alternate layers of p and n material. As indicated in Fig. 11-19, it can be regarded as two separate transistors, an npn and a pnp, that in effect overlap so that the center n and p layers are common to both. The center p section may then be regarded as the base of the npn transistor, and the center n section may be regarded as the base of the pnp transistor. When forward voltage is applied to the SCR, conduction is blocked by the center junction, which acts as a back-biased diode.

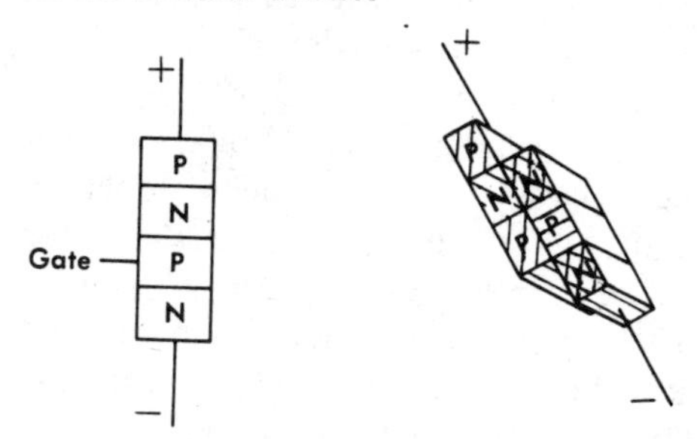

Fig. 11-19. The four-layer SCR is analogous to two transistors sharing common layers.

If current is injected at either center section (positive at the p layer, or negative at the n layer), the SCR will conduct and the center section (the *gate*) loses control. This means that current through the SCR, once started, cannot be stopped by changing the voltage at the gate. Current must be interrupted in some other way long enough for the SCR to return to a nonconducting state. One way this can be done is to use an ac supply for the SCR. If the SCR is in a conducting state, it will turn off when the supply voltage reverses polarity.

The "turn-on" action of the SCR is very rapid and complete. Once the SCR is conducting, its resistance is very low. These properties make it particularly useful as an electronically controlled switch, and many practical applications use it for that purpose.

The symbol for an SCR resembles a diode symbol with a third connection added to one end of the cathode bar (see Fig. 11-20).

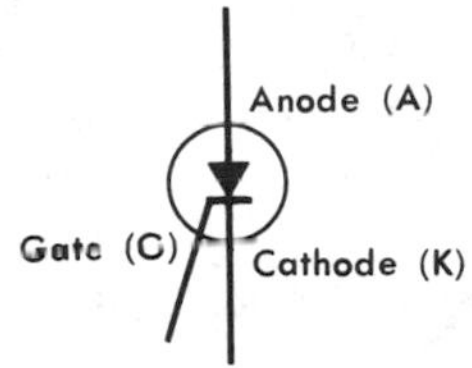

Fig. 11-20. SCR symbol.

ZENER DIODES

When a forward bias is applied to a semiconductor diode, resistance is comparatively low, and several milliamperes, or even amperes, may flow. As voltage is increased, current also increases. But if voltage is applied in reverse polarity, resistance is high and initial current may be only a few microamperes. Then, as the reverse voltage is increased, the current changes very little until finally a point is reached where reverse current increases suddenly. This is called the *zener point* and is shown in Fig. 11-21 as the point labeled "breakdown voltage." When this point is reached, the voltage across the diode remains fixed and only the current increases as more voltage is applied. Fortunately, this "breakdown" is not destructive, and diodes made for this purpose can be operated in or out of this region repeatedly without any change in characteristics.

Because the voltage at the zener point remains constant over a wide range of reverse-current values, the zener diode is well suited for use as a voltage regulator. Zener diodes are designed to emphasize the zener characteristics of the semiconductor doide and are made in a variety of voltage and wattage ratings, from approximately 5 volts and 1 watt to about 200 volts and 150 watts.

Other uses for the zener diode, beside the use as a voltage regulator already mentioned, include surge protection, arc suppression, meter protection, and meter scale expansion.

TUNNEL DIODES

All of the semiconductor devices discussed so far depend for their behavior on the same mechanism, which is merely used in different ways. They all use electrons or holes as current carriers. This process has its limitations, especially when it comes to speed. A finite time is required for a carrier to move from one atom to the next. This time is greatly shortened in the tunnel diode. In this device, current carriers do not go over the junction barrier; they apparently "tunnel" under it. The symbol for a tunnel diode is shown in Fig. 11-22.

With no inherent frequency limitations, tunnel diodes can be switched in intervals as short as 10^{-12} seconds—that is, one millionth of one millionth of a second. Other properties of the tunnel diode are comparative freedom from radiation effects and successful operation over a wide range of temperatures.

The tunnel diode is one of the few devices that exhibit negative-resistance characteristics. In gen-

Fig. 11-22. Tunnel diode symbol.

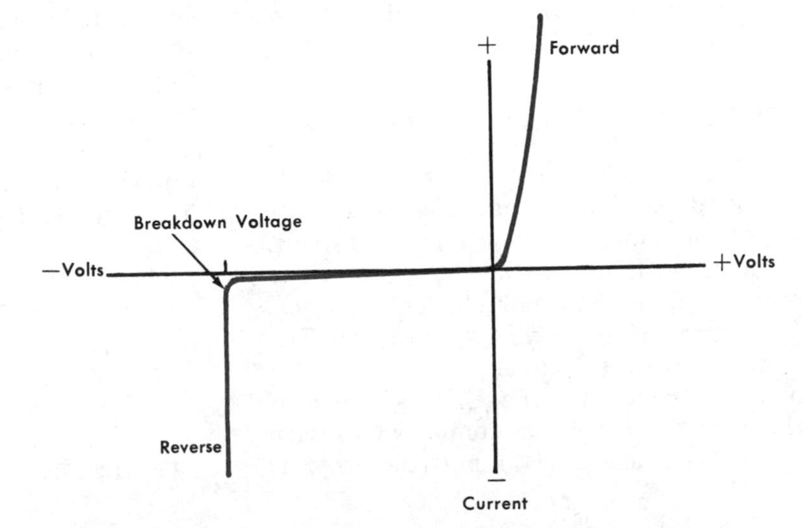

Fig. 11-21. Typical zener-diode characteristic.

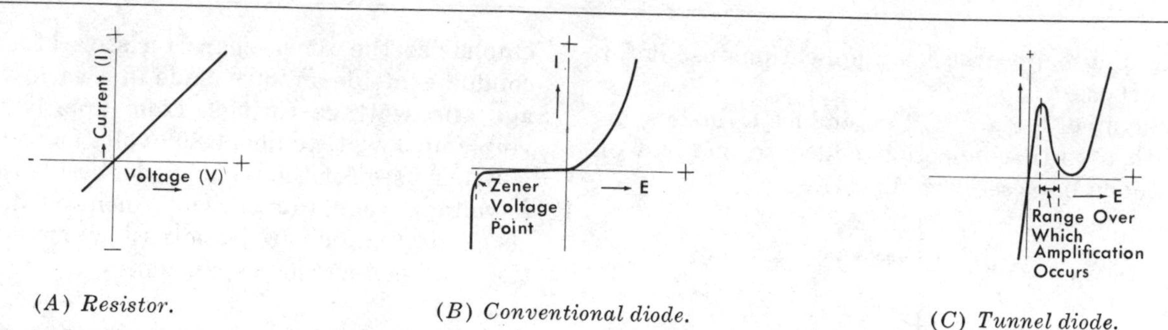

(A) *Resistor.* (B) *Conventional diode.* (C) *Tunnel diode.*

Fig. 11-23. Comparison of voltage/current characteristics of resistor, conventional diode, and tunnel diode.

eral, electronic devices display one of three types of resistance effects when voltage is applied. When the resistance is linear and positive, an increase in voltage results in an increase in current, in direct ratio to the amount of voltage increase. This is illustrated by the graph in Fig. 11-23A. A nonlinear resistance like that of a conventional diode produces the voltage/current characteristic shown in Fig. 11-23B. This is a positive resistance, because the current always increases as the voltage increases. But it is a nonlinear increase, as shown by the curved graph. The graph for a tunnel diode (Fig. 11-23C) shows increasing current at first; then, when a certain point is reached, current begins to decrease and continues to decrease until another reversal point is reached and current increases again. The region of decreasing current is the negative-resistance region of the tunnel diode.

The tunnel diode can be used in many of the same applications as transistors. It requires far less power to operate and, for this reason, is generally smaller than the equivalent transistor. One promising area for tunnel diodes is in the switching circuits of large computers, where they can perform both logic and memory functions. Tunnel diodes offer the advantages of small size, low operating power, high speed, and high reliability.

UNIJUNCTION TRANSISTORS

The unijunction transistor is a three-terminal semiconductor device with characteristics quite different from those of the conventional transistor. It is constructed with an emitter element and two base terminals, but no collector element, as such. The bar structure of the UJT is shown in Fig. 11-24, together with the symbol.

As shown by the figure, the UJT is a bar of n-type silicon with base leads connected to each end. One is called Base 1 (B1) and the other is

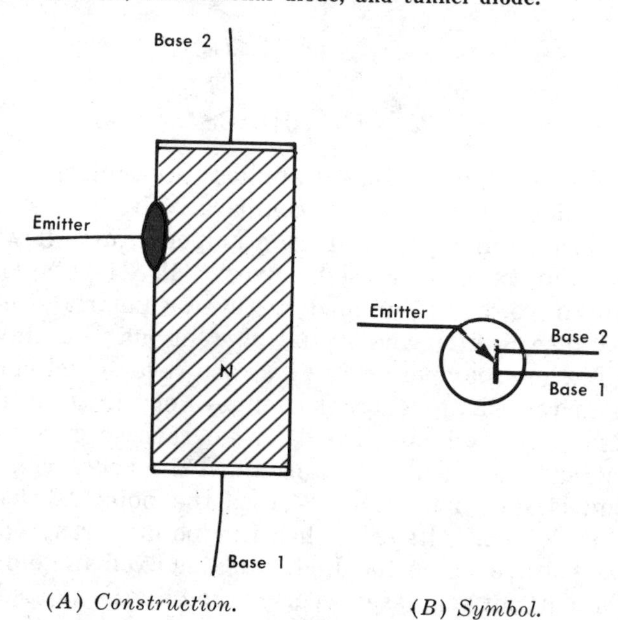

(A) *Construction.* (B) *Symbol.*

Fig. 11-24. Simplified unijunction transistor.

called Base 2 (B2). A pn junction is formed near one end of the bar (the B2 end) by alloying a thin aluminum lead to the top surface of the bar. A circuit representation of the UJT is shown in Fig. 11-25.

In use, B1 is usually grounded and B2 is connected to a positive voltage source. With the emitter circuit open, the resistance between B2 and B1 is between 5000 and 10,000 ohms. The emitter is, in effect, tapped into this resistance a small distance below B2. If the external voltage applied to the emitter is less than that resulting from the IR drop through the crystal from E to

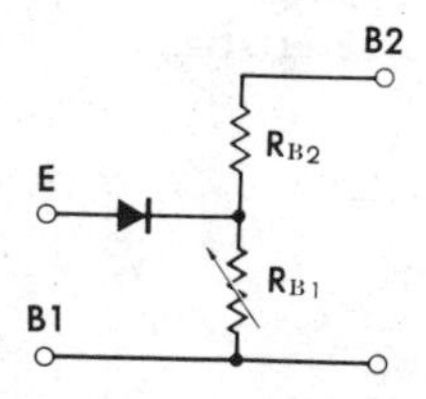

Fig. 11-25. Equivalent circuit of a unijunction transistor.

B1, the emitter will be cut off and very little of its current will flow. On the other hand, if the external emitter voltage exceeds that caused by the internal voltage drop, the emitter will be forward biased and current will flow from emitter to B1. This flow reduces resistance between B1 and the emitter; hence, as current through the emitter increases, the voltage decreases. This gives the unijunction transistor its negative-resistance characteristic.

Some of the applications to which the UJT is particularly suited are pulse generators, sawtooth generators, timing circuits, and trigger circuits.

VARACTORS

A *varactor* is a two-terminal solid-state device that uses the voltage-variable capacitance of a pn junction. Every semiconductor diode has some internal capacitance, but normally this is undesirable. However, in the varactor it is deliberately exploited, and the varactor is designed to increase this effect as much as possible. The capacitance of a varactor is changed by varying the applied voltage.

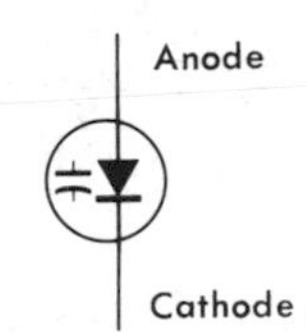

Fig. 11-26. Varactor symbol.

Fig. 11-26 shows just one of several symbols that are used for the varactor diode. It is the symbol for a simple diode with a capacitor symbol added. Varactors are not available in high capacitance values, but added capacitance can be obtained by paralleling several units. They follow the same laws as simple capacitors in respect to series, parallel, and voltage combinations.

The varactor is used in many applications where it replaces the variable capacitor (with movable plates) in tuned circuits such as rf tuned circuits and oscillators. Other uses are in automatic frequency control, frequency multipliers and dividers, modulators, sweep generators, and light-controlled capacitors. In fact, it can be useful anywhere a controlled voltage can be adapted to controlling capacity.

THERMISTORS

The *thermistor* is a relatively simple, two-terminal semiconductor device, whose useful characteristic is that its resistance changes with a change in its temperature. If its resistance increases with an increase in temperature, it is said to have a *positive temperature coefficient* (PTC); if its resistance decreases with an increase in temperature, it is said to have a *negative temperature coefficient* (NTC).

The temperature coefficient of a conductor is the value in ohms of the change in resistance of the conductor as it changes in temperature by 1°C. For example, some copper wire has a temperature coefficient of 0.004. If the resistance of a certain length of this wire is 1 ohm at 20°C, and the temperature changes to 22°C, its new resistance will be 1 ohm plus 2×0.004, or 1.008 ohms. Note that the value .008 was added in this case, since copper wire has a positive temperature coefficient, as is true of most pure metals.

Many coefficients used in electronics have temperature coefficients of resistance, although in some the coefficient is almost zero (in constantan, for example, it is 0.00002). In most cases, this property of a conductor is undesirable; it can cause thermal runaway in some transistor circuits if proper measures are not taken to prevent it. The thermistor, however, is purposely designed to enhance this property of temperature coefficient, in controlled amounts.

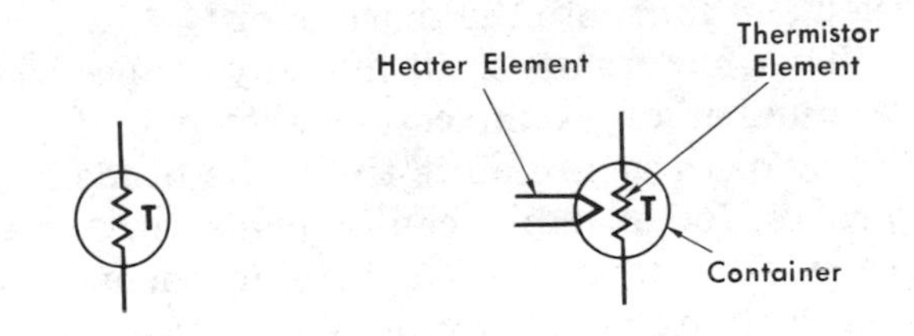

(*A*) *Directly heated.* (*B*) *Indirectly heated.*

Fig. 11-27. Thermistor symbols.

Thermistors are made of special mixtures or compounds, usually semiconductors. Some of the substances used are selected oxides of cobalt, magnesium, manganese, nickel, or uranium. Thermistor circuit symbols are shown in Fig. 11-27. As shown in the figure, thermistors are either directly heated or indirectly heated. In the directly heated type, heating is produced either by ambient temperature or by the current passing through the thermistor, or both. In the indirectly heated type,

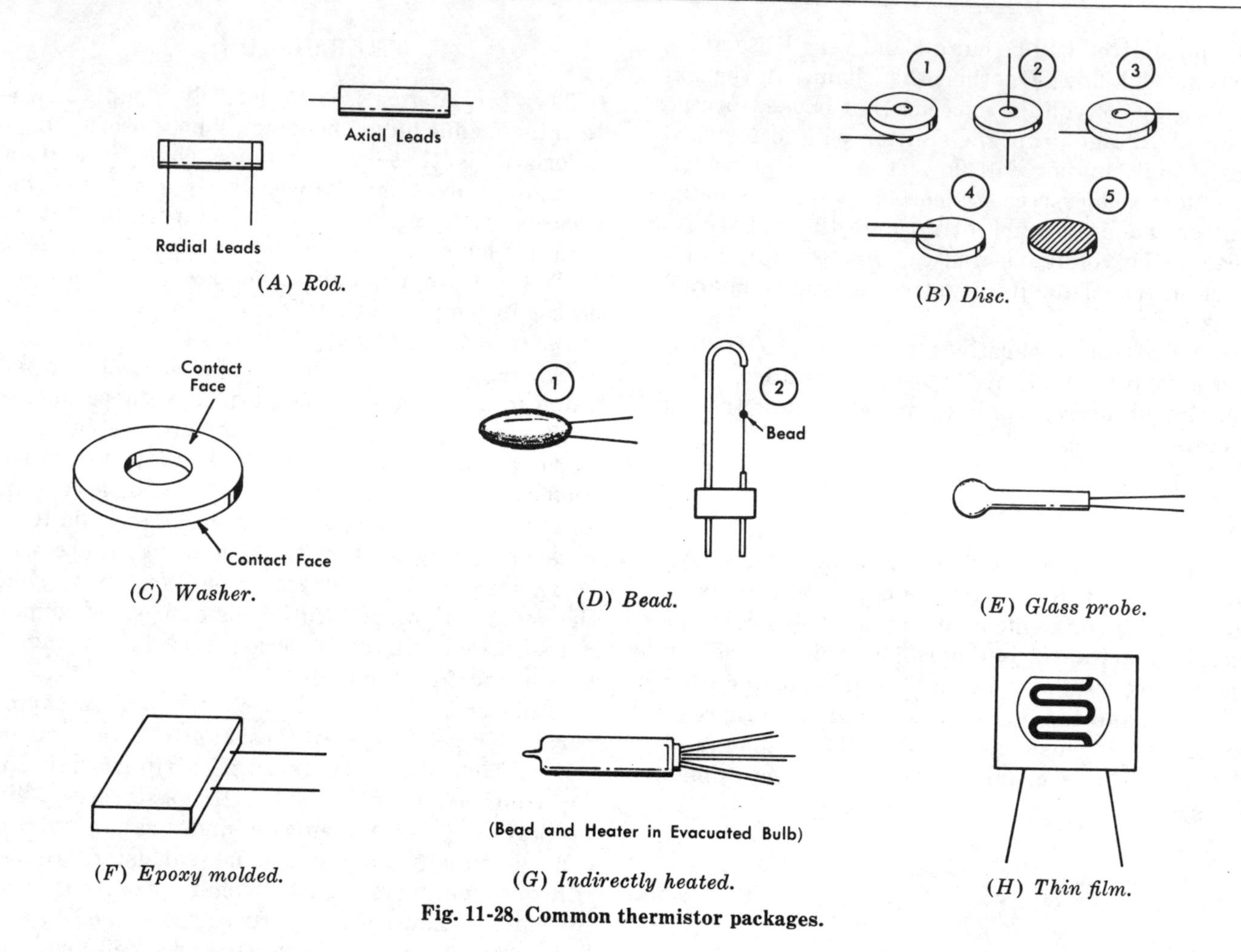

(A) *Rod.* (B) *Disc.* (C) *Washer.* (D) *Bead.* (E) *Glass probe.* (F) *Epoxy molded.* (G) *Indirectly heated.* (H) *Thin film.*

Fig. 11-28. Common thermistor packages.

heat is produced principally by an electric heater element built into the thermistor.

Thermistors are built in many shapes and sizes. A number of examples are shown in Fig. 11-28. Some of these are quite tiny—the beads and glass probes, for example, can be made very small, and this is an advantage in helping them to quickly reach the temperature of their surroundings.

One of the most obvious uses for thermistors is in temperature-measuring instruments. In this application, they have the advantages of high sensitivity, small size, simplicity, and ruggedness. They can be used in a variety of circuits, from a simple 1-thermistor series circuit to a 2-thermistor, 4-arm bridge circuit. They are easily adapted for remote indications.

Other uses, when combined with appropriate electronic circuitry, are dc meter compensation, microwave power measurement, liquid-level sensing, flow meters, vacuum gauges, wind gauges, and altimeters.

VOLTAGE-DEPENDENT RESISTORS (VDRs)

Conventional resistors and many metallic conductors are said to have a *linear* resistance. This means that if a potential of 1 volt is applied to such a resistor a certain current will flow—say, 1 milliampere. Further, it also means that as the potential is doubled or tripled, the current also doubles or triples. All this is in accordance with Ohm's law, $I = E/R$, where R is a constant. In this example, R is equal to 1000 ohms, and this is true no matter what voltage may be applied. A graph of current and voltage for this resistor looks like curve A in Fig. 11-29. The "curve" is a straight line. That is why this resistance is called "linear."

Another type of resistance is called *nonlinear* because its current graph is not a straight line but is curved, as is curve B in Fig. 11-29. In curve B, the value of current in milliamperes at any point is equal to the square of the value of applied volt-

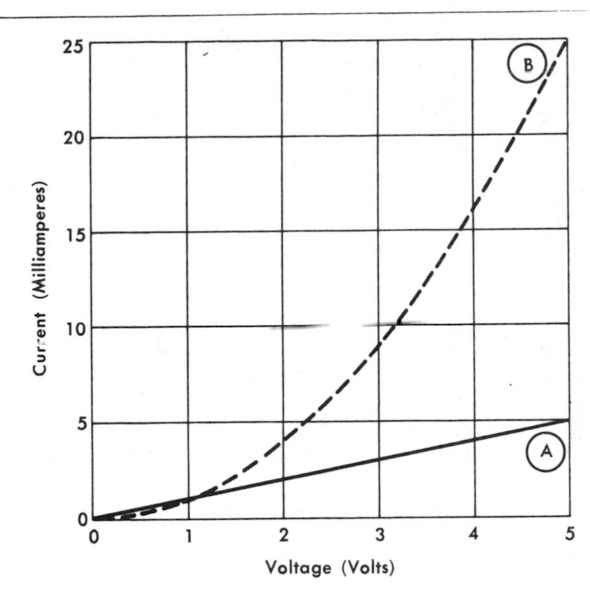

Fig. 11-29. E-I response curves for resistors.

age at that point. Thus at $E = 1$ volt, $I = 1^2 = 1$; at $E = 2$ volts, $I = 2^2 = 4$; at $E = 3$ volts, $I = 3^2 = 9$; etc. This is just one of the many voltage/current characteristics you might find in a study of nonlinear resistance.

If you were to ask "What is the value of the resistor represented by curve B in Fig. 11-29?" we would have to say "That depends." It depends on the applied voltage. In other words, resistor B is a *voltage-dependent resistance.*

As they have done with the varactor and the thermistor, design engineers have taken a normally undesirable characteristic, in this case nonlinear resistance, and emphasized it to the point where it becomes useful in certain electronic applications. In one commercial voltage-dependent resistor, for example, nonlinearity has been increased so much that doubling the voltage increases the current 30 times.

In a nonlinear resistance, the nonlinearity can be one of two types. The type discussed up to this point is that of decreasing resistance as the voltage is increased. This type applies to commercial voltage-dependent resistors (VDRs), forward-biased semiconductor diodes, and positive-plate vacuum-tube diodes. The second type is that of increasing resistance with increasing voltage. This type includes tungsten-filament lamps, ballast-regulating tubes, miniature fuses, and barretters.

VDRs are made from specially processed silicon carbide (carborundum), zinc oxide, or ceramic mixtures. One other type of VDR is made from a pair of matched silicon diodes connected in parallel, cathodes to anodes. Besides the term VDR, some other names by which they are known are: *Globar resistor, Stabistor, Thyrite resistor, Varistite resistor,* and *varistor.* Two of the symbols used for the VDR are shown in Fig. 11-30, and some of the shapes in which it is constructed appear in Fig. 11-31.

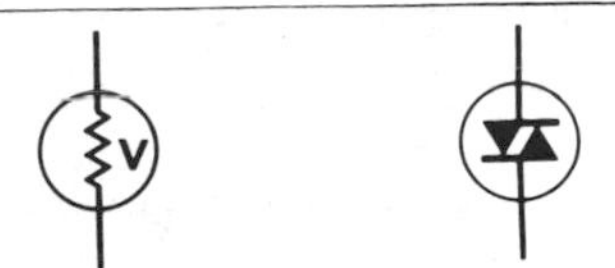

Fig. 11-30. VDR symbols.

The VDR finds use in voltage regulators, voltage dividers, component protection, surge protection, signal expansion and compression, frequency multiplying, filtering, and many other applications.

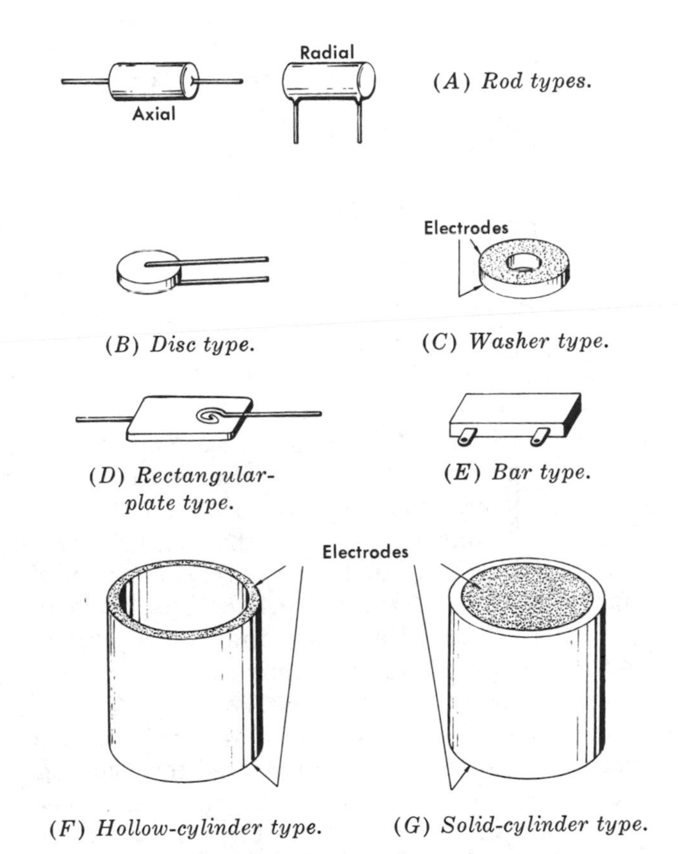

Fig. 11-31. VDR configurations.

INTEGRATED CIRCUITS (ICs)

The construction methods developed in manufacturing diodes, transistors, FETs, MOSFETs, and other semiconductor devices have made it possible to assemble an entire working circuit of transistors, diodes, capacitors, and resistors, all

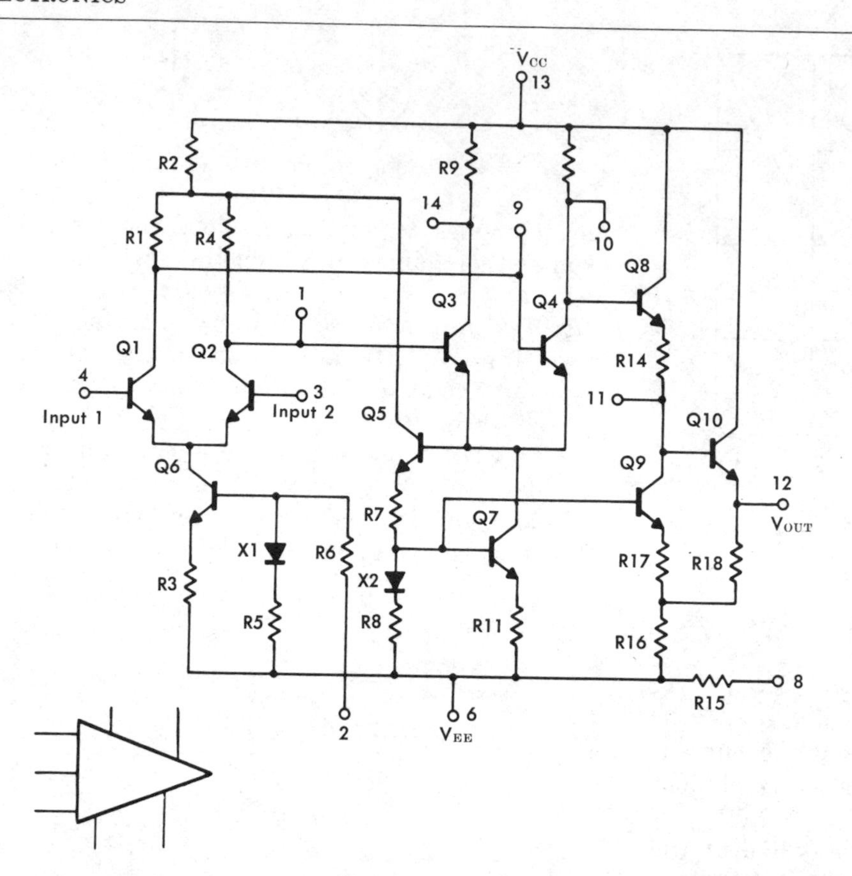

Fig. 11-32. Typical integrated circuit.

formed within a single, tiny chip of semiconductor material. The whole assembly is called an *integrated circuit* or *IC*. A typical integrated circuit and its symbol are shown in Fig. 11-32. This IC contains ten transistors, two diodes, and sixteen resistors, but many more components than this can be assembled in one IC.

Transistors and diodes are fairly easy to produce on an IC chip, but resistors, capacitors, and inductors are much harder to produce, particularly large values of capacitance and inductance. Therefore the trend in ICs is to eliminate resistors, capacitors, and inductors (passive components) wherever possible. Complementary transistors are sometimes used to eliminate the need for coupling devices. An idea of the small size of an integrated-circuit chip can be gained from Fig. 11-33. This is a magnification of about 30 times. This IC contains sixteen transistors, sixteen diodes, and twenty-two resistors.

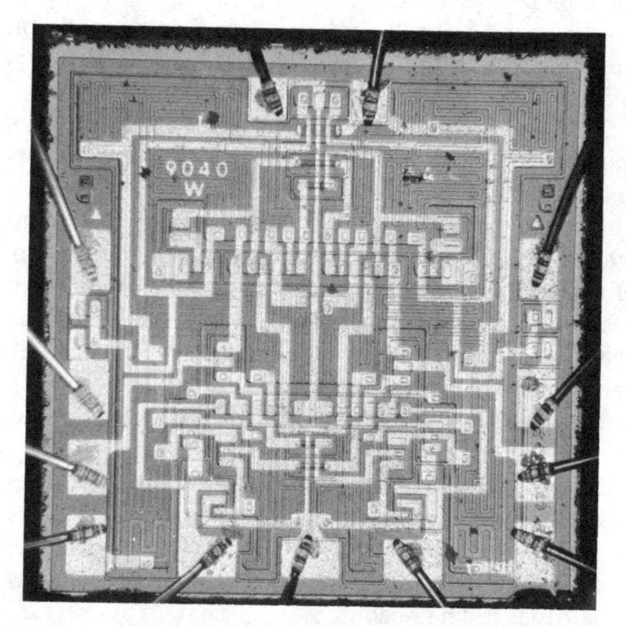

Courtesy Fairchild Semiconductor

Fig. 11-33. A monolith IC chip magnified approximately 30 times.

Digital integrated circuits (flip-flops, gates, inverters, etc.) have been much easier to design and fabricate than linear circuits. The former operate in a discontinuous manner, whereas the latter have a continuous output that must be directly proportional to the input. This has presented several difficulties to manufacturers of integrated circuits. For one thing, they have had trouble fabricating matched pnp and npn elements. It is different to control resistance values precisely.

With earlier ICs, the level of signal that could be handled was low, but that shortcoming is being overcome.

The possible applications of integrated circuits are so great that we can mention only a few of them here. Digital ICs can be used in counting, gating, and frequency multiplying or dividing circuits. This makes them valuable for use in computers and electronic calculators. Linear ICs can perform just about every function now performed by discrete semiconductors or vacuum tubes in combination with other passive components to modify either ac or dc signals, whether continuous or intermittent.

Summary

Semiconductors show many of the properties of insulators and conductors.

The current flow in semiconductors is due to the electrons and the holes.

The electrons in the outer valence ring of the element are the valence electrons.

If two atoms share valence electrons, the atoms are covalent.

N-type semiconductor materials have free electrons, which are their majority current carriers.

P-type semiconductor materials have holes, which are their majority current carriers.

Junction diodes act as circuit rectifiers. They pass a current in one direction, but act as an insulator when the current flow is reversed.

The current flow through a junction transistor (npn or pnp) is controlled by the *bias* of the emitter-to-base junction.

Junction transistors are low-impedance devices; field-effect transistors are high-impedance devices.

The semiconductor controlled rectifier (SCR) can be turned on by varying the gate voltage, but once it is turned on, the gate no longer controls it. To turn the SCR off, the supply current must be interrupted or reversed.

Zener diodes have useful voltage-regulating properties when voltage is applied in reverse-bias polarity.

Tunnel diodes require very little power to operate. They exhibit negative-resistance characteristics and can operate at extremely high frequencies.

Unijunction transistors also have negative-resistance characteristics and are useful in generators and timing circuits.

Varactors are semiconductor diodes especially designed to vary their capacitance as applied voltage is varied.

The thermistor is a semiconductor device whose resistance changes greatly with temperature change.

The voltage-dependent resistor (VDR) is designed to provide nonlinear resistance.

Questions and Problems

12. Sketch the schematic symbols for both an npn and a pnp junction transistor.
13. Name the elements of both an npn and a pnp transistor. Compare the purpose of these elements with those of a triode vacuum tube.
14. In which type of semiconductor are holes the majority current carrier?
15. Sketch the symbol for a junction diode. Indicate the direction of current flow.
16. Explain the terms *valence electrons* and *covalence.*
17. What basic materials are usually used in semi-conductors?
18. What polarity should the base of an npn transistor be with respect to the emitter for collector current to flow?

19. What is the name of the manufacturing step in which an impurity is added to a basic semiconductor material?

20. List two principal advantages of a transistor over a vacuum tube.

21. Which more closely resembles the vacuum tube in operating characteristics, the junction transistor or the field-effect transistor?

22. What characteristics of the SCR make it a good switching device?

23. Is the zener diode usually operated with forward bias or with reverse bias?

24. If the resistance of a conductor increases with increasing temperature, is its temperature coefficient positive or negative?

RECTIFICATION AND DETECTION

Rectification and detection are both methods of changing an alternating current into a pulsating direct current. Rectification is a term applied to the conversion of ac into dc and includes the filtering necessary to smooth the pulsations. Detection is similar to rectification except that selective filtering is used in order to retain a desired component of the pulsating dc.

The rectifier allows current to flow in one direction, but blocks it in the other direction. In the vacuum-tube diode the electrons can flow from the cathode to the anode (plate) only. They cannot go in the opposite direction.

Solid-state devices, such as selenium, germanium, and silicon rectifiers, are usually represented in a schematic by the symbol shown in Fig. 11-34A. The *bar* is the cathode, and the *arrow* is the anode. Such a solid-state rectifier can be used for the same purpose as a vacuum-tube diode.

Half-Wave Rectification

Fig. 11-34B shows a transformer with a diode connected across its output terminals. The electrons in the transformer secondary coil will reverse their direction during each cycle, as indicated by the arrows. During half of the ac cycle (Fig. 11-34C), the plate is negative, and the tube does not conduct. In Fig. 11-34D the electron flow is reversed; the plate becomes positive, and the diode conducts. Fig. 11-34E shows a capacitor in series with the diode. Once during each cycle, the diode conducts, and electrons are transferred from one capacitor plate, through the diode, to the other plate. The diode will not conduct in the opposite direction. Therefore, the capacitor will be charged with the polarity shown. The charge across the capacitor in Fig. 11-34E can be used as a supply source of dc. The action of the transformer and the diode renew the charge as it is drained off. The rectifier circuit in Fig. 11-34E is a *half-wave rectifier* since only half of the alternating cycle is used.

Full-Wave Rectification

The *full-wave rectifier* circuit permits current flow and rectification during both the negative and positive phases of the ac cycle. The most familiar type of full-wave rectifier is shown in Fig. 11-35.

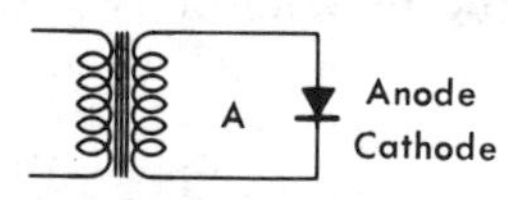

(*A*) *Semiconductor diode rectifier showing cathode and anode.*

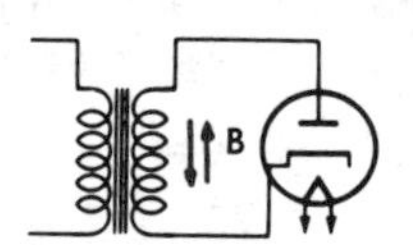

(*B*) *Vicuum-tube circuit showing ac voltage in transformer secondary.*

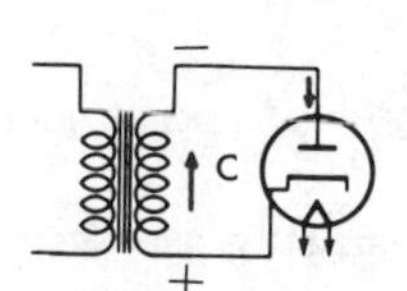

(*C*) *Nonconducting half of cycle.*

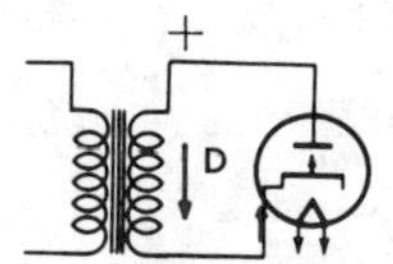

(*D*) *Conducting half of cycle.*

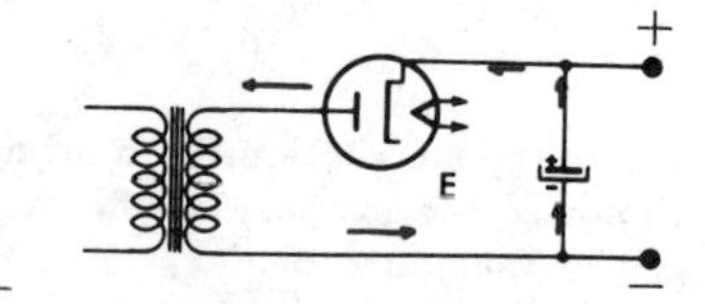

(*E*) *Rectified current charges capacitor.*

Fig. 11-34. Simple rectifier circuits.

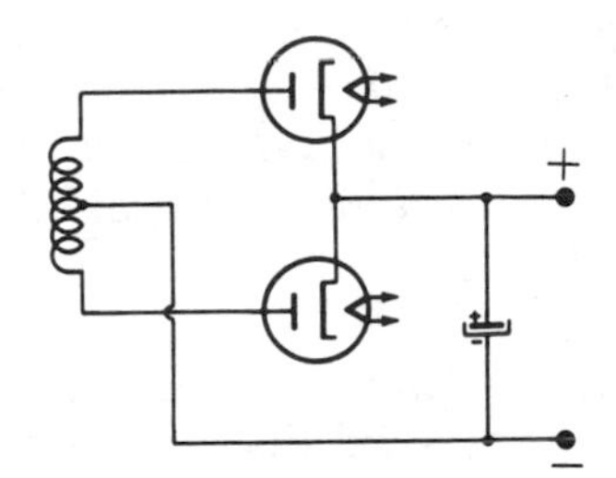

Fig. 11-35. A full-wave rectifier circuit.

This is the same circuit arrangement as a circuit with two half-wave rectifiers in which the two rectifiers alternately charge the same capacitor. One rectifier conducts on one half of the cycle, and the other rectifier conducts on the other half cycle.

Full-Wave Bridge

The full wave *bridge* is a full-wave form of rectifier circuit which does not require that the voltage source be center-tapped (Fig. 11-36). The positive leads of two diodes and the negative leads of two other diodes are connected together to give us the bridge rectifier. The solid arrows show the current path on one-half cycle and the dashed arrows show the flow on the other half cycle.

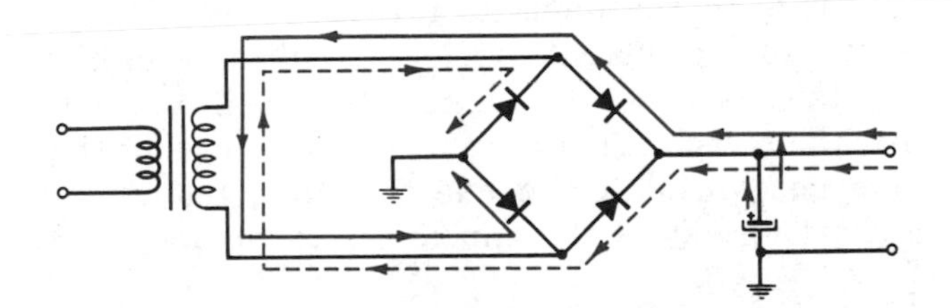

Fig. 11-36. A bridge rectifier circuit.

Voltage Doubler

Two types of *voltage-doubler* circuits are used extensively. These are shown in Figs. 11-37 and 11-38. The circuit in Fig. 11-37 is a *half-wave* doubler.

On one half of the ac cycle, the electrons flow through diode V1 and charge series capacitor C1, as shown by the solid arrow. Capacitor C1 is charged to the peak voltage of the ac line. On the next half of the ac cycle, this capacitor is discharged through diode V2 and charges load capacitor C2. During the following ac alternations, the charge on capacitor C1 is added to the line voltage and charges the load capacitor. The load

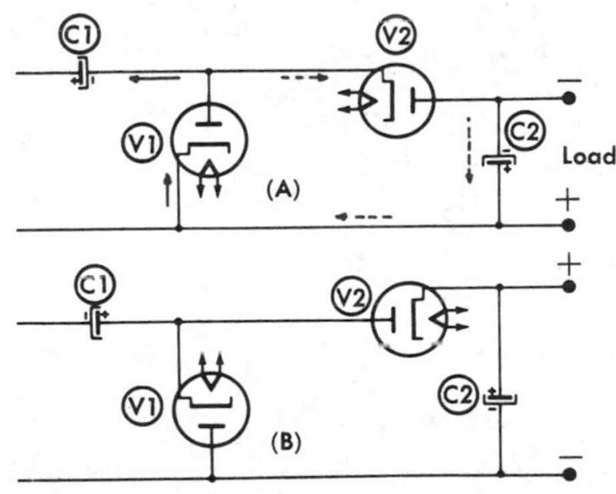

Fig. 11-37. Half-wave doubler circuits.

capacitor then has a voltage equal to the charge of capacitor C1, plus the peak of the line voltage. This charge is equal to twice the peak line voltage. The diodes in Fig. 11-37A are arranged so that the negative voltage is above the ground. Reversing the capacitors and the diodes, as shown in the circuit in Fig. 11-37B, places the positive voltage above the ground. The circuit in Fig. 11-37B is the conventional circuit, since a positive voltage is used for most vacuum-tube applications.

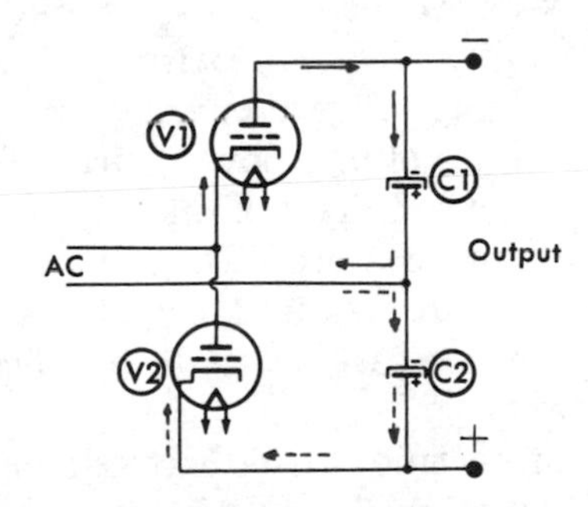

Fig. 11-38. A full-wave doubler circuit.

The *full-wave* doubler shown in Fig. 11-38 is basically a bridge rectifier. It can also be thought of as two half-wave rectifiers arranged with the output voltages in series. Current flow in one direction charges capacitor C1 through diode V1, as shown by the solid arrow. Capacitor C2 is charged by diode V2 when the current flow is reversed. Capacitors C1 and C2 are both charged to the peak voltage of the applied ac. Since the two capacitors are in series, the voltage across the output terminals is equal to the voltage of C1 plus the voltage of C2, or twice the peak voltage of the input.

Detection and Filtering

A detector circuit in a radio receiver separates the high-frequency *carrier* signal from the lower-

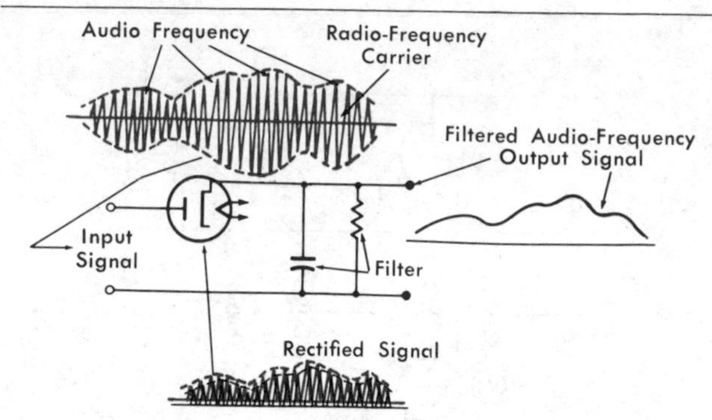

Fig. 11-39. Simple radio-frequency detector circuit.

frequency *audio* signal. More will be said about radio circuits in a later section. All we now need to know is that the higher radio frequencies only *carry* (transmit) the lower audio frequencies. The two frequencies are separated in a receiver by detection. A simple detector circuit containing a diode is shown in Fig. 11-39.

The diode rectifies the radio signal, as shown. The *filter* system, composed of a capacitor and a resistor, then eliminates the higher frequency. This leaves only the audible sound frequencies.

A rectifier—as we have said—permits current to *flow in only one direction.* Alternating current can be changed into direct current in a number of ways. In a high-voltage power circuit, a rotary converter can be used. This is merely a dc generator driven by an ac motor. Dc is produced here by *electromechanical* means. Most of the rectifiers we will consider convert ac to dc by *electrical* or *electronic* means.

A number of rectifiers used extensively in electronics are shown in Fig. 11-40.

Liquid Rectifiers

The liquid rectifier makes use of the fact that certain metals and electrolytes, when in contact with each other, pass current in only one direction. One such rectifier is made of aluminum immersed in a solution of ammonium sulfate. The liquid rectifier was one of the earliest types used.

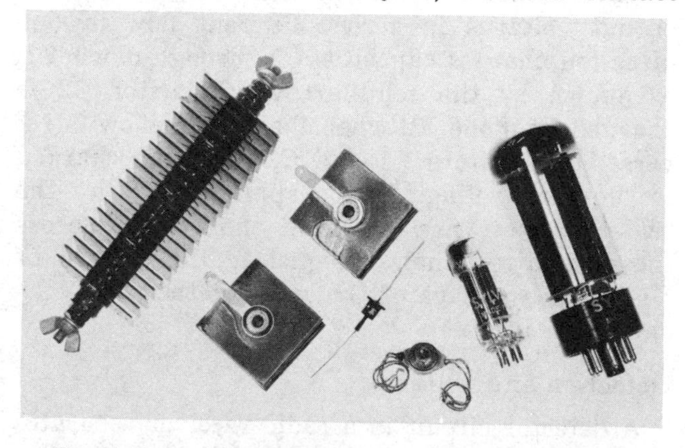

Fig. 11-40. Typical rectifiers used in electronic circuits.

Disc Rectifiers

The *disc rectifier* is one of the more important types. It is made of solid discs or plates of different types of material. The discs are pressed together or are bolted or clamped together.

One of the *dry-disc* rectifiers is the copper-oxide type. It is made of pure copper with an oxide layer formed on the surface of the copper.

Selenium rectifiers are more efficient than the copper-oxide type. They are made by placing a thin coating of selenium on iron. Electrons move easily from the selenium to the iron, but not from the iron to the selenium.

Diffused Type (P and N)

The germanium or silicon diffused rectifier is very efficient. The junction of p-type and n-type material is formed by the *fusion* of the two types of material. (The p and n junction of semiconductors was explained at an earlier point.) The current flow in one direction is *added* to the carriers, lowering the junction resistance. A current flow in the opposite direction causes a *depletion* of the current carriers, resulting in a high resistance across the junction. In the silicon rectifier, the resistance of the junction approaches that of a good insulator in one direction, but almost a short circuit in the opposite direction.

Technical Considerations

Rectifiers and diodes are basically alike. They pass current in only one direction, but not the same amount of current. Nor can they withstand the same amount of voltage or dissipate the same amount of heat. All of these characteristics determine the use to which a diode or rectifier can be put. For example, a *selenium* rectifier will break down if more than 18 to 20 volts per junction is applied in the reverse direction. However, certain *silicon* rectifiers have an inverse voltage rating of as much as 600 volts per junction.

Power rectifiers can conduct large currents, whereas many diodes used as radio-frequency detectors can conduct only a few milliamperes.

Operating temperatures are also important to the operation of detectors and rectifiers. Extremes of heat or cold may cause them to fail completely.

Summary

Rectification is the process of changing an alternating current into a direct current.

Detection is the process of changing an alternating current into a pulsating direct current and retaining a desired component of the pulsating dc by selective filtering.

The half-wave and full-wave rectifiers make up the two basic types of useful rectifier circuits.

A rectifier permits current to flow in one direction only.

Questions and Problems

25. What is a half-wave rectifier?
26. In which direction do the electrons flow in a vacuum-tube diode?
27. Draw a sketch to show the polarity of the voltage on the anode of a rectifier (with respect to its cathode) when the rectifier is conducting electrons. When it is not conducting electrons.
28. Sketch a simple half-wave rectifier circuit in which a vacuum-tube rectifier is used. Also show the direction of electron flow on your sketch.
29. Draw the schematic symbols for a vacuum-tube diode. For a solid-state diode. Label the anode and the cathode of each diode.
30. List the two common types of dry-disc rectifiers.
31. Sketch a full-wave rectifier circuit that includes a center-tapped transformer and a vacuum tube.
32. Sketch a half-wave rectifier circuit (similar to Question 28) with a selenium rectifier.
33. Explain the difference between *rectification* and *detection.*

AMPLIFICATION

The term *amplification,* to most persons, means an increase in the level of a sound signal. However, the idea of amplification is not limited to sound signals. For example, a certain step-up transformer will increase 120 volts of ac to 500 volts. This is a *voltage* amplification. On the other hand, the bell transformer in a home reduces 120 volts to 24 volts, but increases the current almost five times. This is *current* amplification. In both cases, the *power* in the secondary coil of the transformer is always less than the input because of losses in the transformer.

The point to remember is that the term *amplification* denotes a difference—but not always an increase—between the input and the output. Let's consider an amplification of one-half or .5 times. This is an amplification of less than *one,* or actually a *loss* between input and output. But this, too, is amplification. Such gains of less than one are sought in certain circuits.

The vacuum tube and the semiconductor are used to produce and control amplification. Their action can be compared to that of a variable resistor with its center arm connected to one end of the resistor, as shown in Fig. 11-41A. The center arm is the *control,* just as the grid is the control of a vacuum tube and the *base* is the control of a transistor. In this instance, however, the control is mechanical, in the same way a valve in a hydraulic system is a mechanical control.

The variable resistor is not an amplifier; it is only one component of the amplification system. A number of components that might be referred to as amplifiers are shown in Fig. 11-41.

Mechanical control changes the setting of the resistor or of the hydraulic valve shown in Fig. 11-41A. The resistor controls an electrical current flow; the valve, a fluid flow.

Heat control can be accomplished by a furnace thermostat or the automatic sprinkler head shown

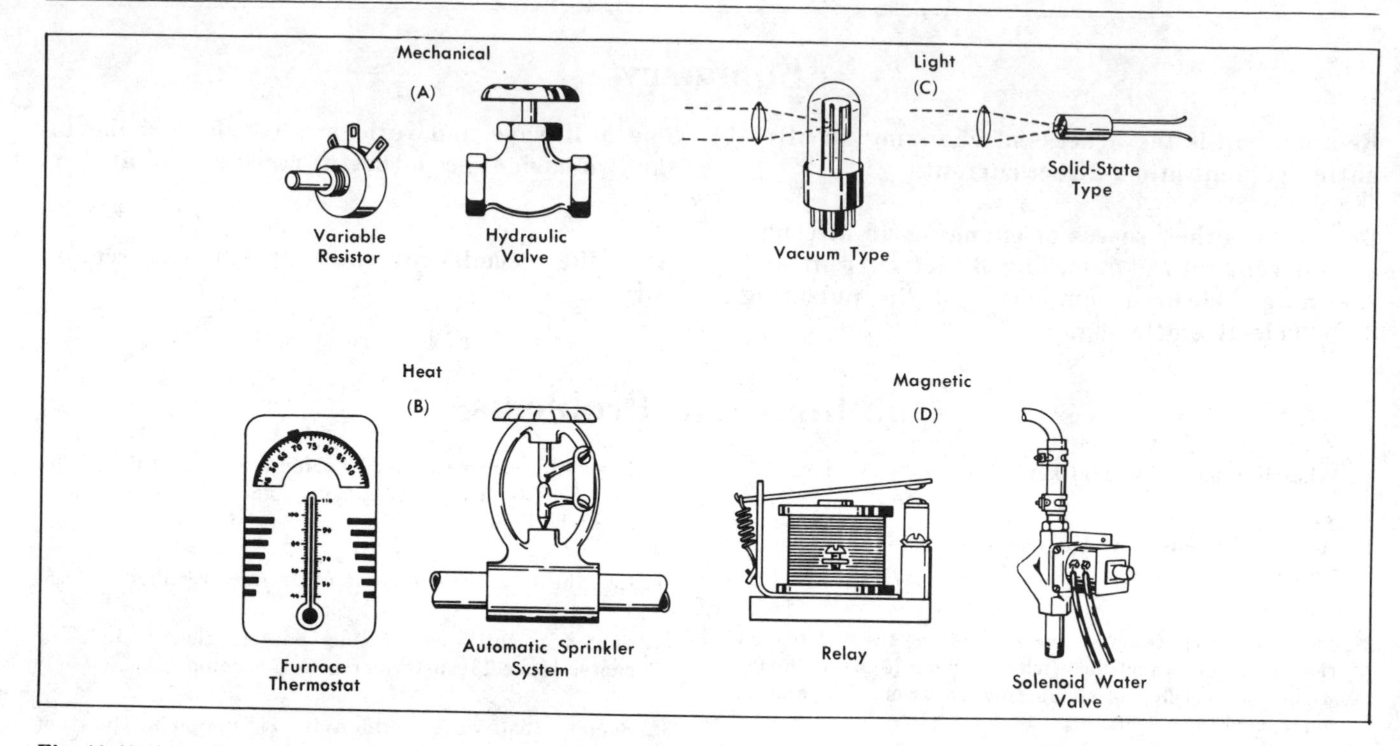

Fig. 11-41. A mechanical control, a method of controlling heat, a method of controlling light, and a magnetic control device.

in Fig. 11-41B. The furnace control uses heat to control the flow of an electrical current. The sprinkler head uses heat to control the flow of a stream of water.

When we think of light as a control, we usually associate it with the photoelectric cell. More recently light control has been associated with small solid-state photosensitive devices, like the ones in Fig. 11-41C.

Electrical power can be controlled magnetically by means of relays. Water pressure can be controlled with a *solenoid* water valve (Fig. 11-41D).

In each of these examples, heat, magnetism, light, or a mechanical force is indirectly applied (through a device) to accomplish a task. This is similar to using a pry bar to move a heavy object.

The vacuum tube and the transistor produce amplification by changing their resistance values, in a manner similar to the variable resistor (potentiometer) in Fig. 11-41A. We can show the action of an amplifier in which a variable resistor is used, by connecting a potentiometer as shown in Fig. 11-42A.

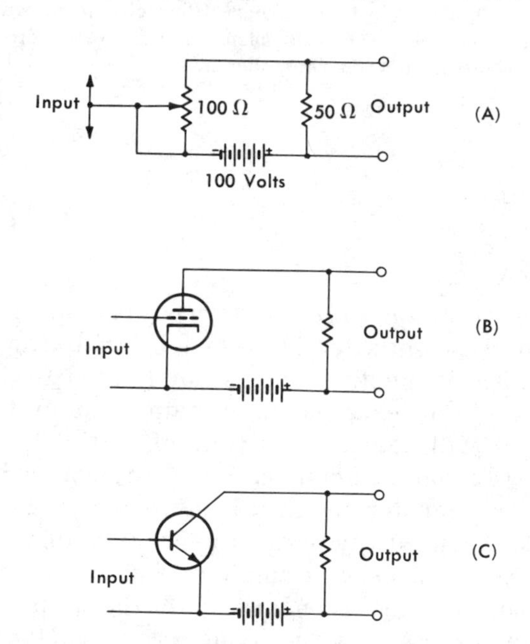

Fig. 11-42. Achieving amplification by using a potentiometer, a vacuum tube, and a transistor.

The control arm of the potentiometer is first positioned at the center for a resistance of 50 ohms. The battery is then connected across a total resistance of 100 ohms. One ampere of current flows. If the potentiometer resistance is reduced to zero, the current increases to 2 amperes. But, if the potentiometer resistance is increased to 100 ohms, the current will only be 0.66 of an ampere. The voltage across the 50-ohm output resistor is changed from 50 ohms times 0.66 ampere to 50 ohms times 2 amperes (from 33 volts to 100 volts). The change in the resistance of the potentiometer produced a corresponding voltage change across the output resistance.

A vacuum tube (Fig. 11-42B) or a transistor (Fig. 11-42C) can be substituted for the potentiometer in Fig. 11-42A to produce similar results. When a signal voltage is applied to the grid of the vacuum tube or to the base of the transistor, a corresponding voltage change occurs across the output resistors.

The Vacuum-Tube Amplifier

A triode vacuum tube consists of a cathode, a grid, and a plate. When electrons "boil off" the cathode, a *space charge* is produced around the cathode. This effect is shown in Fig. 11-43A by the *negative* signs arranged around the cathode. This space charge remains rather constant and stationary unless a voltage of the proper polarity is applied to the plate.

When a positive voltage is applied to the plate, the electrons leave the space-charge area and move toward the plate. There is little resistance to this movement, and the resistance of the *triode* (cathode to plate) is low (See Fig. 11-43B).

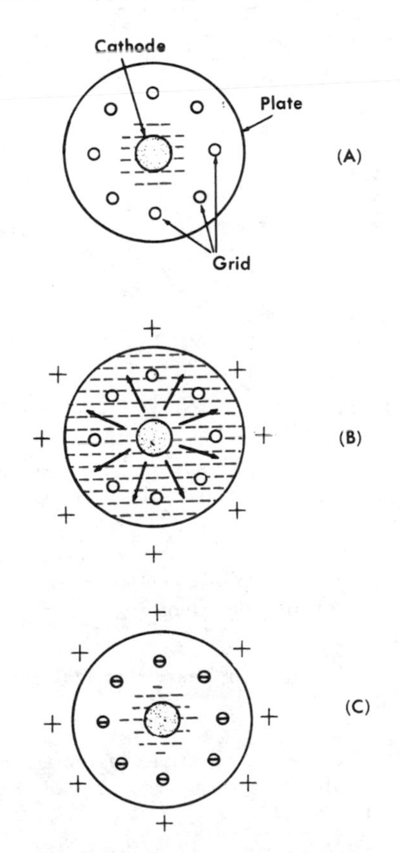

Fig. 11-43. The principle of a vacuum tube.

A high resistance is produced between the cathode and the plate when a negative voltage is applied to the grid (Fig. 11-43C). The electrons are repelled by the charge on the grid and cannot pass the grid to reach the plate.

A change in the amount of charge on the grid permits a varying number of electrons to pass to the plate. Such an action increases or decreases the cathode-to-plate resistance.

It is quite practical to apply 200 volts to the plate of a vacuum tube and to control it from maximum to minimum current with a voltage change of only two or three volts on the grid. Hence, you can easily realize that very high gains in voltage, current, and power are possible from such a vacuum tube.

The Transistor Amplifier

The transistor is a *solid-state* device. Its control action takes place within a solid material. The material (usually germanium or silicon) is made of three doped regions composed of p or n material. We have already learned that these regions are separated by two extremely thin junctions created by the fusion (melting together) of the two types of material.

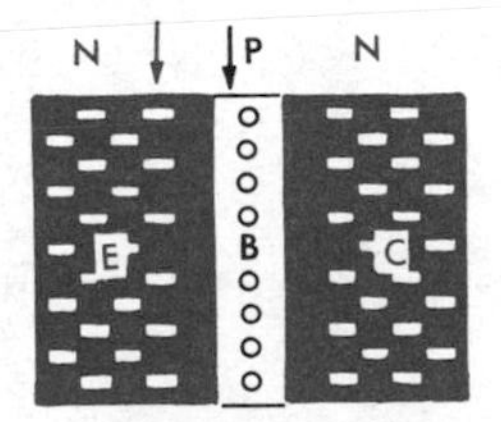

Fig. 11-44. The position of the current carriers in a transistor.

Such a semiconductor can change its resistance between the *emitter* and the *collector* whenever a current is allowed to flow between the *emitter* and the base.

The normal positions of the current carriers in a transistor are shown in Fig. 11-44. The carriers are positioned back from the junction, and an external potential is required to move them up to the junction. We must be able to change the resistance of this unit if we are to use it as an amplifier. Let's observe what takes place when the base and the emitter are shorted together and a voltage of proper polarity is applied from the emitter to the collector. In Fig. 11-45 the battery has caused the electrons to move away from the base region. Only a small current is flowing in the

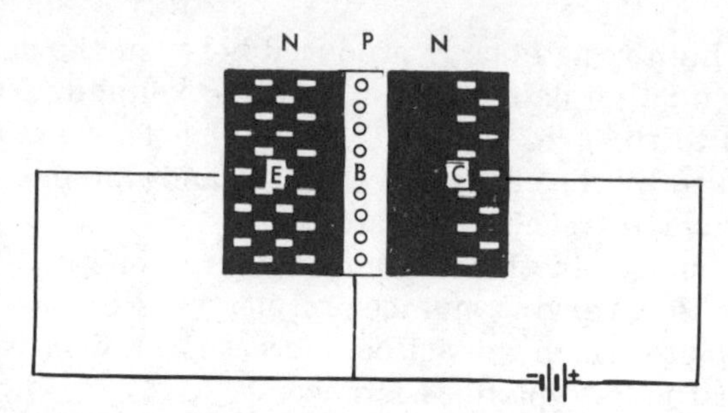

Fig. 11-45. The position of carriers with voltage applied.

circuit. The resistance from the emitter to the collector is extremely *high*.

In Fig. 11-46 a battery is used to place a bias from the base to the emitter. This causes carriers to flow into the base region. These carriers are attracted to the collector, and a current flows between the emitter and the collector. The resistance between the emitter and collector is greatly reduced.

A current flow between the emitter and the base changes the resistance of the transistor to permit amplification. The emitter-to-collector resistance of a transistor can be changed from a few ohms to thousands of ohms by merely changing the amount of current flowing between the emitter and the base.

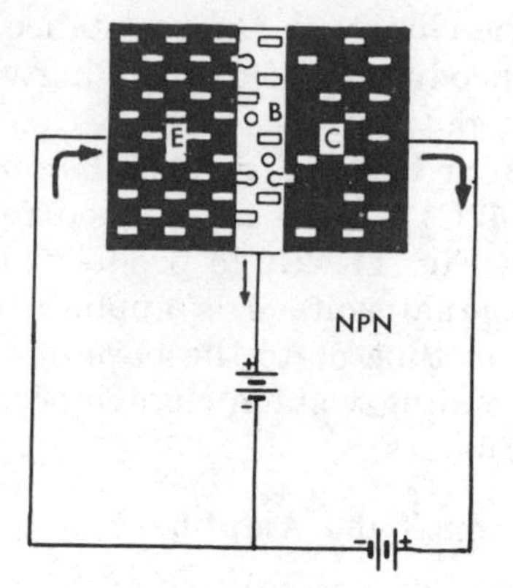

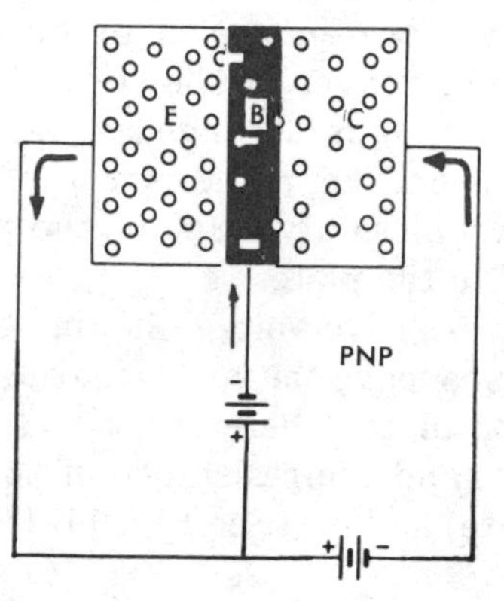

Fig. 11-46. Transistor biased for use as an amplifier.

Summary

Amplification is the ratio of output to input in an amplifier.

A vacuum tube amplifies by controlling a stream of electrons between its cathode and plate.

A transistor amplifies by controlling the flow of carriers from its emitter to its collector.

Questions and Problems

34. Define the term amplification.

35. What polarity of grid voltage is required to halt the flow of electrons in a vacuum tube?

36. What property of a vacuum tube or transistor makes it useful as an amplifier?

37. Which element of a transistor is the input, or control, element?

38. Is the resistance of a vacuum tube with zero grid bias relatively high or low? Explain.

39. Is the resistance of a transistor with a zero bias between its emitter and its base relatively high or low? Explain.

40. Explain the difference between *voltage* amplification and *current* amplification.

41. Will amplification increase the total *power* in a circuit? Explain.

42. Will a decrease in negative bias on the grid of a triode vacuum tube change the resistance of the vacuum tube? If so, how?

ELECTRONIC CIRCUITS

In the early days of radio, batteries were used as the only source of power. There were "A" batteries for filaments, "B" batteries for plate voltage, and "C" batteries for grid bias. Such a radio, with all its batteries, was a large, heavy, clumsy unit. Dry cells (primary cells) had to be replaced often. Secondary cells weren't much better because there was always the problem of recharging them. The great boom in radio came when practical rectifiers became available, so that radios could be operated from ac power lines.

Power Sources

A radio needs power for two basic reasons—filament voltage and plate voltage. Many radio vacuum tubes require 6.3 volts to heat their filaments. Other filament voltages range from approximately .5 volt to 117 volts. This filament voltage can be easily furnished by a step-down transformer connected to an ac source.

Plate-voltage power supplies come in many sizes and they use many different circuits. The most common power-supply circuit consists of a half-wave rectifier connected to an ac line. A typical circuit is shown in Fig. 11-47. One feature of this *ac-dc* circuit is that the sum of all the filament voltages is approximately equal to the 117-volt ac line voltage. The tubes can, therefore, all be connected in series across the line, so that a filament transformer is not needed. (See Fig. 11-47.)

This power-supply circuit will also operate from a dc source if the plug inserted into the outlet is in the proper polarity position. One great disadvantage of this circuit is that the negative side of the circuit is often connected directly to the chassis and to one side of the power plug. If the plug is inserted in the outlet with the negative side of the plug contacting the *hot* side of the ac line, a full 117 volts exists between the chassis of the receiver and *any* grounded object. This, of course, is quite dangerous.

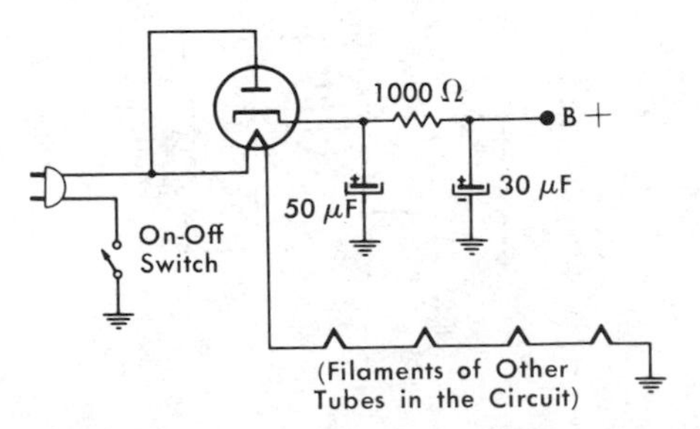

Fig. 11-47. Typical half-wave, ac/dc power supply.

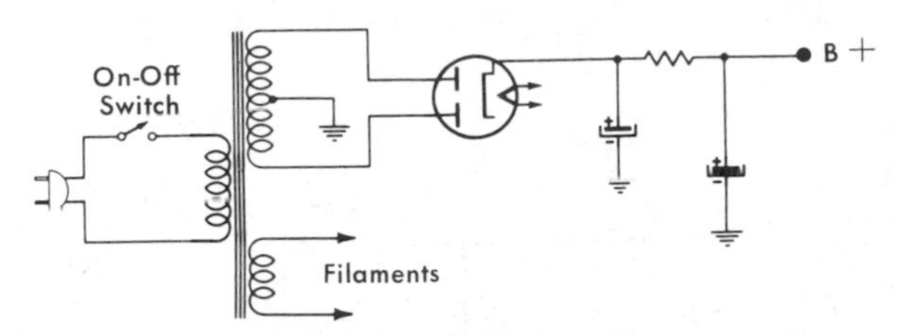

Fig. 11-48. A full-wave power supply.

When plate voltages must be greater than the available line voltage, a step-up transformer is used. Such a circuit is shown in Fig. 11-48. You should be able to recognize this full-wave rectifier circuit. In circuits of this type the filament voltages are often supplied by a low-voltage secondary winding on the same transformer with the high-voltage winding.

In a power supply for solid-state circuits, the output voltage is quite often less than the ac line voltage. Consequently, a step-down transformer is needed, rather than the step-up unit used for vacuum-tube circuits. In addition, the output of the power supply needs to be regulated in many cases to prevent voltage fluctuations from changing the operating point. Three examples of solid-state regulator circuits are shown in Figs. 11-49A, B, and C.

The first example, Fig. 11-49A, is called a series regulator because the regulating element, Q1, is in series with the rectifier output and the load. The power furnished by the output terminals of a power supply may vary for two reasons: changes in power demanded by the load, or fluctuations of voltage in the power source. This circuit responds to changes in load, but not to changes in line voltage. An increase in load demand causes an increase in bias current, reducing the emitter-to-collector resistance of the transistor. The transistor becomes a dynamic resistance in series with the load, maintaining a relatively constant output voltage. The filtering action of C2 is, in effect, multiplied by the beta of the transistor.

Fig. 11-49B shows a very simple application of a zener diode to regulate the output voltage of a

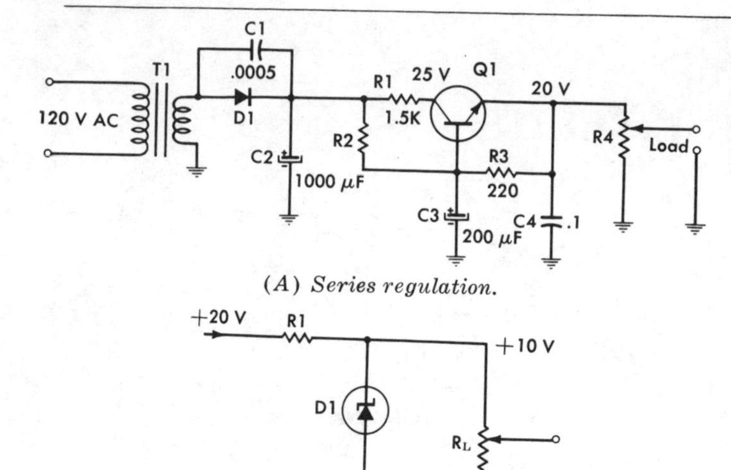

(A) *Series regulation.*

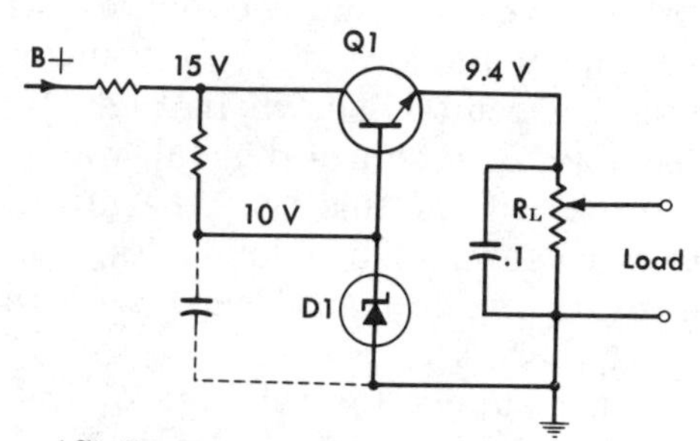

(B) *Simple zener-diode regulator.*

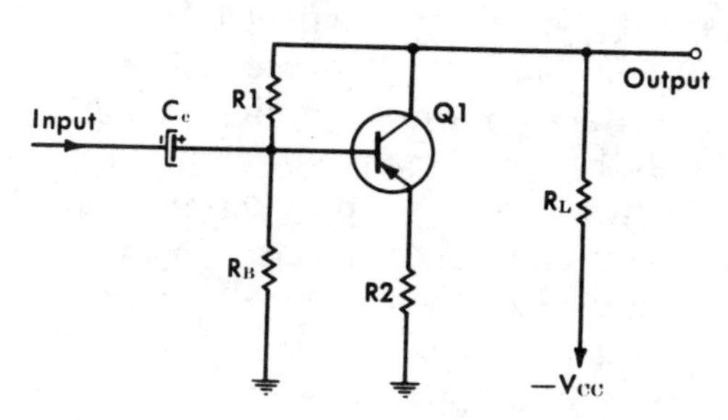

(C) *Improved zener-diode regulator.*

Fig. 11-49. Power-supply regulation.

supply. The zener breakdown point of the diode is 10 volts. The diode and the load, R_L, together draw enough current through R1 to drop 10 volts, leaving 10 volts across the load resistor. If the load demand increases by a certain amount, D1 draws that much less current so that the total through R1 remains unchanged.

The power handling capability of a zener diode can be increased by coupling it to a transistor, as in Fig. 11-49C. The base voltage of the series regulator is fixed by the zener voltage. The emitter voltage will follow the base voltage minus the 0.6-volt diode drop. The effective filtering action of the zener diode is multiplied by the current gain of the transistor. A capacitor can be connected from base to ground (as shown by dotted lines) for extra filtering.

Audio Amplifiers

The circuit in Fig. 11-50 illustrates an elementary audio-frequency amplifier. The input signal to such a circuit is a rather small ac voltage. The vacuum tube amplifies the input signal and produces an amplified version of the signal as an output signal. Such a circuit is known as a *voltage amplifier*. Since this is an audio-frequency amplifier, only those components that will work best at audio frequencies are used.

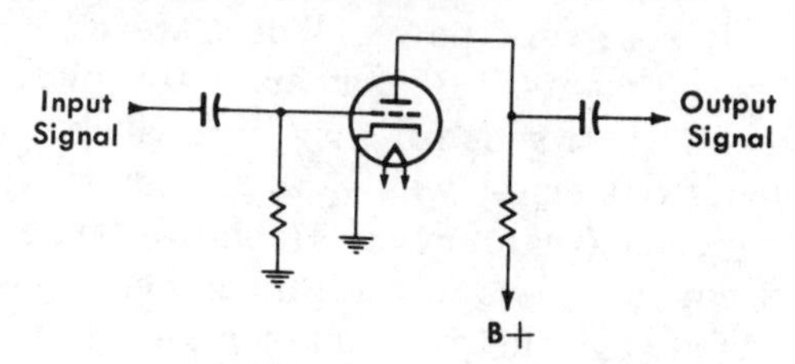

Fig. 11-50. Simple amplifier circuit.

Fig. 11-51 might be considered the transistor version of Fig. 11-50. The collector load resistor, R_L, compares to the plate load resistor, and the base resistor, R_B, compares to the grid resistor in the tube circuit. The coupling capacitor, C_c, is an electrolytic capacitor because larger values of capacitance are needed to match the lower impedance of transistors.

Fig. 11-51. Transistor amplifier stage.

The circuit in Fig. 11-52 represents an *audio-output amplifier* similar to those found in a radio or a television receiver. Again, the input signal is an ac voltage, and the output is an amplified signal. The *input* for this circuit comes from the *output* of a circuit like the one in Fig. 11-50.

Notice that the output of this amplifier tube enters a transformer. The primary coil of this

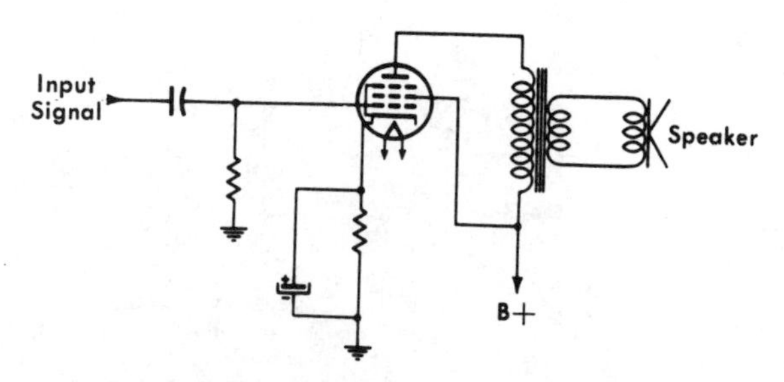

Fig. 11-52. Audio-output amplifier.

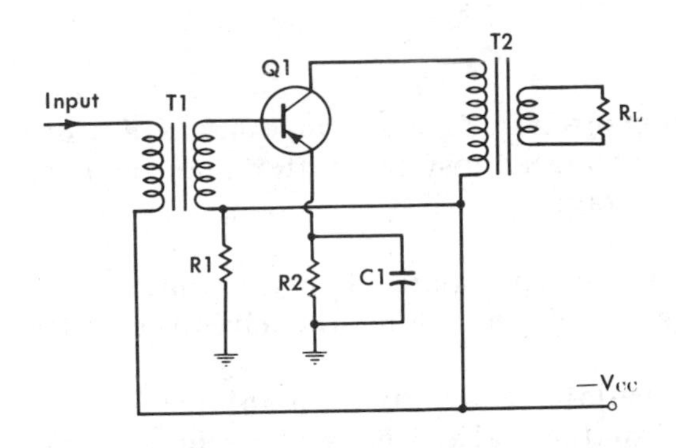

Fig. 11-53. Transistor audio-output amplifier.

transformer is *matched* to the tube output, which is a high-voltage, low-current ac signal voltage.

The matching or output transformer has a primary-coil winding of *many* turns of *small-diameter* wire and a secondary-coil winding of a few turns of a larger-diameter wire. The audio-output amplifier tube "drives" the transformer with a high voltage. The transformer changes the high voltage into a low voltage with the high current required to operate the speaker. The power output of this circuit—the tube and transformer—is measured in watts.

The transistor version of Fig. 11-52 is shown in Fig. 11-53. Here, again, the lower impedance of transistors is taken into account, and an input transformer is used to provide a low-impedance input to the transistor.

Radio-Frequency Amplifiers

A radio-frequency amplifier is similar to the audio amplifier described except that it amplifies the high frequencies of a radio signal instead of the low-frequency audio signal. The circuit in Fig. 11-54 is an example of an rf amplifier.

The input circuit of the rf amplifier consists of a tuned transformer which will transfer a signal from the primary coil to the secondary coil only when the signal is of the same frequency to which the transformer is tuned. At any other frequency,

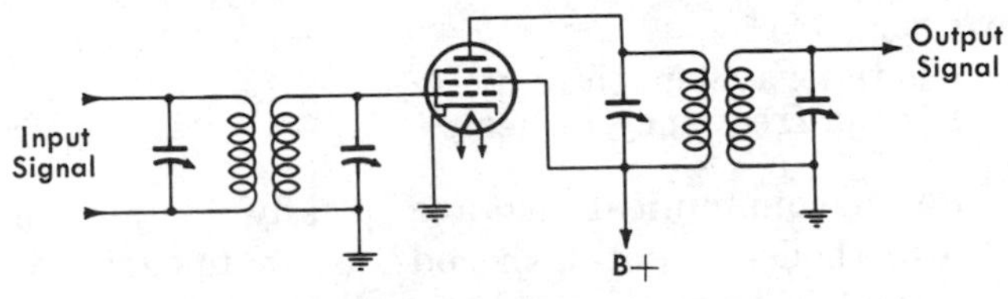

Fig. 11-54. Radio-frequency amplifier.

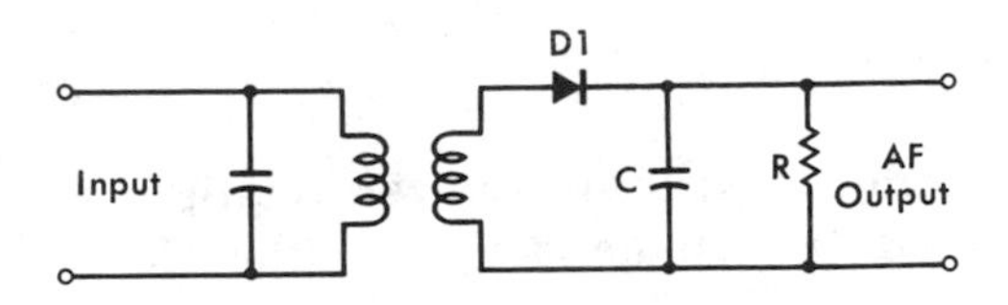

Fig. 11-55. Diode detector circuit.

very little signal will be transferred to the input of the rf-amplifier tube. The transformer in the output circuit is also tuned to a certain frequency and will reject all other frequencies.

Detectors

One of the earliest detectors was a galena crystal with a "cat-whisker" contact. The operator moved the "cat-whisker" around on the surface of the crystal until he found a sensitive position. A galena crystal is like other detectors—it passes current in one direction only.

A simple detector circuit is shown in Fig. 11-55. A semiconductor diode is used as a detector but a vacuum-tube diode could be used, instead, if a filament supply were available.

In the circuit in Fig. 11-55, the transformer circuit selects a radio frequency signal. The diode rectifies the signal, as previously described when detection was discussed.

Oscillators

An oscillator is basically an amplifier which is caused to oscillate by allowing a part of the output signal to re-enter at the input stage in a polarity that aids the input signal. This is called positive feedback. Oscillators can oscillate at almost any frequency if certain components are used in one *feedback* circuit which determines the frequency of oscillation. The feedback circuit in the oscillator shown in Fig. 11-56, is the transformer. The frequency of oscillation is determined by the inductance and capacitance of the transformer.

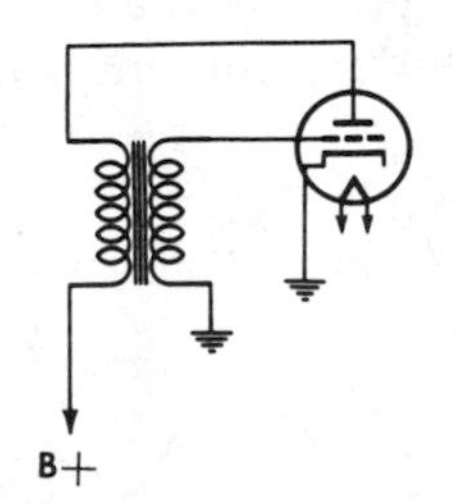

Fig. 11-56. A simple oscillator circuit.

Summary

A vacuum-tube radio receiver requires a power supply for the filament and for the plate.

Vacuum tubes require filament voltage ranging from .5 volt to 117 volts.

The most common plate-voltage power supply is the half-wave rectifier.

A step-up transformer must be used when the plate voltages must be greater than the available line voltage.

The frequency range of any amplifier or oscillator is determined by the circuit components.

An oscillator is simply an amplifier with part of its output signal fed back to its input stage.

Questions and Problems

43. Why were the batteries in early radios replaced by ac power supplies?

44. What filament voltage is commonly used for vacuum tubes?

45. What can change an amplifier into an oscillator?

46. Is it necessary to have a rectifier in an ac-dc receiver connected to a 110-volt dc power line? Explain.

47. Why is an output transformer used in an audio-output amplifier circuit?

48. Could an ac-dc power supply be dangerous? Explain.

49. Does an output transformer of an audio-output amplifier circuit step up or step down the voltage? Why?

50. What is the difference between an audio-frequency amplifier and a radio-frequency amplifier?

WIRELESS COMMUNICATIONS

RF Transmission

High-frequency alternating currents carry the messages sent by wireless transmitting stations. These currents produce both *electrostatic* and *electromagnetic* fields that radiate in all directions from a transmitting antenna at the speed of light (186,000 miles per second), traveling many miles without losing much of their energy. These electrostatic and electromagnetic fields are known as *radio waves.* When a broadcast receiver is tuned to the same frequency as the alternating currents entering the antenna, communication is possible.

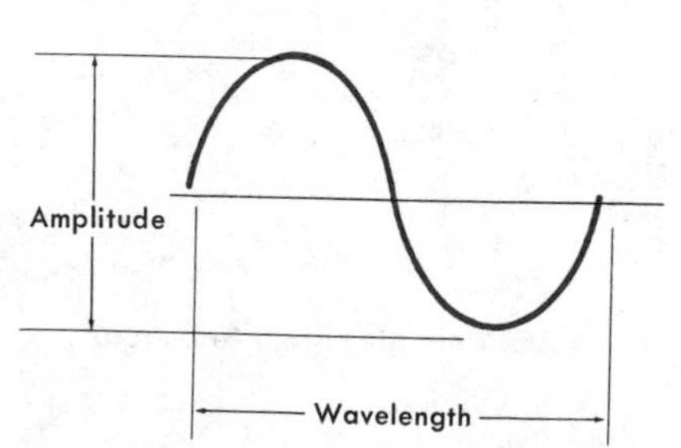

Fig. 11-57. Sine wave of alternating current.

We have learned that alternating current reverses its direction periodically during each cycle (Fig. 11-57). The alternating currents fed to a transmitting antenna follow the same pattern and produce corresponding changes in the strengths of the radiated radio waves.

At any one point in space, the wave passing that point would vary in amplitude during one cycle. The distance a wave travels to complete one cycle is its *wavelength.* Frequency and wavelength are related by the formula:

$$\lambda = \frac{300,000,000}{f}$$

where,

λ is the wavelength in meters,
f is the frequency in hertz.

Radio communications are usually by one of two methods—a dot-dash code, as in telegraphy, or voice, as in telephony. In radiotelegraphy, the transmitter radiates a continuous radio-frequency wave of a constant amplitude. This wave is called a *carrier* wave. The signal will be radiated continuously as long as a sending key at the transmitter is held down. (Fig. 11-58A shows how the current is fed to the antenna.) Whenever the operator moves the key to send the dot-dash code, the antenna current is turned off and on as in Fig. 11-58B. The radiated signal is likewise turned off and on. The receiver reproduces the message as short or long sounds (dots and dashes) which can be heard at the receiver.

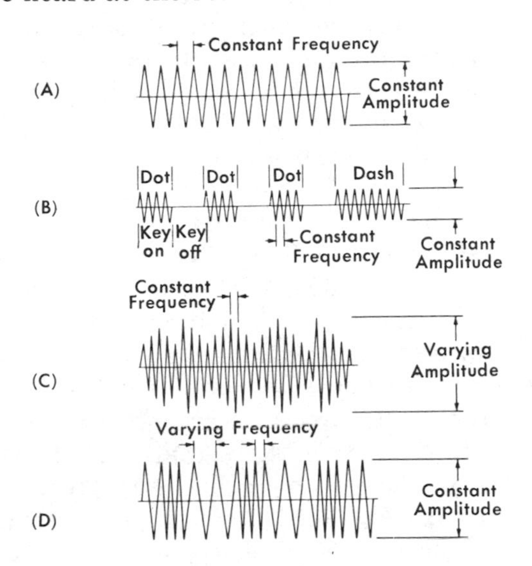

Fig. 11-58. Methods of radio communication.

In radiotelephony, the speaker's voice *modulates* (varies the amplitude of) a constant frequency radio-frequency current (carrier). A typical voice-modulated current is shown in Fig. 11-58C. A receiver can reproduce the message by removing the voice modulation from the constant-frequency carrier wave and presenting the voice to a listener through earphones or a speaker.

This method of transmitting voice is called *amplitude modulation* (abbreviated a-m) because the *amplitude* of the carrier frequency is modulated. A big disadvantage of amplitude modulation is that unwanted noises called static also can modulate the amplitude of the carrier wave, mixing with the voice or other audible input. This fault can be overcome by another method known as *frequency modulation* (fm).

In amplitude modulation, the *wavelength* (time of one cycle) of the carrier wave does not change, only the amplitude. In frequency modulation, the *frequency* (wavelength) changes, but *not* the amplitude. (See Fig. 11-58D.) Without modulation, the carrier wave remains at a *center* frequency. Modulation varies the carrier wave above *and* below this center frequency. The amplitude of the fm carrier is kept at a fixed amplitude. Certain circuits in a frequency-modulation (fm) receiver then receive the transmission, without static.

RF Reception

The reception of a radio transmission begins at the receiver antenna. Extremely small radio-frequency voltages are induced in the antenna as the electrostatic and electromagnetic field pass it. (Remember that voltages are induced whenever a moving field cuts a stationary conductor!) These induced rf voltages are then led to the receiver by a lead-in wire.

The basic radio receiver is a *crystal* set like the one in Fig. 11-59. Signal currents flow through the lead-in wire and induce a voltage (and current) in the secondary coil of the antenna transformer. A variable capacitor connected across the secondary coil acts as a tuning control. When the knob on this capacitor is rotated, the capacitance is changed, permitting different transmitted frequencies to be received. At any one setting of the tuning capacitor, *parallel resonance* exists for one frequency, depending upon the capacitance at that point. When this parallel circuit is tuned to the frequency of a transmitting station, resonance causes maximum current to flow in the secondary circuit at the station frequency.

The crystal diode in Fig. 11-59 rectifies the signal. The audio-modulated current then passes through the earphones and returns to the transformer.

This crystal receiver can be made more sensitive (will pick up more distant stations) if an rf amplifier is added between the receiver and the antenna. Such an amplifier can be tuned by a tuning capacitor, or can remain untuned. In either event, the strength of the received signal is boosted, so that the detected audio signal is louder.

At this point we shall introduce the *block diagram*. It is called a block diagram because each

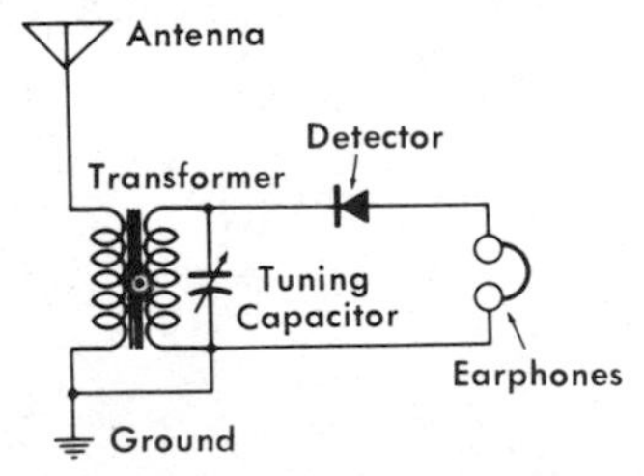

Fig. 11-59. Crystal set (radio receiver with crystal detector).

circuit stage or each portion of electronic equipment is represented by a *block* with the name of that portion in it. A block diagram of the crystal set in Fig. 11-59 is shown in Fig. 11-60. If the rf amplifier is added to the circuit, the block diagram in Fig. 11-61 would represent the receiver. The arrows on the connecting lines indicate the direction of the power flow or the signals, or whatever is transmitted along that path. The block diagram, a form of schematic (drawing) shorthand, is a time-saver when only the overall function of a portion of electronic equipment is being studied or explained.

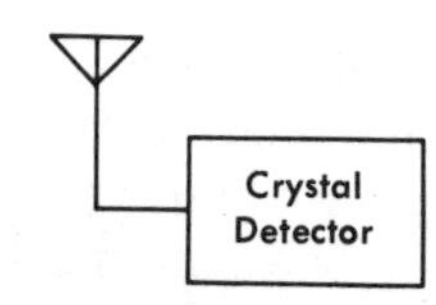

Fig. 11-60. Block diagram of a crystal set.

Earphones are sometimes used for very simple receivers but they are inconvenient. A speaker is more desirable but more power is required for its operation. For instance, a crystal detector would not provide enough power for a speaker. Therefore, an audio-output amplifier must be added to the circuit between the receiver and the speaker. Since this may result in more volume than is needed, it is usual to include a volume control somewhere in the audio circuit.

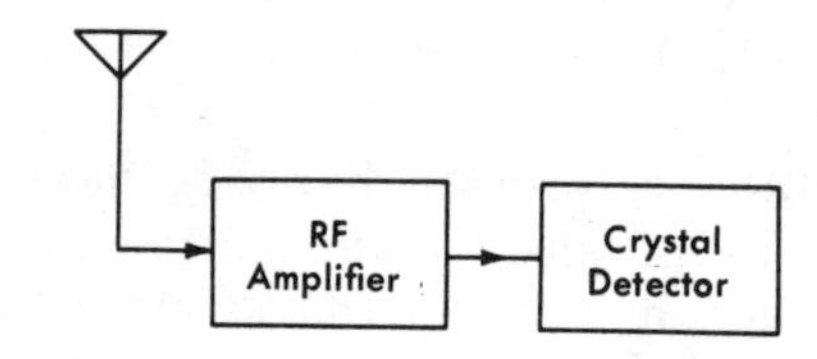

Fig. 11-61. Block diagram of rf amplifier and a crystal detector.

A composite receiver consisting of an rf amplifier, a simple diode detector, and an audio amplifier is commonly known as a trf (tuned radio frequency) receiver. A block diagram of a trf is shown in Fig. 11-62.

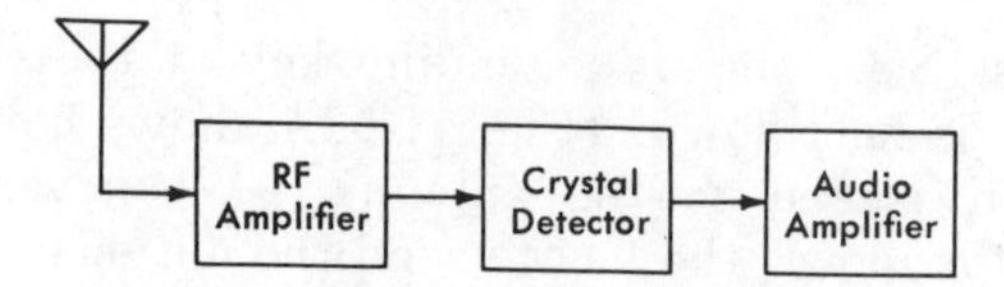

Fig. 11-62. Block diagram of a trf receiver.

SUPERHETERODYNING

The trf principle is not used in modern radio receivers because of its many disadvantages. Instead, modern receivers use a method of reception known as the *superheterodyne* principle. The block diagram in Fig. 11-63 will help us understand the superheterodyne receiver.

The signal voltage from the antenna is at the frequency of the transmitted signal. (If it is a broadcast-band signal, its frequency will be between 550 and 1600 kilohertz.) This signal, together with its audio modulation, enters the frequency-converter stage. Here the signal is changed to a *lower* frequency, but still retains its audio modulation. This change is accomplished by *mixing*, or *heterodyning* the signal (rf) frequency with the output of an oscillator in the receiver. The result is a signal of lower frequency than that of the rf signal, 455 kilohertz in most receivers. This lower frequency is then amplified in a tuned amplifier called an intermediate-frequency (i-f) amplifier. It is so named because the signal which it amplifies is intermediate in frequency, that is, between the rf and audio frequencies.

After leaving the i-f amplifier, the signal is detected and amplified, as it is in any other radio receiver.

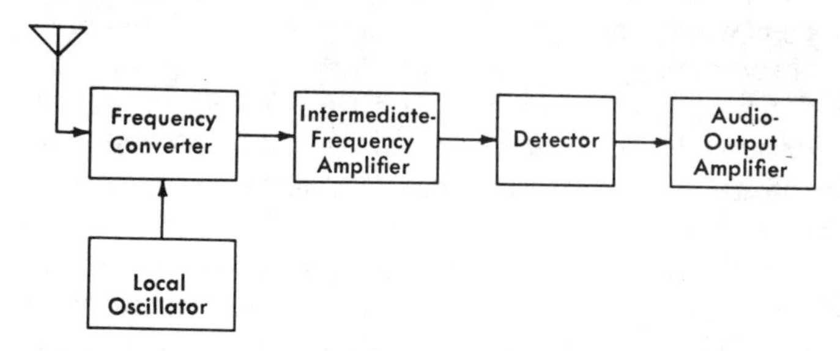

Fig. 11-63. Block diagram of a superheterodyne receiver.

Summary

Radio waves are actually *electromagnetic* and *electrostatic* fields.

Radio waves travel at the speed of light, or 186,000 miles per second.

A radio wave travels the distance of its wavelength during one cycle.

Radio waves induce voltages in a receiving antenna as they move past it.

A block diagram is a form of schematic shorthand.

Questions and Problems

51. State the formula which shows the relationship between radio frequency and wavelength.

52. Explain the superheterodyne principle in your own words.

53. Why are block diagrams used in the study of electronics?

54. What is the wavelength of a signal transmitted by a station operating at a frequency of 1310 kHz?

55. What is the purpose of the tuning capacitor in a receiver circuit?

56. Explain the function of each section of the block diagram in Fig. 11-62.

57. Explain the function of each section of the block diagram shown in Fig. 11-63.

58. What is the difference between an a-m and an fm radio signal?

TELEVISION

The television receiver is as much a part of our everyday living as the automobile, the radio, and the movies. The silent movie was an entertainment medium for some time before sound was added. In radio and electronics it was just the opposite: we had sound long before we had pictures!

The Television Signal

The television signal is made up of two carriers—a frequency-modulated (fm) carrier and an amplitude-modulated (a-m) carrier. The fm carrier contains only the *sound* portion of the telecast. The a-m carrier contains the picture information (video) plus the synchronizing (sync) information.

The television rf spectrum is composed of 82 *channels*. Three groupings of channels are in the *vhf* (very high frequency) range; the remainder are in the uhf (ultrahigh frequency) range. Each channel is a *band* of frequencies six megahertz (MHz) wide. All the information needed to reproduce the picture and sound is transmitted within this band.

Channels 2 to 4
are in the 54-MHz to 72-MHz range (vhf)

Channels 5 and 6
are in the 76-MHz to 88-MHz range (vhf)

Channels 7 to 13
are in the 174-MHz to 216-MHz range (vhf)

Channels 14 to 83
are in the 470-MHz to 890-MHz range (uhf)

The location of the picture and sound carriers for Channel 7 is shown in Fig. 11-64. The a-m picture carrier is sent at a frequency of 1.25 MHz

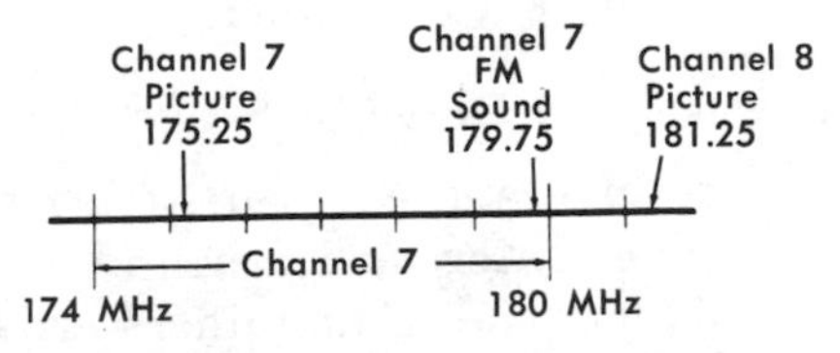

Fig. 11-64. Location of picture and sound carriers for Channel 7.

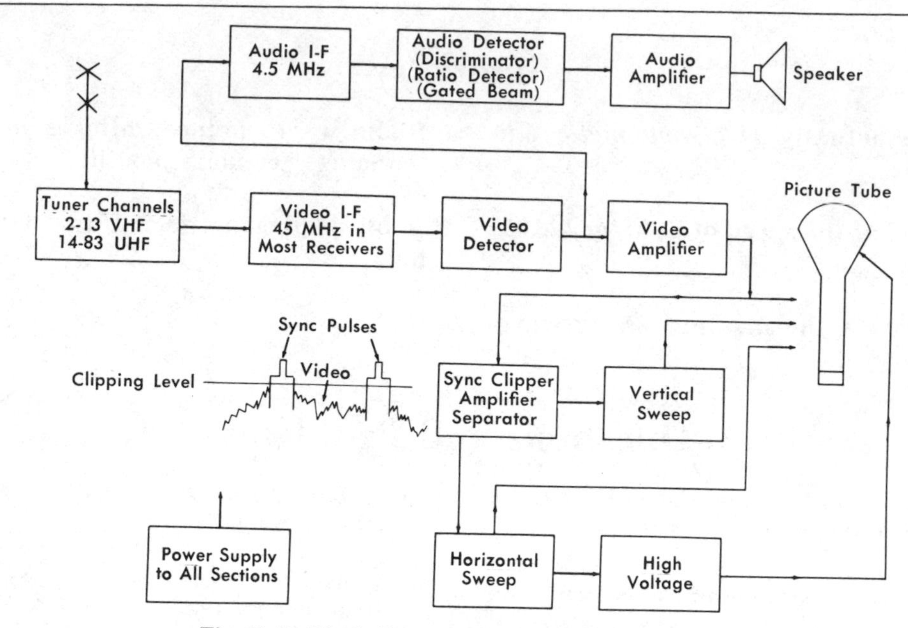

Fig. 11-65. Block diagram of a television receiver.

above the lower limit of the assigned channel. The fm sound carrier is sent at a frequency of exactly 4.5 MHz greater than that of the picture carrier. The 4.5-MHz separation of the picture carrier and the sound carrier makes possible the *intercarrier* sound system used in almost all television receivers.

THE BLACK-AND-WHITE TELEVISION RECEIVER

Fig. 11-65 shows a block diagram of the main section of a television receiver.

The Tuner

The rf signal received from the antenna is first amplified by an rf amplifier. This signal is then fed to a *mixer* (or frequency converter) where it is heterodyned with an unmodulated rf signal from a local oscillator. The output of the *tuner* (rf amplifier and mixer) is then sent to the video i-f amplifiers. The tuners of most home receivers select any one of twelve channels (2-13), plus a uhf position that allows uhf Channels 14 to 83 to be tuned with a uhf tuner.

The Video I-F and Detector Stages

In most receivers the video i-f strip is a 45-megahertz type. Intercarrier sound is also used in most receivers. The i-f amplifiers are so aligned that the sound carrier can be amplified along with the video signal but at a reduced level.

The video and sound carriers are then detected and *mixed* (heterodyned). Since the transmitted sound and video are separated by a frequency of 4.5 MHz, a *beat-frequency difference* of 4.5 MHz is produced which also contains the fm modulation. The detector output consists of a 4.5-MHz sound signal, which is fed to the sound i-f amplifier, and a video signal, which is fed to the video amplifier, before going to the picture tube.

The Sound I-F Detector and Audio Stages

The 4.5-MHz intercarrier signal is amplified by the sound i-f amplifier to raise the signal strength high enough to be detected. The fm detector can be one of many types, such as a discriminator, a *ratio detector,* or a *gated-beam detector*. The audio signal from the detector is applied to a conventional audio-amplifier and output stage. It, in turn, drives the speaker. Certain types of fm detectors, such as the gated-beam type, have an output level high enough to drive the output stage directly.

The Video-Amplifier Stage

The signal from the detector is sent to the video amplifier. There, the signal is amplified to a level sufficient to drive the grid or cathode of the picture tube. The video amplifier also amplifies the vertical and horizontal synchronizing pulses.

Although the 4.5-MHz sound signal is present at the input to the video amplifier, it must not be allowed to reach the picture tube. Therefore, one or more 4.5-MHz *sound traps* are used in the video amplifier to eliminate sound interference (*sound bars*) in the picture.

The Sync-Amplifier, Clipper, and Separator Stages

This section consists usually of one or two high-gain stages. The necessary *clipping* (separation) and amplification are done by the cutoff and *saturation* characteristics of a tube or transistor. The synchronizing pulses are clipped (separated) from the video signal because the tube amplifies only that part of the signal above the line marked *clipping level* in Fig. 11-65. The video portion of the signal is not amplified. The output of the sync amplifier is sent to the vertical- and horizontal-sweep sections.

The Vertical- and Horizontal-Sweep Stages

If you look closely at the face of a picture tube, you will see a *scanning raster* made up of a series of horizontal lines. These lines are make by a single electron beam that moves from left to right and then quickly returns to the left. This rapid return to the left usually cannot be seen on the screen because the beam is *cut off,* or *blanked out,* by the sync pulses from the video amplifier. The electron beam also travels from top to bottom. Again the retrace is blanked out. The dark bar seen when the picture rolls *vertically* is the vertical-blanking signal. Vertical retrace takes place during this interval.

The horizontal- and vertical-sweep oscillators are kept in step (synchronized) by the vertical- and horizontal-sync pulses. The outputs of these oscillators are amplified and fed to a set of coils (the deflection yoke) on the neck of the picture tube. The magnetic field produced by these sweep signals causes the electron beam to move back and forth across the screen in step with the video signal to form a picture on the fluorescent screen of the picture tube.

The High-Voltage Stage

The high voltage necessary to operate the cathode-ray tube is produced by using the flyback voltage of the horizontal-output stage. The horizontal-output transformer has a high-voltage winding. The rapid collapse of its magnetic field during flyback time produces a very high voltage pulse (as high as 25,000 volts in some receivers). This pulse is rectified and filtered, and then applied to the high-voltage anode of the picture tube.

The Power Supply

A low-voltage power supply supplies the plate and filament voltages for the various tubes. In receivers with the tube filaments in series, the power supply may be just a selenium or silicon rectifier and a filter capacitor. Other receivers may contain a power transformer—with two, three, or four sets of windings—placed in a circuit with one or more rectifiers, filter capacitors, and filter chokes.

COLOR TELEVISION

When color is added to a television picture, more receiver parts and circuits are needed. The basic sections of a color receiver are the same as the ones in a black-and-white receiver except for added sections to receive and process the color information. A three-color picture tube must also be used, as illustrated in Fig. 11-66.

The color receiver will reproduce color only when the program is in color. Otherwise, an ordinary black-and-white picture will be seen.

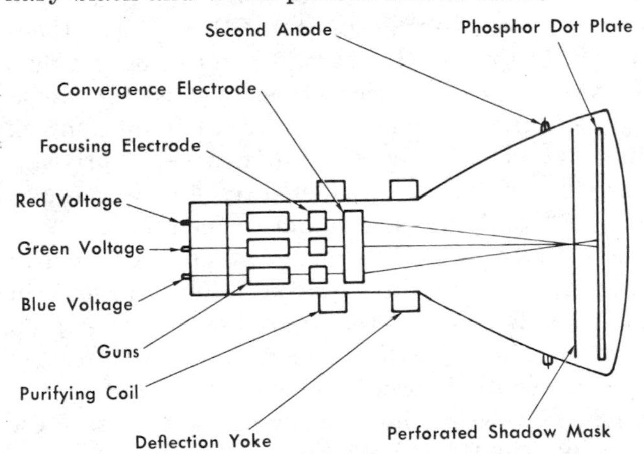

Fig. 11-66. Elements of a color television picture tube.

Summary

The television spectrum is composed of 82 channels. Channels 2 to 13 are in the vhf range; Channels 14 to 83, in the uhf range.

The television signal is both an amplitude-modulated and a frequency-modulated carrier wave. The sound part of the signal is frequency modulated; the video and sync parts of the signal are amplitude modulated.

The fm carrier is transmitted at a frequency exactly 4.5 megahertz above that of the a-m carrier.

Questions and Problems

59. How many channels are assigned for home television reception?

60. What does vhf mean? uhf?

61. What channel numbers are included in the vhf and uhf frequencies?

62. What characteristic of a television transmission makes it possible for a receiver to have an intercarrier sound system?

63. Where does the 4.5-MHz intercarrier sound originate?

64. What is the purpose of the yoke on the neck of a picture tube?

65. What kind of carrier signal is transmitted for the video television signal?

66. What kind of carrier signal is transmitted for the sound signal?

67. What does the term *raster* mean? How is a raster produced?

INDUSTRIAL ELECTRONICS

Because almost every home has a radio and a television set, we may think that these are all there is to electronics. However, we must not neglect the vast field of industrial electronics. Industrial and military equipment is often far more complex than the more familiar radio or tv set. Moreover, industry finds much wider uses for electronics than as merely an entertainment medium.

For this reason, this last chapter is devoted to some of the ways electronics is put to work in industry. Because of the limited space, we can do no more than skim the surface. Readers interested in a career in electronics should look further into industrial electronics, for some of the greatest opportunities exist here.

Relays

Relays have been used for many years to control or limit all kinds of mechanisms. In fact, relays are still a very important control device and are becoming even more important. They act as switches or actuating mechanisms in electronic controls, sensing devices, measuring equipment, and computers. Since we have already studied how relays operate, we will just note, in this section, how they are used with other electronic components for timing, heat regulation, light control, and other control operations.

Timers

Many industrial operations must be timed to a split second. For example, resistance (spot or seam) welders must be precisely timed so that an exact number of cycles of ac current pass through the electrodes for each weld. This means that the alternations—occurring at 1/120th of a second—must be counted and controlled!

Most such timing circuits contain what is called an *RC circuit*. The principal parts of this circuit are a capacitor, a resistor, a triode tube, and, of course, a power supply. You will notice in Fig. 11-67 that resistor R and capacitor C are in the grid circuit of the triode tube. The plate circuit of the tube controls the relay shown. When a plate current flows (when the tube conducts) there is a current in the relay winding.

The triode tube is biased so that, with little or no voltage (bias) on the grid, the tube will conduct, and enough current will flow to close the relay to the ON position. A larger negative grid bias will repel or "block" the electrons emitted by the cathode so that none can reach the plate. No current will flow in the plate circuit and through the relay windings. Therefore, the relay will open to an OFF position.

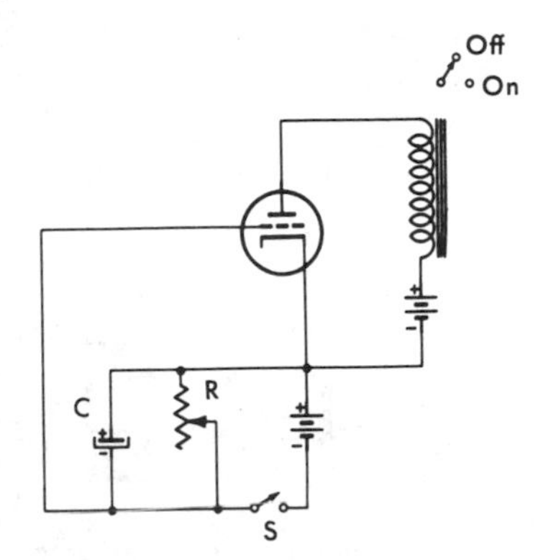

Fig. 11-67. RC time-delay circuit.

The negative grid bias is obtained from capacitor C (Fig. 11-67). As switch S is closed, the capacitor becomes charged. The grid becomes negative and cuts off the tube.

As switch S is opened, the timing cycle starts. The capacitor must discharge through resistor R. The resistance holds back the flow of electrons so that, with a certain size capacitor and a certain value resistor, this discharge will take a certain length of time.

The timing period may be from a small fraction of a second to several minutes, depending on the value of the parts. The time interval in seconds is equal to the capacitance in microfarads multiplied by the resistance in megohms. For example, an 8-μF capacitor and a 7-megohm resistor would give 8 μF times 7 megohms or 56 seconds. This is an approximate calculation because the voltage used to charge the capacitor and the amount of negative bias needed to cut off the tube must also be considered.

Notice in Fig. 11-67 that R is a variable resistor and that the timing period can be changed by adjusting the value of R.

Electronic timers are used in a wide variety of applications. Welding equipment has been mentioned as an example. However, in any operation which requires precise timing, or timing for a sequence of several operations, a timing circuit similar to the one just described will probably be used.

Heat Sensing and Controlling

Heat is used in some way in the manufacturing processes of almost all industries today. Controlled heat is necessary in chemical processing, in the production and fabrication of metals, in the plastics industry, and even in the manufacture of electronic components.

One very common heat control, the *bimetallic thermostat*, is shown in Fig. 11-68. It is used for control of moderate temperatures and is accurate to within 2° or 3° F. Its operation is based upon two metals (shown as *A* and *B*) having different expansion rates. As the two metal strips are heated, *B* will expand more than *A*. The strips will bend and open the contact points in the relay circuit.

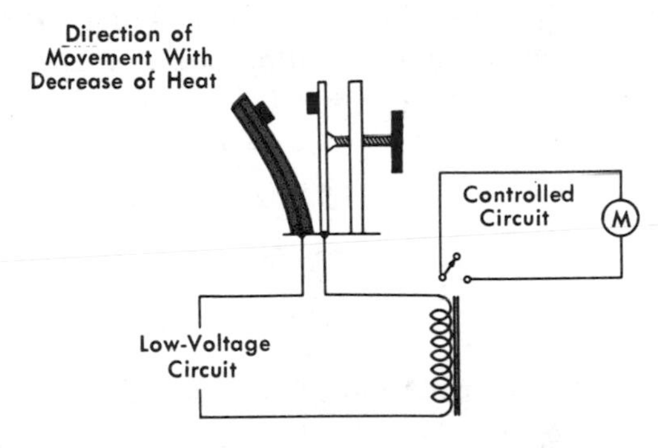

Fig. 11-68. Thermostat control.

Another type of control is a *thermocouple*. A simplified unit is illustrated in Fig. 11-69. Two different metals (*A* and *B*) are heated at their junction. An electrical potential is generated in proportion to the amount of heat, as shown by the meter.

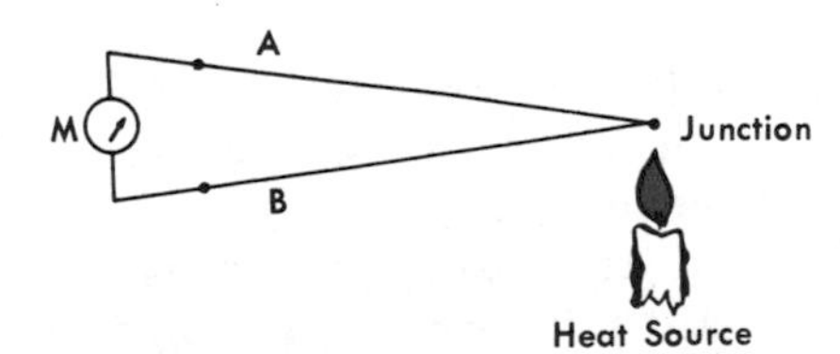

Fig. 11-69. Simplified thermocouple.

Some metals, when heated, will give off (emit) electrons. Other materials, when heated, will emit very few or no electrons. You will remember that the filament or cathode of vacuum tubes will emit electrons. This is the reason they are called *thermionic* materials.

When heated, one metal of the thermocouple will emit electrons, but the other will not. This creates the electrical potential used as the control voltage for the unit—which will adjust for the desired heat setting.

In another rather new method of heat control, the property of some metal oxides is used. These materials, called *thermistors,* change their resistance values with changes of heat. The change in resistance is used to sense and control the heat variation. Thermistors can accurately detect *infrared radiation* from the material being heated.

In both the thermistor and the thermocouple, the output (amount of electrical change) would not be large enough to directly control other electrical devices, such as relays or motor controls. For this reason, an amplifier circuit (Fig. 11-70) is needed. A small change on the grid of the tube will make a larger, amplified change in the output circuit of the tube. Many of the circuits in industrial electronics include amplifiers to increase the accuracy and power needed for control circuits.

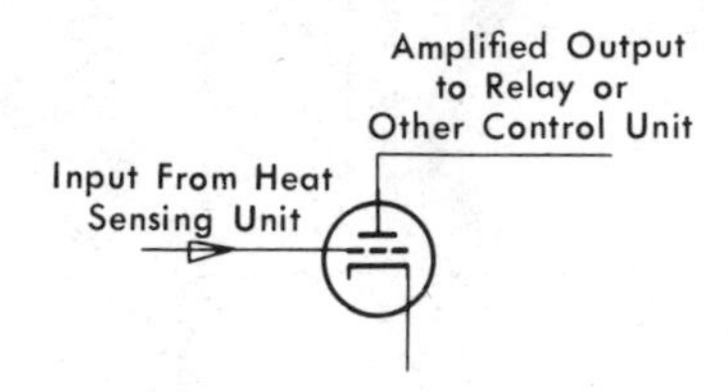

Fig. 11-70. Amplifier circuit for heat-sensing unit.

Phototubes

In an earlier section we described the phototube, but said little of its industrial application. You will remember that the cathode (Fig. 11-71) is coated with a material that will emit electrons when struck by light. The number of electrons emitted is proportional to the amount of light striking the cathode.

A very common use of the phototube is in sensing when it is dark enough for the city street lights to be turned on.

In chemical processing, the density of a fluid can be constantly inspected by a unit like the one in Fig. 11-72. A similar unit is used as a smoke-detecting fire alarm. Smoke also changes light intensity, so that an alarm system is turned on by the relay.

The amplified output of two phototubes is used for motor control (Fig. 11-73) in a metal-cutting

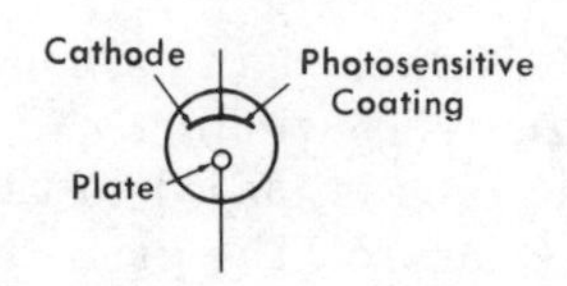

Fig. 11-71. Phototube symbol.

operation. The cutting torch will 'follow" the pattern block by means of two motors, which drive the cutter head assembly in the direction shown by *Plane "A"* and *Plane "B."* Both phototubes are set for zero output when half of each cathode is shaded by the pattern. When more or less than half of the photocell is shaded, a signal is transmitted to one of the drive motors. The motor will drive until the signal is corrected to a normal, or zero, value.

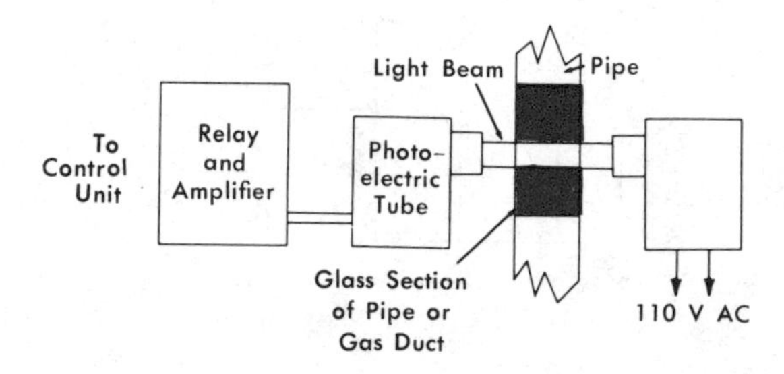

Fig. 11-72. Variations in density alter the amount of light reaching phototube.

Another application of phototubes is in the printing industry. When many copies of a small item, such as a label, are printed, the printing is done on a larger roll of paper. Later, the individual items are cut out. To ensure that the labels will be cut at the edge, instead of in the middle, a little

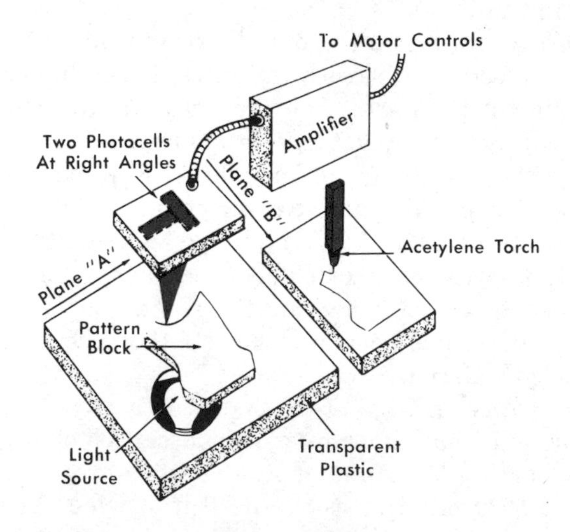

Fig. 11-73. A pair of phototubes guide acetylene torch in cutting a precise pattern.

"reference" mark is printed. A phototube scans these marks, as shown in Fig. 11-74. Each time a mark is detected, a cutter trims one of the labels.

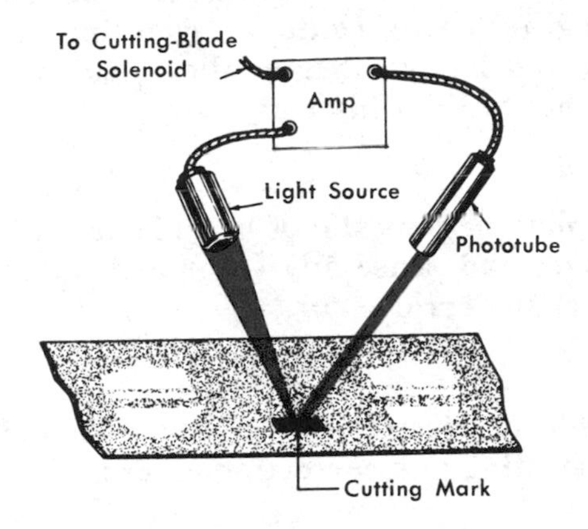

Fig. 11-74. Precise cutting of labels is accomplished with phototube control.

Automation

In the past few years we all have heard the term *automation*. It has been spoken of as some super-automatic operation. In fact there are machines, and sometimes a whole series of operations, which are automatic and *self-correcting*. This ability to be self-correcting is commonly thought of as automation. An automatic tool or device which can "inspect" its own product, detect an error, and then correct the error, is an automated device.

The inspection and correction are usually done by an electronic circuit called a feedback circuit. When an error is detected, a corrective signal is fed back and the amount of error is automatically corrected. Equipment as complex as the automated devices is beyond the scope of this book. However, we will use simplified examples to illustrate the principle of feedback and self-correcting units. Fig. 11-75 is a block diagram of such a unit. The sensing and measuring is often done by an electronic circuit similar to some described earlier in this section. These electronic measurements are often "mixed," or compared with a standard of

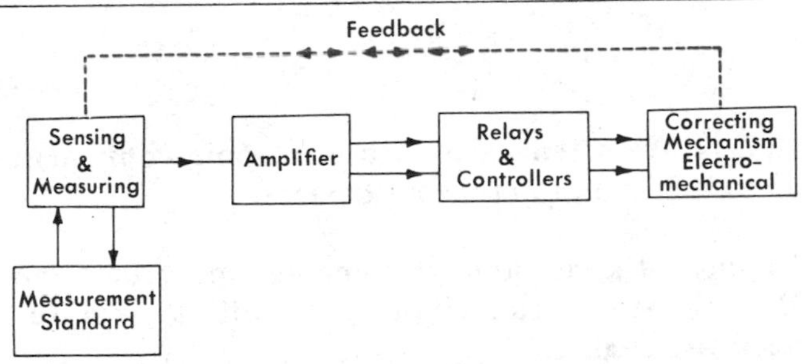

Fig. 11-75. Block diagram of an automatic operation.

measurement, to show the amount and direction of an error. These signals from the sensing and standard units are too small to be used by the correcting mechanism. Therefore, they must be amplified before being sent to the controller unit. The relay, motor controllers, and other control devices automatically operate the *electromechanical* correcting mechanism. The amount of correction is "read" by the sensing unit, and the whole process of sensing and correcting continues.

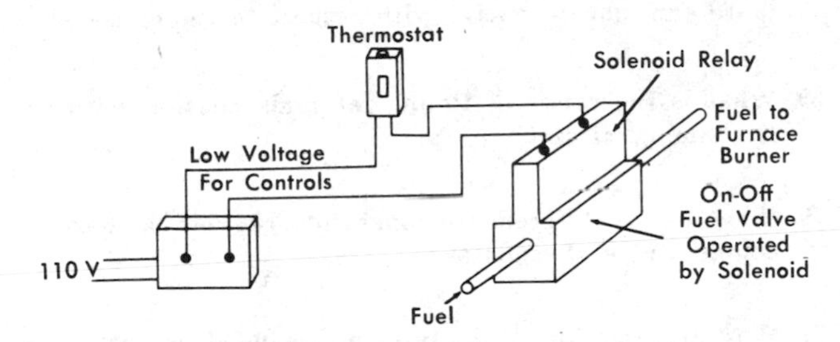

Fig. 11-76. Automated furnace control.

Compare the diagram in Fig. 11-75 with the furnace control in Fig. 11-76. The bimetallic thermostat is a sensing device. The "error" from the desired room temperature is sent to the relay (solenoid). This *error signal* does not represent the value of error, only an open or closed circuit. The solenoid is an electromechanical control which opens or closes a fuel valve. The furnace here would be the *correcting mechanism*. The amount of its correction (heat produced) would be fed back to the thermostat (sensing unit), and a continuous measurement and correction would take place.

Summary

Relays are often used with electronic equipment as switches and actuating devices.

In most timing and sequencing operations, an RC (resistance-capacitance) circuit is used for accurate timing.

The length of the time delay of an RC circuit depends on the values of the capacitance and the resistor.

A bimetallic thermostat, a thermocouple, or a thermistor is often used as the heat-sensing element in heat-sensing equipment.

Phototube controls are used for measuring, indexing, following contours, counting, alarm systems, and many other applications by detecting a change in light intensity.

Automation consists of a machine which will measure and sense any error and automatically correct the error.

A feedback circuit is used in automated devices for detecting and correcting errors.

Questions and Problems

68. What happens when the grid of a vacuum tube is made more and more negative with respect to the cathode?

69. What is the range of timing intervals possible with an RC timing circuit?

70. What is the difference in operation between a thermocouple and a thermistor?

71. Why are amplifier circuits often included as part of a sensing and control device?

72. Approximately what time interval can be obtained with a 4-μF capacitor and a 6-megohm resistor?

73. What does the term *electromechanical* mean? How are electromechanical devices used?

74. Will plate current flow in a tube which is biased to cutoff? (See Question No. 68.)

75. Compare the operating principles of a bimetallic thermostat and a thermocouple, in which two dissimilar metals are used.

ANSWERS TO QUESTIONS

Chapter 1

1. The nucleus.
2. Hydrogen.
3. Static electricity is either stationary or at rest. Dynamic electricity is electrons in motion.
4. (a) volt (b) ampere.
5. Silver—conductor
   Mica—insulator
   Platinum—conductor
   Pure iron—conductor
   Bakelite—insulator
   Rubber—insulator
   Tungsten—conductor
   Air—insulator
   Gold—conductor
   Germanium—semiconductor
   Mineral oil—insulator
   Lead—conductor
   Silicon—semiconductor
6. Heating (iron, lamp, range, toaster).
   Chemical (batteries).
   Magnetic (doorbell, meter, clock).
7. The most common appliance would be an electric iron. Most electric irons are rated at 1000 watts and since one horsepower equals 746 watts the rating would be 1000 divided by 746, or approximately 1.304 horsepower. Other appliances found in the home which use electricity at a fast rate are electric heaters, toasters, and ranges.
8. Rate. It is coulombs per second.
9. If an object has more of one kind of fluid than the other, it is charged, but if it contains equal amounts of fluid or none at all it is neutral.
10. If an object contains too much fluid it is positive, and if it has lost fluid or does not have enough it is negative.
11. It is a gas-filled, dual-electrode discharge tube which glows when electric current passes through the gas.
12. Electrons, protons, neutrons.
13. Like charges repel, unlike charges attract.
14. All matter is made up of three kinds of particles within the atom, which are called electrons, protons, and neutrons. Niels Bohr.
15. Static means standing still or at rest. Static charges are found on the outer surfaces of all charged objects.
16. Current electricity moves and does work; static electricity stands still and does little work.
17. Positive.
18. Electroscope.
19. Induction.
20. The force (either attraction or repulsion) between two charges (either like or unlike) is directly proportional to the product of the charges, and inversely proportional to the square of the distance between them.
21. Capacitors.
22. Friction, contact, or by induction.
23.

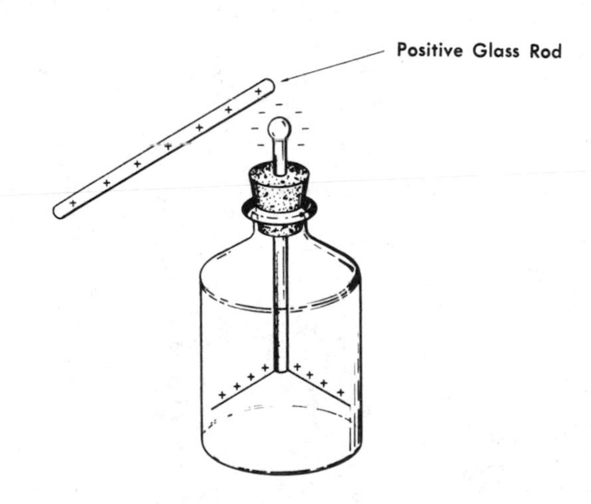

24. The amount of divergence is a rough measurement of the amount of charge. Statcoulomb.
25. Two or more plates of conducting material mounted very close together and separated by an insulator.
26. Air, oil, paper, wax, mica, ceramic, various oxides.
27. Yes. A higher dielectric constant allows the capacitor to store a greater charge per volt applied.
28. Doubled.
29. (a) No. (b) No. (c) Current appears to pass through but only the signal passes through because of the insulation of the dielectric.
30. No. Yes. If a polarized capacitor is connected backwards it will have excessive leakage.
31. Electrolytic capacitors.

Chapter 2

1. Current will be limited to a low value.
2. Battery; generator.
3. The other two lamps will not burn; the circuit is open.
4. Equal current will flow through each resistor.
5. Yes—R3 and R4 will form a series circuit.
6. No—the current path is broken.
7. (a) I, (b) E, (c) P.
8. (a) volts, (b) ohms, (c) amperes.
9. Ammeter.
10. 2 seconds.
11. 1 volt.
12. Voltmeter.
13. The ohm.
14.

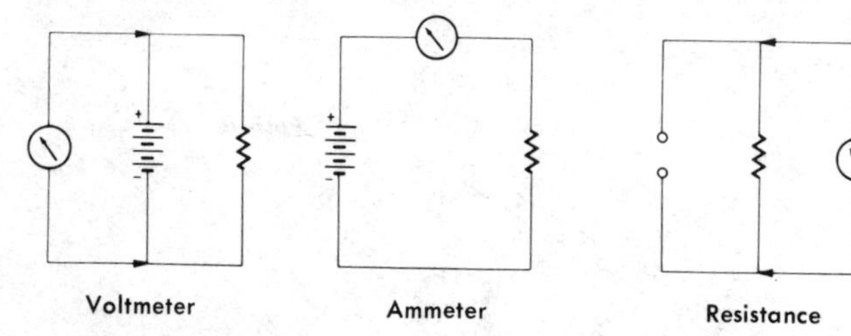

15. The ampere.
16. The volt.
17. An ohmmeter.
18. $I = \frac{E}{R}$
19. 108 volts.
20. 64 volts.
21. 40.8 amps.
22. 3.14 ohms.
23. 105 ohms.
24. $R = \frac{E}{I}$
25. Any one of the quantities can be obtained if the other two are known.
26. 1.1 amperes.
27. Wirewound, carbon, fusible, variable, temperature sensitive.
28. (a) 10.4 r, (b) 9.8 r, (c) 17.2 r.
29. Directly.
30. Reduced by a factor of 4.
31. 87 ohms.
32. Material which readily carries electrical current at normal temperatures.
33. 100 psi.
34. (a) A, (b) I, (c) B, (d) A.
35. (a) 1 psi, (b) 3 psi, (c) 90 psi.
36. Yes.
37. Yes.
38. (a) A, (b) I, (c) B, (d) A.
39. (a) 2 volts, (b) 93 volts, (c) 90 volts.
40. Yes.
41. 1 ampere.
42. 1 volt.
43. 100 volts.
44. 90 volts.
45. Yes.
46.

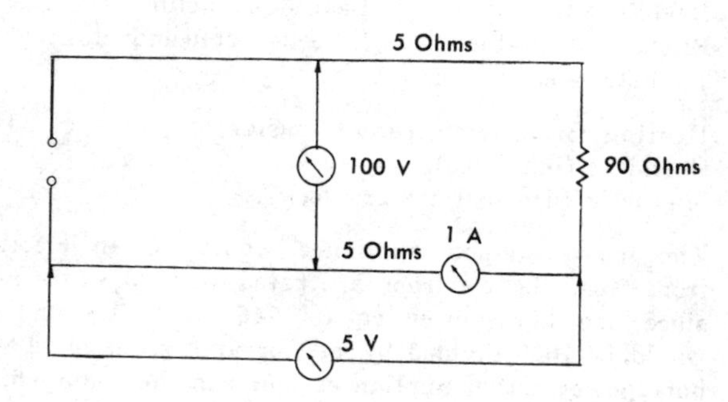

47. 5 ohms.
48. 2.73 ohms.
49. 5063 ohms.
50. 1015 ohms.
51. 3 ohms.
52. (a) 1.5 amperes, (b) 2.25 amperes.
53. 137.5 watts.
54. 1.66 amperes.
55. 72 ohms.
56. 480 watts.
57. 3.4 amperes.
58. 100 volts.
59. 72 watts.

Chapter 3

1. Six cells in series. Potential of each cell is 2.1 volts when charged.

2. (a) In a primary cell, chemical action depletes the materials, which cannot be restored. A secondary cell has plates that change their chemical structure as the cell discharges. The original chemical state is restored by charging. (b) A cell is a single unit containing one cathode and one anode in an electrolyte. A battery is a combination of two or more cells.

3. Positive terminal. Charging the battery in reverse would ruin it.

4. Zinc will be used faster by the group in series.

5. Six cells.

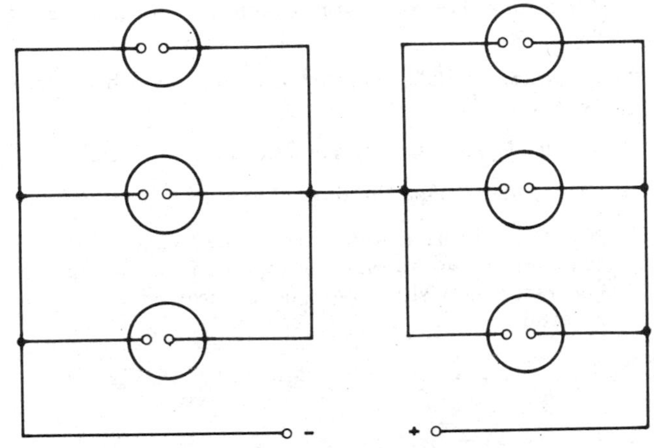

6. Neither. They supply the same potential.

7. Lighter weight and less danger of spilling.

8. With full load some voltage is lost across the internal resistance.

9. No. It serves only to raise the temperature of the cell.

10. No, because that is considered when rated limits are set. However, useful life may be reduced.

11. Temperature and condition of electrolyte and plate material.

12. Because a no-load voltage reading means that no current is being drawn.

13. When the stove is first turned on and the coils are cool.

14. Efficiency will not increase.

15. 1 ampere.

16. .625 ampere.

17. 1.1 volts.

Chapter 4

1. Use iron filings sprinkled on a cardboard.

2. Low permeability means that it is difficult to align the magnetic domains. Once aligned, the domains tend to hold the alignment for the same reason.

3. Substances with a permeability slightly greater than one are paramagnetic. Ferromagnetism is attributed to substances similar to iron and its alloys with a permeability much higher than unity (one). Diamagnetism is attributed to substances similar to antimony and bismuth with a permeability of less than one.

4. Because of shielding problems. The compass would be shielded from the magnetic field of the earth by the surrounding metal of the submarine.

5. By induction, placing (in contact with) a magnet on the blade. By heating or heavy blows from a hammer.

6. They enter—although this magnetic pole is located near the earth's geographic north pole it is really a magnetic south pole, since it attracts the north pole of a compass.

7. Magnetic flux—reluctance.

8. Yes. An electromagnet is a coil of wire with an iron core in the center. The coil is called a solenoid.

9. Grasp the wire with the left hand so that the thumb points in the direction of the current and the fingers will then point in the direction of the magnetic lines of force.

10. Yes. The iron core merely affords an easy path for the magnetic lines of force to follow.

11. No. The electromagnet would be stronger because of the core which strengthens the lines of force.

12. No.

13. To concentrate magnetic lines of force.

14. Magnetomotive force is the force that drives magnetic lines of force through a magnetic circuit.

15. No. Moving the wire in a magnetic field will induce current, but it can also be induced by moving the field or changing the strength of the field.

16. The changing magnetic lines of force induce a current in the secondary.

17. 5 volts.

18. There is no current when no lines are cut.

19. Current is produced when the conductor cuts the magnetic lines of force.

20. The winding to which voltage is applied is considered to be the primary winding.

21. Increase the speed of the conductor cutting the magnetic field, the number of magnetic lines of force, or the strength of the magnetic field.

22. Yes—current would be induced if the axis of rotation of the loop was at right angles to the earth's magnetic field. It would not be induced if the axis was parallel to the earth's magnetic field.

23. Because of the increased number of magnetic lines of force.

24. A force produced as a result of self-induction.

25. One volt will be produced by a single conductor cutting through 100,000,000 lines of magnetic flux in one second.

Chapter 5

1. 100 hertz.
2. (a) Change the strength of field (number of lines of force), (b) Change the rate at which the conductor moves across the lines of force, (c) Change the direction the conductor moves across the field.
3. Two times.
4. Economical transmission over long distances. It is easily transformed to higher or lower voltages.
5. The maximum voltage equals 110×1.414 or 155.5 volts—the average voltage equals $.637 \times$ maximum or 99.05 volts.
6. Because zero would result if both halves were used.
7. $\pi/4, \pi/2, \pi, 5\pi/4, 3\pi$.
8. 220 volts.
9. 21.21 amps.
10. 300 volts.
11. $X_C = \frac{1}{2\pi fC} = \frac{1}{6.28 \times 60 \times 5 \times 10^{-9}}$

    $I = \frac{E}{X_C} = 57$ microamperes.
12. $X_L = 2\pi fL = 6.28 \times 60 \times 6 = 2260.80$ ohms

    $I = \frac{E}{X_L} = \frac{110}{2260.80} = .05$ amp.
13. A change in current produces a magnetic field that induces a voltage which opposes the current change producing the field.
14. $I = \frac{E}{Z} = \frac{110}{\sqrt{(X_L - X_C)^2}} = \frac{110}{\sqrt{(2653 - 755)^2}} = .058$ ampere.
15. 200,000,000.
16. Lead.
17. Lag.
18. The formulas for inductive and capacitive reactance both contain the term f, for frequency. In the formula for X_L, f is a multiplier. Therefore, if the frequency f increases, X_L increases. In the formula for X_C, f is a divisor. Therefore X_C decreases as f increases.
19. Impedance is the total opposition to current flow in an ac circuit.
20. 94,250 ohms.
21. Voltage and current are in phase. Voltage and current are 90° out of phase.
22. No—they are vector voltages and cannot be added directly.
23. Lag.
24. Yes.
25. 14.136 ohms.
26. 183.67 ohms.
27. 4.4 amperes.
28. 15.71 amperes.

Chapter 6

1. A machine that converts mechanical power into electrical power.
2. A special kind of generator or dynamo which produces ac voltage.
3. The field provides the magnetic lines of force.
4. The armature cuts the lines of force.
5. No—the only basic difference in the two generators is the method used to take the current from the unit. Ac generators use slip rings; dc generators use a commutator.
6.

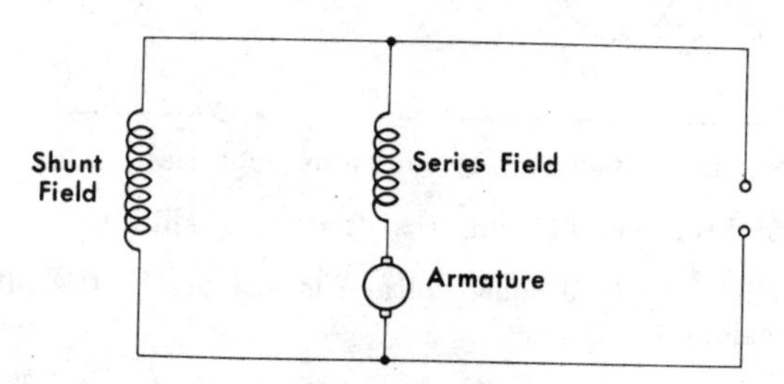

7. By lifting the brushes from the commutator and sending a surge of current through the field coils in the proper direction.
8. Dc generator.
9. Power for the field magnets of self-excited generators is supplied by the generators themselves. Power for the field magnets of alternators comes from an outside source, such as a battery.
10. To convert electrical energy into mechanical energy. In horsepower.
11. It completes the magnetic circuit for the field poles.
12. Iron has good magnetic properties.
13. To reduce eddy-currents.
14. Induction, repulsion, repulsion-induction.
15. Series-wound (universal) motor.
16. Induction.
17. Repulsion.
18. The armature fields are induced by the alternating fields of the field coils.
19. Brushes, armature, commutator, and field.

20. Electrical energy is converted into mechanical energy.
21. Yes. The same current flows through both field and armature windings so reversal of current in ac does not affect operation.
22. Eddy-currents are reduced.
23. No.
24. 500 volts.
25. 0.4 ampere.
26. Step-down.
27. 5 to 1 ratio.
28. Primary 50 milliamperes—secondary 0.5 ampere.

Chapter 7

1. It depends upon the force between two charged bodies.
2. 0.00117 ohms.
3. 149,972 ohms.
4. Yes.
5. 1000 ohms-per-volt.
6. Because the meter might be damaged by current from the circuit.
7. The rectifier type.
8. It draws very little current—the tube amplifier makes the meter more sensitive.
9. 29,972 ohms. 299,972 ohms.
10. Three-way switch.
11. Indicating.
12. It breaks connections at two points simultaneously.
13. As many as desired.
14. A switch.
15. It should be mounted so that gravity cannot close it.
16. The mercury switch.
17. A change in temperature.
18. Ac relays have laminated cores and a shading coil.
19. Because they can control high voltage or current safely from a distance.
20. A pilot relay is used to control another relay.
21. A control relay switches one or more power circuits.
22. Protective relays protect machinery by opening overloaded circuits.
23. Regulating relays automatically limit inputs to prevent overloaded circuits.
24. Its action is single-pole, single-throw, normally closed, double-break.
25. Switching circuits, regulating voltage, telephone equipment.
26. By gradually cutting starting resistance on the armature.
27. The armature may burn out.
28. It increases.
29. Yes. It has an 1800-watt limit.
30. Its purpose is to protect electrical circuits from dangerous overloads.
31. No.
32. Excess heat is developed and melts the fuse link.
33. It will withstand harmless, short-time overloads.
34. The nonrenewable type.
35. Time-lag fuses. They withstand starting surges without blowing.
36. Yes. Appearance of blown fuse sometimes indicates nature of trouble.

Chapter 8

1. One circular mil equals .7854 square mil.
2. Length, diameter, and type of material.
3. A conductor is a material that passes electric current easily.
4. No. 12.
5. No. 3.
6. 52634.
7. They have more free electrons and offer less resistance to current.
8. In various electrical codes (National Electrical Code).
9. Yes. It is a material with no free electrons.
10. Glass.
11. 6.0.
12. No.
13. 470.
14. Since ice is a conductor, it forms a path for the high-voltage current.
15. A dielectric has few free electrons and obstructs current flow. A conductor has many free electrons and passes current.
16. Yes. Spark-plug wire requires heavy insulation; doorbell wiring handles low voltage and does not require much insulation.
17. Yes. The dielectric strength of glass is 760 volts per mil; rubber is 470.

18. Yes. The glass would be a good insulator and the steel frame would act as an electrostatic shield.

19. It doubles the service voltage.

20. A service head protects the insulation and keeps rain or snow out of the entrance conduit.

21. It controls the ceiling light.

22. Both wires.

23. A splice must be mechanically secure and soldered and taped.

24. In outlet boxes and other junctions where there is no strain on the wires.

25. The fixture splice.

26. Stranded wire can be flexed many times without breaking. Solid wire would break after a few flexings.

27. It provides two different supply voltages and handles larger loads.

28. S3.

29. Swp.

30. Higher.

31. Yes.

32. By a white metal screw.

33. If the two remaining wires supply unequal loads, the smaller load will receive too much voltage. See Question 30.

34. Five tons, for a balanced load.

35. Underwriter's Laboratories.

36. No. There are three other codes which govern—state code, local (city) code, and central station code.

37. No wire smaller than No. 14.

38. 3-pole type designed for grounding for personal protection.

39. No. 12.

40. Yes.

41. Proper size of wire, proper type of insulator, protection for runs of wire, secure and permanent splices and connections, protection against shock and fire.

42. Conduit, armored cable, and metal raceway.

43. No. 12.

Chapter 9

1. Radiation.

2. Induced eddy currents.

3. Conduction, convection, and radiation.

4. Metals.

5. Insulators.

6. BTUs (British thermal units).

7. Easily controlled and can be produced by several methods.

8. (a) electric blanket, (b) certain types of cooking devices, (c) industrial applications.

9. The joule is a unit of work equal to one watt-second. The calorie is equal to 0.239 joule. One BTU is equal to 252 calories.

10. 2400 watts or 2.4 kW.

11. 12.2 minutes.

12. Ceiling light will develop indirect lighting which will light a room in general. Direct lighting would be best for reading or close work.

13. 4 lumens.

14. To glow when exposed to ultraviolet light.

15. To eliminate flicker and to produce a surge in starting voltage.

16.

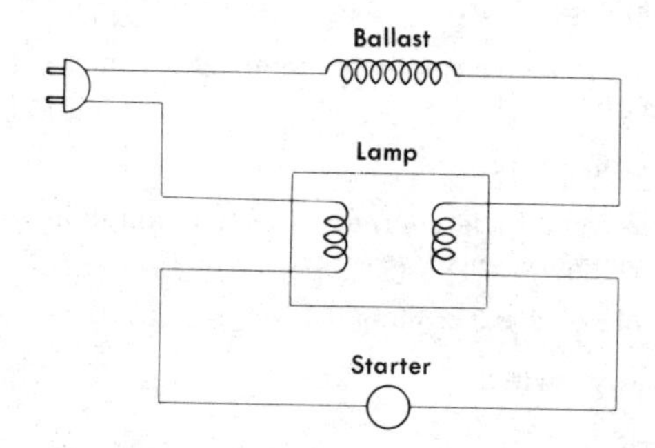

17. Intensity of source and distance from source. Increase or decrease wattage of lamp. Change distance.

18. 5 foot-candles.

19. Operate in pairs on two-phase circuit or sets of three on three-phase circuit. On single-phase circuits operate in pairs with a ballast.

20. 62.5 foot-candles.

21. 2.738 feet.

Chapter 10

1. Scientists describe electromagnetic radiation as disturbances in electric and magnetic fields of force.

2. No.

3. 186,000 miles or 300 million meters per second.

4. 45 meters.

5. No. They require a medium to travel in.

6. Visible light, ultraviolet light, and infrared.

7. By projecting a high-voltage electron stream against a metal target in a vacuum.
8. No. Because of their extremely short wavelength they are able to penetrate solid objects that reflect longer wavelengths.
9. Yes. They may kill the cells or affect their growth or structure.
10. By ionizing them.
11. Remove or adds electrons.
11. An electrical potential causes a movement of electrons and ions.
13. Concentration of charges on sharp points and ionization by heat.
14. Yes. Thermal agitation moves the atoms about and electrons are being lost by collision and then regained.
15. If heat is applied, the speed of the gas molecules will be increased, helping ionization to take place.
16. A high-velocity stream of electrons and ions traveling between the cathode and anode in a partial vacuum.
17. The properties of being deflected by electric or magnetic fields and the ability to excite a phosphorescent screen.
18. Current flow through a metal conductor is a movement of free electrons—current flow through an electrolyte is a movement of ions.
19. In a voltaic cell a chemical action produces an electric current. In an electrolytic cell an electrical current produces a chemical action.
20. Cu^{++} and SO_4^{--}. Copper sulphate.
21. A complete electrical path from one terminal of the cell, through a load, to the other cell terminal.
22. A complete electrical circuit from one terminal of the cell, through a dc power source, to the other cell terminal.
23. The electrolytic solution is a salt of the plating metal.
24. A piece of the pure metal to be used for the plating.
25. Removing aluminum from the ore, plating metal surfaces, and refining other metals to a purer state.
26. Toward the cathode. They are attracted by the negative charge.
27. Negative—the anode is more positive.

Chapter 11

1. The surplus electrons gathered around the cathode. The screen grid.
2. High-speed electrons strike the metal plate and knock additional electrons loose.
3. The plate, suppressor grid, screen grid, control grid, and cathode (or filament).
4. By controlling cathode temperature, plate voltage, and grid voltage.
5. Field emission, photoelectric emission, thermionic emission, and secondary emission.
6. Because it is closer to the emitting element, the cathode.
7. Negative.
8. Two. Electron emitter and electron collector.
9. It suppresses secondary emission effects by repelling the secondary electrons back to the plate.
10. It moves the space charge electrons away from the cathode by attracting them toward the plate.
11. 15,750 scans per second.
12.

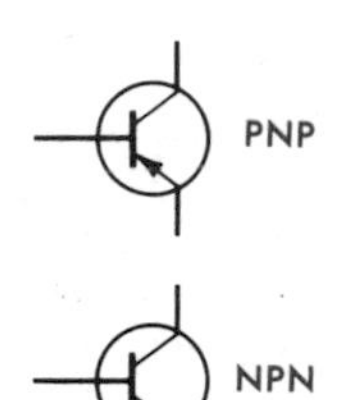

13. Emitter, base, and collector. The emitter compares to cathode, base compares to grid, and collector compares to plate.
14. P type.
15.

16. Valence electrons are the electrons in the outer ring of an atom. Covalence electrons are electrons shared by two atoms of the same element.
17. Germanium or silicon.
18. Positive.
19. Doping.
20. Smaller size and lower power requirements.
21. The field-effect transistor; its impedance is high.
22. Fast turn-on and low forward resistance.
23. Reverse bias.
24. Positive.
25. A rectifier that rectifies only one half of the ac waveform.
26. From cathode to anode.

27.

Conducting

Nonconducting

28.

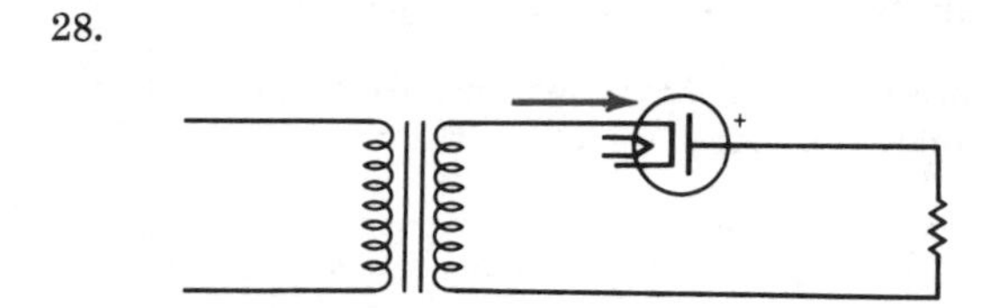

29.

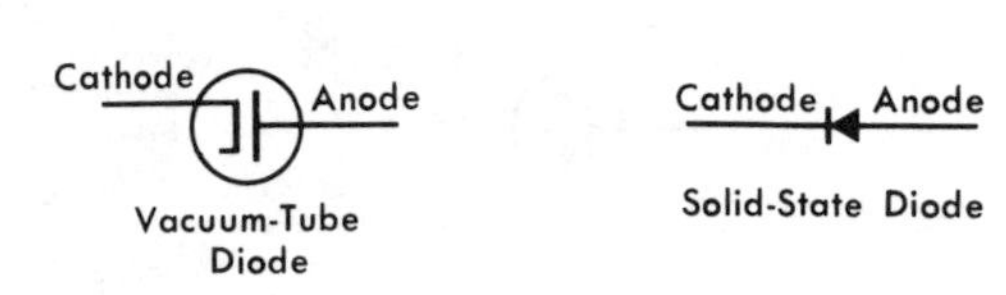

30. Copper-oxide and selenium.

31.

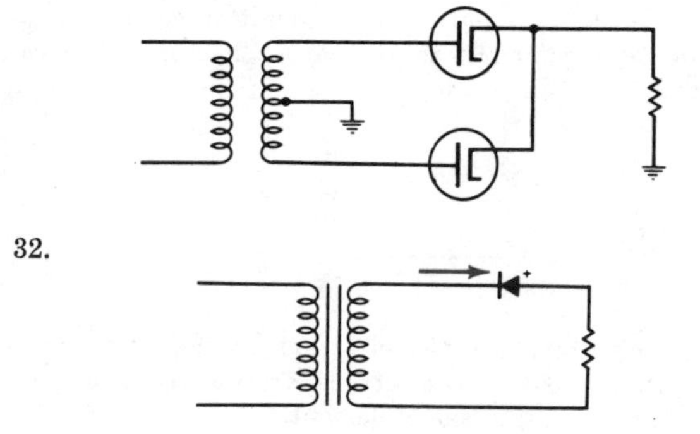

32.

33. Rectification changes ac to dc. Detection changes ac to dc but keeps a desired component of the pulsating dc by selective filtering.

34. Amplification is the ratio of output to input in an amplifier.

35. Negative.

36. Small inputs can control large outputs.

37. The base.

38. Low. The grid does not oppose the flow of electrons to the plate.

39. High. It takes base current to cause collector current.

40. Voltage amplification increases voltage with little increase in current. Current amplification increases current with little increase in voltage.

41. No. The increased power comes from the power supply, not from the amplifier. The amplifier is like a valve that opens to let more power out of the supply.

42. Yes. It lets more plate current flow. This is the same in effect as a lower tube resistance.

43. Batteries were large, expensive, and needed frequent replacement.

44. A common filament voltage is 6.3 volts.

45. Positive feedback from output to input.

46. No. It can be connected to an ac line.

47. The speaker needs high current at low voltage and the output tube supplies low current at high voltage. The output transformer changes one to the other.

48. Yes. In some units one lead of the line cord is connected to the chassis. The radio could be plugged in so that the chassis is hot.

49. Step-down. See Answer 47.

50. One amplifies the comparatively low audio frequencies; the other amplifies the radio frequencies. Otherwise, they are similar.

51. $\lambda = \frac{300,000,000}{f}$

where,

λ is wavelength in meters,
f is frequency in hertz.

52. High radio frequencies are beat or heterodyned with an oscillator signal of another frequency. The difference between the two frequencies is the i-f or intermediate frequency. This is lower than the original radio frequency and easier to amplify.

53. Because they can present ideas in a compact, simplified form.

54. Approximately 229 meters.

55. The tuning capacitor resonates with the inductor at the frequency of the desired station.

56. Rf amplifier amplifies the rf signal of the selected station. Crystal detector removes audio signal from the rf signal. Audio amplifier increases strength of audio signal so that it will operate speaker.

57. Local oscillator furnishes signal to heterodyne with incoming rf signal. Frequency converter mixes oscillator signal and rf signal to get i-f signal. Intermediate frequency amplifier amplifies i-f signal. Detector removes audio signal from i-f signal. Audio output amplifier amplifies audio signal and drives speaker.

58. The a-m signal is an rf carrier signal whose *amplitude* is changed at an audio rate. The fm signal is an rf carrier whose *frequency* is changed at an audio rate.

59. 82 channels.

60. Very high frequency (vhf). Ultrahigh frequency (uhf).

61. Vhf: Channels 2 through 13. Uhf: Channels 14 through 83.

62. The 4.5-MHz separation of the picture carrier and the sound carrier.

63. The 4.5-MHz sound signal is produced in the sound detector.

64. It causes the electron beam to move back and forth across the face of the picture tube.

65. An amplitude-modulated signal.

66. A frequency-modulated signal.

67. The series of horizontal lines traced on the face of the picture tube by the electron beam.

68. Plate current is reduced and finally stopped altogether.

69. From a small fraction of a second to several minutes.

70. The thermocouple produces an electric potential between two junctions of different temperatures. The thermistor changes resistance as its temperature is changed.

71. Because a signal produced by a sensor may not be strong enough to be useful.

72. 24 seconds.

73. A device using electrical energy to produce mechanical movement.

74. No. See answer 68.

75. The bimetallic thermostat uses a change in temperature to produce mechanical movement; the thermocouple uses a difference in temperature between two points to produce an electrical potential.

INDEX